CAMBRIDGE TRACTS IN MATHEMATICS

General Editors



**187 Convexity: An Analytic Viewpoint**

CAMBRIDGE TRACTS IN MATHEMATICS



A complete list of books in the series can be found at www.cambridge.org/mathematics.
Recent titles include the following:

150. Harmonic Maps, Conservation Laws and Moving Frames (2nd Edition). By F. Hélein
151. Frobenius Manifolds and Moduli Spaces for Singularities. By C. Hertling
152. Permutation Group Algorithms. By A. Seress
153. Abelian Varieties, Theta Functions and the Fourier Transform. By A. Polishchuk
154. Finite Packing and Covering. By K. Böröczky, Jr
155. The Direct Method in Soliton Theory. By R. Hirota. Edited and translated by A. Nagai, J. Nimmo, and C. Gilson
156. Harmonic Mappings in the Plane. By P. Duren
157. Affine Hecke Algebras and Orthogonal Polynomials. By I. G. Macdonald
158. Quasi-Frobenius Rings. By W. K. Nicholson and M. F. Yousif
159. The Geometry of Total Curvature on Complete Open Surfaces. By K. Shiohama, T. Shioya, and M. Tanaka
160. Approximation by Algebraic Numbers. By Y. Bugeaud
161. Equivalence and Duality for Module Categories. By R. R. Colby and K. R. Fuller
162. Lévy Processes in Lie Groups. By M. Liao
163. Linear and Projective Representations of Symmetric Groups. By A. Kleshchev
164. The Covering Property Axiom, CPA. By K. Ciesielski and J. Pawlikowski
165. Projective Differential Geometry Old and New. By V. Ovsienko and S. Tabachnikov
166. The Lévy Laplacian. By M. N. Feller
167. Poincaré Duality Algebras, Macaulay's Dual Systems, and Steenrod Operations. By D. Meyer and L. Smith
168. The Cube-A Window to Convex and Discrete Geometry. By C. Zong
169. Quantum Stochastic Processes and Noncommutative Geometry. By K. B. Sinha and D. Goswami
170. Polynomials and Vanishing Cycles. By M. Tibăr
171. Orbifolds and Stringy Topology. By A. Adem, J. Leida, and Y. Ruan
172. Rigid Cohomology. By B. Le Stum
173. Enumeration of Finite Groups. By S. R. Blackburn, P. M. Neumann, and G. Venkataraman
174. Forcing Idealized. By J. Zapletal
175. The Large Sieve and its Applications. By E. Kowalski
176. The Monster Group and Majorana Involutions. By A. A. Ivanov
177. A Higher-Dimensional Sieve Method. By H. G. Diamond, H. Halberstam, and W. F. Galway
178. Analysis in Positive Characteristic. By A. N. Kochubei
179. Dynamics of Linear Operators. By F. Bayart and É. Matheron
180. Synthetic Geometry of Manifolds. By A. Kock
181. Totally Positive Matrices. By A. Pinkus
182. Nonlinear Markov Processes and Kinetic Equations. By V. N. Kolokoltsov
183. Period Domains over Finite and *p*-adic Fields. By J.-F. Dat, S. Orlik, and M. Rapoport
184. Algebraic Theories. By J. Adámek, J. Rosický, and E. M. Vitale
185. Rigidity in Higher Rank Abelian Group Actions I: Introduction and Cocycle Problem. By A. Katok and V. Nitica
186. Dimensions, Embeddings, and Attractors. By J. C. Robinson
187. Convexity: An Analytic Viewpoint. By B. Simon

# Convexity
## An Analytic Viewpoint

BARRY SIMON
*California Institute of Technology*



CAMBRIDGE
UNIVERSITY PRESS

CAMBRIDGE UNIVERSITY PRESS
Cambridge, New York, Melbourne, Madrid, Cape Town,
Singapore, São Paulo, Delhi, Tokyo, Mexico City

Cambridge University Press
The Edinburgh Building, Cambridge CB2 8RU, UK

Published in the United States of America by Cambridge University Press, New York

www.cambridge.org
Information on this title: www.cambridge.org/9781107007314



First published 2011

Printed in the United Kingdom at the University Press, Cambridge

*A catalog record for this publication is available from the British Library*

*Library of Congress Cataloging in Publication data*
Simon, Barry, 1946–
Convexity : an analytic viewpoint / Barry Simon.
p. cm.
Includes index.
ISBN 978-1-107-00731-4 (hardback)
1. Convex domains. 2. Mathematical analysis. I. Title.
QA639.5.S56 2011
516′.08 – dc22 2011008617

ISBN 978-1-107-00731-4 Hardback

# Contents

# Preface

Convexity of sets and functions are extremely simple notions to define, so it may be somewhat surprising the depth and breadth of ideas that these notions give rise to. It turns out that convexity is central to a vast number of applied areas, including Statistical Mechanics, Thermodynamics, Mathematical Economics, and Statistics, and that many inequalities, including Hölder's and Minkowski's inequalities, are related to convexity.

An introductory chapter (1) includes a study of regularity properties of convex functions, some inequalities (Hölder, Minkowski, and Jensen), the Hahn–Banach theorem as a statement about extending tangents to convex functions, and the introduction of two constructions that will play major roles later in this book: the Minkowski gauge of a convex set and the Legendre transform of a function.

The remainder of the book is roughly in four parts: convexity and topology on infinite-dimensional spaces (Chapters 2–5); Loewner's theorem (Chapters 6–7); extreme points of convex sets and related issues, including the Krein–Milman theorem and Choquet theory (Chapters 8–11); and a discussion of convexity and inequalities (Chapters 12–16).

The first part begins with a study of Orlicz spaces in Chapter 2, a notion that extends $L^p$. The most interesting new example is $L^1 \log L$ but the theory also illustrates parts of $L^p$ theory. Chapter 3 introduces the notion of locally convex spaces and includes a discussion of $L^p$ and $H^p$ for $0 < p < 1$ to illustrate what can happen in nonlocally convex spaces. Among the issues discussed are uniqueness of topologies on $\mathbb{R}^n$ as a topological vector space, the fact that infinite-dimensional spaces are never locally compact, Kolmogorov's theorem that a topological vector space has a topology given by a norm if and only if 0 has a bounded convex neighborhood, Fréchet and barreled spaces. Chapter 4 deals with finding hyperplanes to slip between disjoint convex sets. It is an appealing geometric notion, mainly important for technical reasons. Chapter 5 discusses dual topologies and the Mackey–Arens theorem which describes all topologies in which $Y$ is the dual of $X$ where $Y$ is a rich family of linear functionals on $X$. We also discuss Legendre transforms in great generality and prove Fenchel's theorem on when the double Legendre transform of a function is the function itself. Polar sets are a key technical tool.

The second part discusses Loewner's theorem and related ideas: when does a function, $f$, on $(a, b)$ preserve matrix inequalities between matrices with eigenvalues in $(a, b)$ and when is $f$ convex applied to matrices. The answer is given by a deep theorem of Loewner that describes the set of such $f$: for the monotonicity question, they must be analytic on $(a, b)$ and have an analytic continuation to all of $\mathbb{C}_+$ with $\operatorname{Im} f > 0$ if $\operatorname{Im} z > 0$! We describe the framework in Chapter 6 and the first proof of Loewner's theorem in Chapter 7 (there will be another proof in Chapter 9).

The third part focuses on geometric ideas, especially extreme points. Chapter 8 proves several basic results in this area, most notably the result that combines theorems of Minkowski and Carathéodory that any point $x \in K$, a compact convex subset of $\mathbb{R}^\nu$, is a convex combination of at most $\nu + 1$ extreme points, and the Krein–Milman theorem that a compact convex subset of a locally convex space is the closed convex hull of its extreme points. We begin the discussion of ergodic theory continued in the next chapter. Chapter 9 shows that if the set of extreme points is closed (often true, but it can even fail in the finite-dimensional case), then any point is an integral of extreme points. Applications include Bernstein's theorem on totally positive functions, Bochner's theorem, and a second proof of Loewner's theorem. There are several examples presented where the extreme points are dense rather than closed, showing the need for extending the representation theory to situations where the extreme points are not closed: that is the subject known as Choquet theory – existence is the topic of Chapter 10 and uniqueness Chapter 11. Uniqueness turns out to be associated to vector order, so that subject is partially discussed in Chapter 11.

The fourth and final part continues the discussion of convexity and inequalities. Chapter 12 discusses Hadamard's three-circle and three-line bounds in the theory of analytic functions, and applies it to the Riesz–Thorin and Stein interpolation theorems. Applications to Young's inequality and analyticity of $L^p$ semigroups follow. Chapter 13 details a remarkable inequality of Prékopa about integrals of log concave functions. Applications include the Brunn–Minkowksi inequality for convex sets, the classical isoperimetric inequality, and an isoperimetric inequality for Dirichlet ground states. We also give a proof of the general Brunn–Minkowski inequality. Chapters 14 and 15 deal with two threads in rearrangement, both going back to work of Hardy, Littlewood, and Pólya. Chapter 14 focuses on the Brascamp–Lieb–Luttinger inequality, its proof including the study of Steiner symmetrization and applications including additional isoperimetric inequalities. Chapter 15 studies the issue of when $\int \varphi(|g(x)|)\, d\mu(x) \leq \int \varphi(|f(x)|)\, d\mu(x)$ for all even convex functions, $\varphi$, on $\mathbb{R}$. Along the way, we will prove Birkhoff's theorem identifying the extreme points in the set on $n \times n$ matrices, $A$, with $a_{ij} \geq 0$, and so that each row and each column sums to 1. Chapter 16 provides a variational principle for entropy based on Legendre transforms and uses it to prove a semicontinuity result of importance in spectral analysis.

While this book is extensive, there are numerous topics in convexity left out – some of them are indicated in the Notes (Chapter 17).

I'd like to thank Almut Burchard, Brian Davies, Leonid Golinskii, Helge Krüger, Elliott Lieb, Michael Loss, and especially Derek Robinson for useful comments about this book. As always, the love and support of my family, especially my wife Martha, was invaluable.

Barry Simon
Los Angeles

# 1

# Convex functions and sets

This chapter has the fundamental definitions and some of the basics concerning differentiability and Jensen's inequality that will play central roles throughout the book. We'll also define the gauge of a convex set and Legendre transforms of functions, two notions central to Chapters 2–5. And we'll phrase the Hahn–Banach theorem as essentially a statement about tangents to convex functions.

A function, $f$, from an interval, $I \subset \mathbb{R}$, to $\mathbb{R}$ is called *convex* if and only if for all $x, y \in I$ and $\theta \in [0, 1]$,

$$f(\theta x + (1 - \theta)y) \leq \theta f(x) + (1 - \theta) f(y) \tag{1.1}$$

Geometrically (see Figure 1.1), (1.1) says for $z$ in the interval $[x, y]$, the pairs $(z, f(z))$ lie below the straight line from $(x, f(x))$ to $(y, f(y))$. It is remarkable that such a simple definition is so useful and rich. In particular, we will see that both Hölder's and Minkowski's inequalities are consequences of convex machinery – so much so that we will provide four distinct proofs of Hölder's inequality, three in this chapter and one in the next.

An equivalent formula to (1.1) is that for $x, y, z \in I$ with $x < y < z$, we have the determinant

$$\begin{vmatrix} x & f(x) & 1 \\ y & f(y) & 1 \\ z & f(z) & 1 \end{vmatrix} \geq 0$$

This is a geometric statement about a triangle being positively oriented.

To extend the definition from $\mathbb{R}$ to $\mathbb{R}^\nu$ (and beyond), we need to begin with domains of definition that generalize the role of intervals.

**Definition** A subset $K$ of a real vector space, $V$, is called *convex* if and only if for any $x, y \in K$ and $\theta \in [0, 1]$, $\theta x + (1 - \theta)y \in K$.

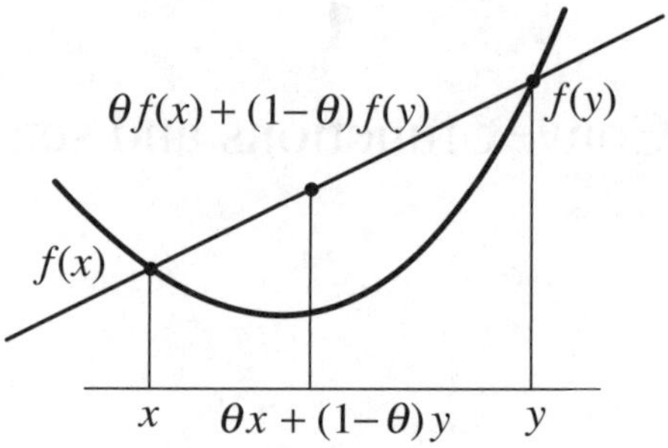


Figure 1.1 The meaning of a convex function

Thus, $K$ is convex if it contains the line segment between any pair of points in $K$.

**Definition** Let $K$ be a convex subset of a vector space $V$. A function $f\colon K \to \mathbb{R}$ is called

(i) *convex* if (1.1) holds for all $x, y \in K$ and $\theta \in [0, 1]$,
(ii) *concave* if $-f$ is convex,
(iii) *affine* if $f$ is convex and concave,
(iv) *strictly convex* if $f$ is convex and strict inequality holds in (1.1) whenever $x \neq y$ and $\theta \in (0, 1)$.

Thus, concavity means that

$$f(\theta x + (1-\theta)y) \geq \theta f(x) + (1-\theta)f(y)$$

and affine means

$$f(\theta x + (1-\theta)y) = \theta f(x) + (1-\theta)f(y)$$

If $K = V$ and $f$ is affine with $f(0) = 0$, then first

$$f(\theta x) = f(\theta x + (1-\theta)0) = \theta f(x)$$

and then

$$f(x+y) = 2f(\tfrac{1}{2}\,x + \tfrac{1}{2}\,y) = f(x) + f(y)$$

so $f$ is linear. In general, if $K = V$, every affine function $f$ is of the form $f(x) = f(0) + \ell(x)$ with $\ell$ linear.

For some purposes in later chapters, it is convenient to allow a convex function to take the value $+\infty$ and to extend $f$ from $K \subset V$ to all of $V$ by setting it to $\infty$ on $V \backslash K$. Since $K$ is convex, (1.1) then still holds for all $x, y \in V$. In this chapter though, we will suppose $f < \infty$ at all points of definition.

The following connection between convex sets and functions is easy to check:

**Proposition 1.1** *Let $K$ be a convex subset of $V$ and $f\colon K \to \mathbb{R}$. Define*

$$\tilde{\Gamma}(f) = \{(x, \lambda) \in V \times \mathbb{R} \mid x \in K,\, \lambda > f(x)\} \tag{1.2}$$

*Then $f$ is a convex function if and only if $\tilde{\Gamma}(f)$ is a convex subset of $V \times \mathbb{R}$.*

Moreover, a simple induction together with

$$\sum_{j=1}^{m} \theta_j x_j = \theta_m x_m + (1 - \theta_m) \sum_{j=1}^{m-1} \varphi_j x_j$$

where $\varphi_j = \theta_j (1 - \theta_m)^{-1}$ shows that

**Proposition 1.2** (First Form of Jensen's Inequality) *If $K$ is a convex subset of $V$ and $x, \dots, x_m \in K$ and $\theta_1, \dots, \theta_m \in [0, 1]$ with $\sum_{j=1}^{m} \theta_j = 1$, then $\sum_{j=1}^{m} \theta_j x_j \in K$. If $f\colon K \to \mathbb{R}$ is convex, then*

$$f\biggl(\sum_{j=1}^{m} \theta_j x_j\biggr) \le \sum_{j=1}^{m} \theta_j f(x_j) \tag{1.3}$$

The following is sometimes useful:

**Proposition 1.3** *Let $f\colon I \subset \mathbb{R} \to \mathbb{R}$ be continuous. Then $f$ is convex if and only if for all $x, y \in I$,*

$$f(\tfrac{1}{2}\, x + \tfrac{1}{2}\, y) \le \tfrac{1}{2}\, f(x) + \tfrac{1}{2}\, f(y) \tag{1.4}$$

*Remarks* 1. (1.4) is called *midpoint convexity*.

2. Since convexity is a statement about $f$ restricted to straight lines, the result immediately extends to $f$ defined on $K \subset \mathbb{R}^\nu$ and on $K \subset V$, any vector space with a topology in which scalar multiplication and addition are continuous functions.

*Proof* Obviously, convexity implies (1.4). Suppose conversely that (1.4) holds. Write $\frac{1}{4}x + \frac{3}{4}y = \frac{1}{2}(\frac{1}{2}x + \frac{1}{2}y) + \frac{1}{2}y$ and conclude that

$$\begin{aligned} f(\tfrac{1}{4}\, x + \tfrac{3}{4}\, y) &\le \tfrac{1}{2}\, f(\tfrac{1}{2}\, x + \tfrac{1}{2}\, y) + \tfrac{1}{2}\, f(y) \\ &\le \tfrac{1}{4}\, f(x) + \tfrac{3}{4}\, f(y) \end{aligned}$$

By a simple induction, (1.1) holds for all dyadic rationals $\theta = j/2^n$, $j = 0, 1, 2, \dots, 2^n$. Then, by continuity, it holds for all $\theta$. □

The following more complicated result will be convenient in Chapter 13:

**Proposition 1.4** *Let $f\colon I \subset \mathbb{R} \to \mathbb{R}$ be lsc. Then $f$ is convex if and only if for all $x, y \in I$, (1.4) holds.*

*Remark* lsc is short for lower semicontinuous, that is, if $x_n \to x$, then $f(x) \le \liminf f(x_n)$.

*Proof* Let $[a,b] \subset I$ be a bounded interval. Since $f$ is lsc, it is bounded below and takes its minimum value at a point $c \in [a,b]$ (for let $\alpha = \inf_{x\in[a,b]} f(x)$, let $x_n$ be a sequence with $f(x_n) \to \alpha$, and let $c$ be a limit point of the $x_n$). We will prove continuity on $[c,b]$. The proof for $[a,c]$ is similar, once one has that Proposition 1.3 applies. For notational simplicity, suppose $c = 0$ and $b = 1$.

By (1.4) and $f(0) \le f(1)$, $f(\frac{1}{2}) \le f(1)$, and since $f(0)$ is the minimum, $f(0) \le f(\frac{1}{2})$. Once one has this, $f(0) \le f(\frac{1}{4}) \le f(\frac{1}{2})$ and then since $f(\frac{1}{2}) \le \frac{1}{2}[f(\frac{1}{4}) + f(\frac{3}{4})]$, we conclude $f(\frac{1}{2}) \le f(\frac{3}{4})$, and then by (1.4), $f(\frac{3}{4}) \le f(1)$. By induction for any pair of $x, y$ in $\mathbb{D}$, the dyadic rationals in $[0,1]$, $x < y \Rightarrow f(x) \le f(y)$.

It follows for any $y \in [0,1]$,

$$\tilde{f}(y) = \lim_{\substack{x\in\mathbb{D}\\ x\uparrow y}} f(x) = \sup_{\substack{x\in\mathbb{D}\\ x\uparrow y}} f(x)$$

exists. Notice $\tilde{f}$ is monotone and

$$\tilde{f}(y) = \lim_{x\uparrow y} \tilde{f}(x) \tag{1.5}$$

We claim $\tilde{f}(y)$ is continuous, for pick $x_n \uparrow y$ and $z_n \downarrow y$ with $x_n, z_n \in \mathbb{D}$ and arrange that $\frac{1}{2}(x_n + z_n) \ge y$. Then, with $\tilde{f}(y+) \equiv \lim_{x\downarrow y} \tilde{f}(x) = \lim_{x\downarrow y,\, x\in\mathbb{D}} f(x)$, we have by (1.4) that

$$\begin{aligned}
\tilde{f}(y+) &= \lim_{n\to\infty} \tilde{f}(\tfrac{1}{2}\, x_n + \tfrac{1}{2}\, z_n) \\
&\le \lim_{n\to\infty} \tfrac{1}{2}\, \tilde{f}(x_n) + \tfrac{1}{2}\, \tilde{f}(z_n) \\
&= \tfrac{1}{2}\, \tilde{f}(y) + \tfrac{1}{2}\, \tilde{f}(y+)
\end{aligned}$$

so $\tilde{f}(y+) \le \tilde{f}(y)$ which means, by monotonicity and (1.5), that $\tilde{f}$ is continuous.

By the lsc hypothesis,

$$f(y) \le \tilde{f}(y) \tag{1.6}$$

On the other hand, pick $x_n \uparrow y$ with $x_n \in \mathbb{D}$ and let $z_n = y + 2(x_n - y)$ so $x_n = \frac{1}{2} z_n + \frac{1}{2} y$ and by (1.4),

$$f(x_n) \le \tfrac{1}{2}\, f(z_n) + \tfrac{1}{2}\, f(y) \le \tfrac{1}{2}\, \tilde{f}(z_n) + \tfrac{1}{2}\, f(y)$$

by (1.6). Taking $n \to \infty$ and using the continuity of $\tilde{f}$, we see that

$$\tilde{f}(y) \le f(y)$$

Thus, $f = \tilde{f}$ is continuous and so convex. □

*Remark* In the proof of Proposition 9.15, we will see another situation where midpoint convexity implies convexity, namely, if $f$ is monotone.

In the following, the use of Proposition 1.3 is of purely notational simplicity; one can directly deal with general $\theta$.

**Theorem 1.5** *Let $f: I \to \mathbb{R}$ with $I$ an open interval and let $f$ be $C^2$. Then $f$ is convex if and only if*

$$f''(x) \geq 0 \tag{1.7}$$

*for all $x \in I$. If $K$ is an open convex subset of $\mathbb{R}^\nu$ and $f$ is $C^2$ on $K$, then $f$ is convex if and only if the Hessian $\partial^2 f/\partial x^i \partial x^j$ is positive definite at each point.*

*Remark* This result is extended to $f$'s which are not $C^2$ in Theorem 1.29.

*Proof* Consider first the case $\nu = 1$. Taylor's theorem with remainder says that for $\delta x > 0$,

$$f(x \pm \delta x) = f(x) \pm \delta x f'(x) + \int_0^{\delta x} (\delta x - y) f''(x \pm y)\, dy \tag{1.8}$$

and thus,

$$\tfrac{1}{2}\,[f(x+\delta x)+f(x-\delta x)]-f(x) = \tfrac{1}{2}\int_0^{\delta x} (\delta x-y)[f''(x+y)+f''(x-y)]\, dy \tag{1.9}$$

It follows that if (1.7) holds, then $f$ obeys (1.4), and so $f$ is convex by Proposition 1.3. Conversely, if $f$ is convex, the left side of (1.9) is nonnegative for each $x$ and each sufficiently small $\delta x$. Thus, taking the right side of (1.9) and dividing by $\frac{1}{2}(\delta x)^2$ and taking $\delta x \downarrow 0$, we see that (1.7) holds. We have thus proven the result if $\nu = 1$.

For general $\nu$, we note that convexity is a statement about the values of $f$ restricted to line segments in $K$. Thus, $f$ is convex on $K$ if and only if for all $x_0 \in K$ and $e \in \mathbb{R}^\nu$, $e \neq 0$, if we define $I_e(x_0) = \{\lambda \in \mathbb{R} \mid x_0 + \lambda e \in K\}$ and $F(\lambda; x_0, e) = f(x_0 + \lambda e)$, then $F$ is convex as a function on $I_e(x_0)$. From the one-dimensional case, we see that $F$ is convex if and only if $F''(\lambda) \geq 0$ for such $\lambda$. Since $F(\lambda; x_0, e) = F(\lambda - \lambda_0; x_0 + \lambda_0 e, e)$, we see $f$ is convex if and only if for each $x_0$ and $e \neq 0$,

$$F''(0; x_0, e) \geq 0 \tag{1.10}$$

Since

$$F''(0; x_0, e) = \sum_{i,j=1}^{\nu} e_i e_j \frac{\partial^2 f}{\partial x_i \partial x_j}(x_0)$$

(1.10) is equivalent to the positive definiteness of the Hessian, as claimed. □

*Remark* The proof shows if $f''(x) > 0$ (indeed, if $f''$ is a.e. strictly positive), then $f$ is strictly convex. The example $f(x) = x^4$, which is strictly convex but has $f''(0) = 0$, shows the converse is not true as a pointwise statement.

**Example 1.6** Let $f(x) = e^x$ on $(-\infty, \infty)$. Then $f''(x) = e^x > 0$ so $f$ is convex. Midpoint convexity

$$e^{\frac{1}{2}(x+y)} \leq \tfrac{1}{2}\, e^x + \tfrac{1}{2}\, e^y$$

is (if $a = e^x$, $b = e^y$ so $a, b$ are arbitrary numbers on $(0, \infty)$)

$$\sqrt{ab} \leq \tfrac{1}{2}\,(a + b) \tag{1.11}$$

the arithmetic-geometric mean inequality. Thus, convexity of $e^x$ generalizes this inequality. Since midpoint convexity implies convexity, (1.11) actually implies convexity of $x \to e^x$.

Using Proposition 1.2 with $\theta_1 = \cdots = \theta_m = 1/m$ and this $f$, we see that if $a_1, \ldots, a_m > 0$, then

$$(a_1 \ldots a_m)^{1/m} \leq \frac{1}{m}\,(a_1 + \cdots + a_m) \tag{1.12}$$

The function $g(x) = \log x$ for $x \in (0, \infty)$ obeys $g''(x) = -1/x^2 < 0$ so $g$ is concave. By the above remark, $f$ is strictly convex and $g$ is strictly concave.

It is no coincidence that $\log x$, the inverse of $e^x$ is concave. If $f$ is strictly monotone and $f$ is convex, then $f^{-1}$, the inverse function, is concave. For let $x, y$ be given in $\operatorname{Ran} f$ and let $a, b$ be chosen so that $x = f(a)$, $y = f(b)$. Then, convexity of $f$ implies that

$$\theta x + (1 - \theta)y \geq f((1 - \theta)a + \theta b) \tag{1.13}$$

Since $f$ is monotone, so is $f^{-1}$, so applying $f^{-1}$ to (1.13), inequalities are preserved and thus,

$$\begin{aligned} f^{-1}(\theta x + (1 - \theta)y) &\geq (1 - \theta)a + \theta b \\ &= (1 - \theta)f^{-1}(x) + \theta f^{-1}(y) \end{aligned}$$

that is, $f^{-1}$ is concave. □

**Proposition 1.7** *Let $f\colon [0, a] \to \mathbb{R}$ be convex and monotone increasing. Then $g\colon \{x \in \mathbb{R}^\nu \mid |x| \leq a\} \to \mathbb{R}$ by*

$$g(x) = f(|x|)$$

*is a convex function.*

*Proof*

$$\begin{aligned} g(\theta x + (1 - \theta)y) &= f(|\theta x + (1 - \theta y)|) \\ &\leq f(\theta|x| + (1 - \theta)|y|) \\ &\leq \theta g(x) + (1 - \theta)g(y) \end{aligned} \tag{1.14}$$

(1.14) follows from the assumed monotonicity of $f$ and the triangle inequality

$$|\theta x + (1-\theta)y| \le \theta|x| + (1-\theta)|y| \qquad \square$$

*Remarks* 1. The proof shows that $x \mapsto f(\|x\|)$ is convex on the ball of radius $a$ in any normed linear space.

2. The proof also shows if $f$ on $\mathbb{R}$ is even and convex, then $f$ is monotone increasing on $[0,\infty)$.

**Example 1.8** Let $p \ge 1$ and let $f(x) = |x|^p$ on $\mathbb{R}$. By the last proposition, for $f$ to be convex, we only need that $f$ is convex on $[0,\infty)$. By continuity, convexity on $(0,\infty)$ suffices and for that, we need only note that $f''(x) = p(p-1)x^{p-2} \ge 0$ so $f$ is convex if $p \ge 1$ and strictly convex if $p > 1$. $\square$

Convexity and the triangle inequality are intimately related:

**Theorem 1.9** *Let $V$ be a vector space and let $F\colon V \to [0,\infty)$ be homogeneous of degree* 1, *that is,*

$$F(\lambda x) = \lambda F(x) \tag{1.15}$$

*for all $x \in V$ and $\lambda \in [0,\infty)$. Then the following are equivalent:*

(i) *$F$ is convex.*
(ii) *$\{x \mid F(x) \le 1\}$ is a convex set.*
(iii)

$$F(x+y) \le F(x) + F(y) \tag{1.16}$$

*In particular, $F$ obeying* (1.15) *is a seminorm if and only if $F$ is convex with*

$$F(-x) = F(x) \tag{1.17}$$

*and a norm if and only if $F$ is convex, strictly positive on $V\backslash\{0\}$, and* (1.17) *holds.*

*Proof* (i) $\Rightarrow$ (ii) If $F$ is convex and $x, y \in K \equiv \{z \mid F(z) \le 1\}$, then $F(\theta x + (1-\theta)y) \le \theta F(x) + (1-\theta)F(y) \le 1$ so $\theta x + (1-\theta)y \in K$, that is, $K$ is convex.

(ii) $\Rightarrow$ (iii) Consider first the case $F(x) \ne 0 \ne F(y)$. Let $\theta = F(x)/[F(x) + F(y)]$. Since $x/F(x), y/F(y) \in K \equiv \{z \mid F(z) \le 1\}$, (ii) implies $\theta x + (1-\theta)y = (x+y)/[F(x)+F(y)]$ lies in $K$, that is, $F(x+y)/[F(x)+F(y)] \le 1$, that is, (1.16) holds.

If $F(x) = 0$ and $F(y) = 1$, then for any $\lambda > 0$, $\lambda x, y \in K$ so $\alpha_\lambda(x+y) = \theta_\lambda(\lambda x) + (1-\theta_\lambda)y$ with $\theta_\lambda = 1/(1+\lambda)$ and $\alpha_\lambda = \lambda/(1+\lambda)$. Thus,

$$\frac{\lambda}{1+\lambda} F(x+y) \le 1$$

so taking $\lambda \to \infty$, $F(x+y) \le 1$, so (1.16) holds. If $F(x) = 0$ and $F(y) \ne 0$, repeat the argument with $x$ replaced by $x/F(y)$ and $y$ by $y/F(y)$.

If $F(x) = F(y) = 0$, then for any $\lambda > 0$, $\lambda x, \lambda y \in K$ so $\frac{1}{2}\lambda(x+y) \in K$ so $F(x+y) \le 2/\lambda$. Taking $\lambda \to \infty$, $F(x+y) = 0$, so (1.16) again holds.

(iii) $\Rightarrow$ (i) If (iii) holds, then

$$\begin{aligned} F(\theta x + (1-\theta)y) &\le F(\theta x) + F((1-\theta)y) && \text{(by (1.16))} \\ &= \theta F(x) + (1-\theta)F(y) && \text{(by (1.15))} \end{aligned}$$

□

**Corollary 1.10** *Let $K$ be a convex subset of $V$ which obeys*

(i) *For any $x \in V$, $\lambda x \in K$ for some $\lambda > 0$.*

(ii) *If $x \in K$, then $-x \in K$.*

*Define*

$$\|x\| = \inf\{\lambda \in (0,\infty) \mid \lambda^{-1}x \in K\} \tag{1.18}$$

*Then $\|\cdot\|$ is a seminorm on $V$. Moreover,*

$$\{x \mid \|x\| \le 1\} = \bigcap_{\lambda > 1} \lambda K \tag{1.19}$$

$$\{x \mid \|x\| < 1\} = \bigcup_{\lambda < 1} \lambda K \tag{1.20}$$

*Remarks* 1. When (i) holds, we say that $K$ is *absorbing*. When (ii) holds, we say $K$ is *balanced*.

2. Thus, $\{x \mid \|x\| < 1\} \subset K \subset \{x \mid \|x\| \le 1\}$. But see Remark 5 below.

3. For any convex set $K$ with $0 \in K$, the function of the right side of (1.18) is called the *gauge* of $K$.

4. By (ii), $0 = \frac{1}{2}(x - x) \in K$, so $\{\mu \mid \mu x \in K\}$ is a symmetric interval $I$. $\|x\|$ is defined so $\sup_{\mu \in I} \mu = \|x\|^{-1}$.

5. If $V$ is $\mathbb{R}^\nu$ and $K$ is open (resp. closed), then $K = \{x \mid \|x\| < 1\}$ (resp. $\{x \mid \|x\| \le 1\}$). More generally, if for all $x$, $\{\lambda \mid \lambda x \in K\} \subset \mathbb{R}$ is open, $K = \{x \mid \|x\| < 1\}$, and if the set is closed, then $K = \{x \mid \|x\| \le 1\}$.

6. If $V$ is a complex vector space and (ii) is replaced by $x \in K$ and $|\zeta| = 1$ (for $\zeta \in \mathbb{C}$), then $\zeta x \in K$, then $\|\cdot\|$ is a complex seminorm.

*Proof* By (i), $\{\lambda \mid \lambda^{-1}x \in K\}$ is nonempty so $\|\cdot\|$ is everywhere defined. Clearly, if $\mu > 0$,

$$\begin{aligned} \|\mu x\| &= \inf\{\lambda \mid \lambda^{-1}\mu x \in K\} \\ &= \inf\{\lambda\mu \mid (\lambda\mu)^{-1}\mu x \in K\} \\ &= \inf\{\lambda\mu \mid \lambda^{-1}x \in K\} = \mu\|x\| \end{aligned}$$

so $\|\cdot\|$ is homogeneous of degree 1. Moreover, by (ii), $\|-x\| = \|x\|$.

Now

$$\begin{aligned}\{x \mid \|x\| \leq 1\} &= \{x \mid \inf\{\lambda \mid \lambda^{-1}x \in K\} \leq 1\}\\ &= \{x \mid \lambda^{-1}x \in K \text{ for all } \lambda > 1\}\\ &= \bigcap_{\lambda>1} \lambda K\end{aligned}$$

proving (1.19). Since this set is convex, Theorem 1.9 shows $\|\cdot\|$ is a seminorm.

Similarly,

$$\begin{aligned}\{x \mid \|x\| < 1\} &= \{x \mid \inf\{\lambda \mid \lambda^{-1}x \in K\} < 1\}\\ &= \{x \mid \lambda^{-1}x \in K \text{ for some } \lambda < 1\}\\ &= \bigcup_{\lambda<1} \lambda K\end{aligned}$$ □

Gauges of balanced, absorbing, convex sets will play a key role in the theory of locally convex spaces; see Chapter 3.

Theorem 1.9 shows that there is an interplay between subadditive functions, convex functions, and homogeneous functions of degree one. There is an analog for sets.

**Definition** Let $V$ be a vector space. A *cone* in $V$ is a subset $K \subset V$ so that $x \in K$ and $\lambda \geq 0$ implies $\lambda x \in K$. $K$ is called additive if and only if $x, y \in K$ implies $x + y \in K$.

The following analog of Theorem 1.9 is immediate:

**Theorem 1.11** *Let $K$ be a cone in $V$. Then $K$ is convex if and only if $K$ is additive.*

It is also easy to see that if $K$ is both convex and additive and $0 \in K$, then $K$ is a cone.

Convex cones are often easier to deal with than convex sets. For this reason, given any convex $K \subset V$, we define the set

$$K_{\text{sus}} = \{(\lambda x, \lambda) \mid x \in K,\ \lambda \geq 0\} \subset V \times \mathbb{R} \tag{1.21}$$

called the *suspension* of $K$. It is easy to see that $K_{\text{sus}}$ is a convex cone if and only if $K$ is convex.

**Example 1.12** (Orlicz Spaces and Minkowski's Inequality) Let $(M, d\mu)$ be a measure space with $\mu(M) \equiv 1$. ($\mu(M)$ finite is easily handled, $\mu$ $\sigma$-finite is harder, but many of the results in the next chapter – suitably modified – hold.) Let $F$ be a convex function on $[0, \infty)$ with $F(0) = 0$ and $F(y) > 0$ for all $y > 0$. We suppose that $\lim_{y \downarrow 0} F(y) = 0$. We will use the fact proven below (see Theorem 1.19) that

$F$ is continuous. For any measurable function $f$ on $M$, define

$$Q_F(f) = \int F(|f(x)|)\, d\mu(x) \tag{1.22}$$

where $Q_F$ may be $+\infty$. Then, because $F$ is convex, $Q_F(\cdot)$, where finite, is convex and thus,

$$K = \{f \mid Q_F(f) \leq 1\}$$

is a convex set which clearly obeys $-K = K$ since $Q_F(-f) = Q_F(f)$.

Define $\tilde{L}^{(F)}(M, d\mu)$ by

$$\begin{aligned}&\tilde{L}^{(F)}(M, d\mu)\\ &\quad = \{f \text{ measurable from } M \text{ to } R \mid Q_F(\alpha f) < \infty \text{ for some } \alpha > 0\}\end{aligned} \tag{1.23}$$

Clearly, $\tilde{L}^{(F)}$ is closed under scalar multiplication. Moreover, if $\gamma = (\alpha^{-1} + \beta^{-1})^{-1}$, we have

$$Q_F(\gamma(f+g)) \leq \gamma\alpha^{-1} Q_F(\alpha f) + \gamma\beta^{-1} Q_F(\beta g) \tag{1.24}$$

since $\gamma(f+g) = \gamma\alpha^{-1}(\alpha f) + \gamma\beta^{-1}(\beta g)$, $F$ is convex, and $\gamma\alpha^{-1} + \gamma\beta^{-1} = 1$. By (1.24), $\tilde{L}^{(F)}$ is closed under sums, so $\tilde{L}^{(F)}$ is a vector space.

Note that if $Q_F(\alpha f) < \infty$ for some $\alpha$, then by the monotone convergence theorem ($F(x)$ is monotone on $[0, \infty)$ by hypothesis),

$$\lim_{\alpha \downarrow 0} Q_F(\alpha f) = 0 \tag{1.25}$$

Moreover, if $f$ is not a.e. 0,

$$\lim_{\alpha \to \infty} Q_F(\alpha f) = \infty \tag{1.26}$$

(It may happen $Q_F(\alpha F) = \infty$ for some $\alpha < \infty$.) By (1.25), hypothesis (i) of Corollary 1.10 holds.

Thus, Corollary 1.10 lets us construct a seminorm

$$\|f\|_F = \inf\{\lambda > 0 \mid Q_F(\lambda^{-1} f) \leq 1\} \tag{1.27}$$

called the *Luxemburg norm* associated to $F$. By (1.23), we get a norm by taking equivalence classes of functions equal a.e. We call this space the Orlicz space associated to $F$ and denote it by $L^{(F)}(M, d\mu)$.

If $F(x) = |x|^p$, then

$$Q_F(f) = \int |f(x)|^p \, d\mu(x)$$

and $Q_F(\lambda^{-1} f) = \lambda^{-p} Q_F(f)$. Therefore, $Q_F(\lambda^{-1} f) \leq 1$ if and only if

$\lambda^{-p} Q_F(f) \leq 1$ or $\lambda \geq Q_F(f)^{1/p}$. Thus,

$$\|f\|_{F=x^p} = \left( \int |f(x)|^p \, d\mu(x) \right)^{1/p} \tag{1.28}$$

so Luxemburg norms and Orlicz spaces generalize $L^p$ norms and spaces. Moreover, we have proven that the object in (1.28) obeys the triangle inequality, that is, we have proven Minkowski's inequality

$$\|f + g\|_p \leq \|f\|_p + \|g\|_p \tag{1.29}$$

If one goes through this proof – essentially, the proof of (ii) $\Rightarrow$ (iii) in Theorem 1.9 – we get Minkowski's inequality by noting that

$$Q_p(f) = \int |f(x)|^p \, d\mu(x)$$

is convex. So, with $\theta = \|f\|_p / (\|f\|_p + \|g\|_p)$,

$$\begin{aligned} Q_p\left( \frac{f+g}{\|f\|_p + \|g\|_p} \right) &= Q_p\left( \frac{\theta f}{\|f\|_p} + \frac{(1-\theta)g}{\|g\|_p} \right) \\ &\leq \theta Q_p\left( \frac{f}{\|f\|_p} \right) + (1-\theta) Q_p\left( \frac{g}{\|g\|_p} \right) \leq 1 \end{aligned}$$

which implies Minkowski's inequality. The theory of Orlicz spaces will be presented in the next chapter. □

We next want to begin our discussion of Hölder's inequality. Let us state the general form:

**Theorem 1.13** *Suppose* $1 \leq p, q, r \leq \infty$ *and*

$$\frac{1}{p} + \frac{1}{q} = \frac{1}{r} \tag{1.30}$$

*Let* $M, d\mu$ *be a* $\sigma$*-finite measure space. If* $f \in L^p(M, d\mu)$ *and* $g \in L^p(M, d\mu)$*, then* $fg \in L^r(M, d\mu)$ *and*

$$\|fg\|_r \leq \|f\|_p \|g\|_q \tag{1.31}$$

We begin with some preliminaries:

(i) Since all norms involve $|\cdot|$, we can, without loss, suppose $f, g \geq 0$.

(ii) If $r = \infty$, then $p = q = \infty$ and the result is obvious. So suppose $r < \infty$. $\|fg\|_r^r = \|f^r g^r\|_1$ and $\|f\|_p^r = \|f^r\|_{p/r}$, $\|g\|_q^r = \|g^r\|_{q/r}$. Since (1.30) is equivalent to $r/p + r/q = 1$, we can suppose without loss that $r = 1$.

(iii) By a limiting argument, we can suppose that $f$ and $g$ have supports of finite measure, and so we need only consider the case $\mu(M) < \infty$.

(iv) Once $\mu(M) < \infty$, we can suppose by a limiting argument for some $\varepsilon$, $\varepsilon < f < \varepsilon^{-1}$ and $\varepsilon < g < \varepsilon^{-1}$, or equivalently that $f = e^{F/p}$, $g = e^{G/q}$ with $F, G$ bounded, so (1.31) becomes

$$\int \exp\left(\frac{F}{p} + \frac{G}{q}\right) d\mu \le \left(\int \exp(F)\, d\mu\right)^{1/p} \left(\int \exp(G)\, d\mu\right)^{1/q} \tag{1.32}$$

Letting $\theta = p^{-1}$ so $q^{-1} = (1-\theta)$, we rewrite (1.32) as

$$\mathcal{F}(\theta F + (1-\theta)G) \le \theta \mathcal{F}(F) + (1-\theta)\mathcal{F}(G)$$

where

$$\mathcal{F}(F) = \log\left[\int \exp(F(x))\, d\mu(x)\right] \tag{1.33}$$

We have thus shown

**Proposition 1.14** *Hölder's inequality is equivalent to the convexity of the function $\mathcal{F}$ given by* (1.33) *defined on bounded functions $F$ on a measure space $(M, d\mu)$ with $\mu(M) < \infty$.*

Our first two proofs of Hölder's inequality are thus

**Theorem 1.15** *Let $(M, d\mu)$ be a finite measure space. The function, defined for $F \in L^\infty(M, d\mu)$ by*

$$F \mapsto \log\left[\int \exp(F(x))\, d\mu(x)\right] = \mathcal{F}(F)$$

*is convex.*

*First Proof* By Proposition 1.3, we need only note that $t \mapsto \mathcal{F}(tF + F_0)$ is continuous in $t$ (by the dominated convergence theorem) and then prove midpoint convexity, that is,

$$\mathcal{F}(\tfrac{1}{2} F + \tfrac{1}{2} G) \le \tfrac{1}{2} \mathcal{F}(F) + \tfrac{1}{2} \mathcal{F}(G) \tag{1.34}$$

Let $f = \exp(\frac{1}{2}F)$, $g = \exp(\frac{1}{2}G)$. Then (1.34) is equivalent to

$$\int fg\, d\mu \le \left(\int f^2\, d\mu\right)^{1/2} \left(\int g^2\, d\mu\right)^{1/2} \tag{1.35}$$

which is the Schwarz inequality (which can be proven by noting that $\lambda \mapsto \int (f + \lambda g)^2\, d\mu$ is a nonnegative quadratic polynomial, $a\lambda^2 + b\lambda + c$, so its discriminant $4ac - b^2$ is nonnegative, which is (1.35)). □

*Second Proof* It is easy to see that $t \mapsto \mathcal{F}(F_0 + tF_1) \equiv \mathcal{F}(t; F_0, F_1, d\mu)$ is $C^\infty$ in $t$ so, by Theorem 1.5, we need only prove the second derivative is nonnegative. Since $\mathcal{F}(t + t_0; F_0, F_1, d\mu) = \mathcal{F}(t; F_0 + t_0 F_1, F_1, d\mu)$, we need only show that the derivative is positive at $t = 0$. Since $\mathcal{F}(t; F_0, F_1, d\mu) = \mathcal{F}(t; 0, F_1, e^{F_0}\, d\mu)$, we can suppose $F_0 = 0$. Finally, since $\mathcal{F}(t; F_0, F_1, c\, d\mu) = \mathcal{F}(t; F_0, F_1, d\mu) + \log c$, we can suppose $\int d\mu = 1$.

Thus, we have reduced to the case $\mu(M) = 1$ and

$$f(t) = \log \int \exp(tF)\, d\mu$$

and proving $f''(0) \geq 0$. Let $g(t) = \exp(f(t))$. Then $g'(0) = f'(0)$ since $f(0) = 1$ and $g''(0) = f''(0) + f'(0)^2$ or

$$\begin{aligned} f''(0) &= g''(0) - (g'(0))^2 \\ &= \int F^2\, d\mu - \left( \int F\, d\mu \right)^2 \end{aligned}$$

which is positive by the Schwarz inequality since $\int 1\, d\mu = 1$. □

*Remark* The first proof says that the general Hölder inequality can be derived from the special case $p = q = 2$ (which is easy to prove). It is remarkable that while the initial steps in the two proofs are very different, the critical final step in each case is the Schwarz inequality.

We now return to the general theory of convex functions. Our next theme is to look at regularity – initially at boundedness and continuity.

As a preliminary, we note:

**Definition** Let $C$ be a *hypercube* in $\mathbb{R}^\nu$, that is, for some $a \in \mathbb{R}$ and $x \in \mathbb{R}^\nu$,

$$C = \{y \mid |y_i - x_i| \leq \tfrac{a}{2},\ i = 1, \dots, \nu\} \equiv C_a(x)$$

Call $x$ the *center*, $x_0(C)$, of $C$ and the $2^\nu$ points $(x_1 \pm a/2, x_2 \pm a/2, \dots)$ for the $2^\nu$ choices of $\pm$ in each coordinate the *corners* of $C$ and denote this set by $\delta C$. It is easy to see that any point in $C$ is a convex combination of corners.

**Lemma 1.16** *Let $F$ be a convex function on $C$. Then $F$ is bounded; indeed,*

$$\sup_{x \in C} F(x) = \max_{x \in \delta C} F(x) \tag{1.36}$$

$$\inf_{x \in C} F(x) \geq 2F(x_0(C)) - \max_{x \in \delta C} F(x) \tag{1.37}$$

*Remark* $F$ is defined and a priori finite at each point, so its max over any finite set is finite.

*Proof* Since any $x \in C$ can be written

$$x = \sum_{y \in \delta C} \theta_y(x) y$$

with $\theta_y(x) \geq 0$ and $\sum_{y\in\delta C}\theta_y(x) = 1$, and $f$ is convex,

$$\begin{aligned} F(x) &\leq \sum_{y\in\delta C} \theta_y(x)F(y) \\ &\leq \Big(\sup_{y\in\delta C} F(y)\Big) \sum_{y\in\delta C} \theta_y(x) \\ &= \sup_{y\in\delta C} F(y) \end{aligned}$$

proving (1.36).

On the other hand, if $x \in C$, $\tilde{x} \equiv 2x_0(C) - x \in C$ also and $\frac{1}{2}(x+\tilde{x}) = x_0(C)$. Thus,

$$F(x) \geq 2F(x_0(C)) - F(\tilde{x})$$

and (1.37) follows from (1.36). □

**Theorem 1.17** *Let $U \subset \mathbb{R}^\nu$ be an open convex set on $\mathbb{R}^\nu$ and let $F$ be a convex function on $U$. Then $F$ is bounded on any compact subset $K$ of $U$.*

*Proof* By a standard compactness argument, we need only show for any $x \in U$, there is a neighborhood, $N_x$, of $x$ on which $F$ is bounded. Since $U$ is open, we can find a hypercube $C_a(x)$ with $C_a(x) \subset U$. By the lemma, $F$ is bounded on $C_a(x)$, so we can take $N_x$ to be the interior of $C_a(x)$. □

See Proposition 1.21 below for a strong version of this boundedness.

Next, we turn to continuity.

**Proposition 1.18** *Let $V$ be a normed vector space and let $F$ be a bounded convex function on $B = \{x \mid \|x\| \leq 1\}$. Then $F$ is continuous at $x = 0$.*

*Remark* Since $B$ contains $\{x \mid \|x - x_0\| \leq 1 - \|x_0\|\}$, this shows $F$ is continuous on $\{x \mid \|x\| < 1\}$ and the proof shows uniform continuity on $\{x \mid \|x\| < 1 - \varepsilon\}$ for each $\varepsilon > 0$.

*Proof* Given $x \in B$, let $\tilde{x} = x/\|x\|$. Then $x = (1 - \|x\|)0 + \|x\|(\tilde{x})$ so

$$F(x) - F(0) \leq \|x\|[F(\tilde{x}) - F(0)] \tag{1.38}$$

Moreover,

$$0 = \frac{1}{1+\|x\|}\,x + \frac{\|x\|}{1+\|x\|}\,(-\tilde{x})$$

so

$$F(0) - F(x) \leq \frac{\|x\|}{1+\|x\|}\,[F(-\tilde{x}) - F(x)] \tag{1.39}$$

Thus,

$$|F(x) - F(0)| \leq \|x\| \sup_{z,y \in B} |F(z) - F(y)| \tag{1.40}$$

proving continuity. □

*Note* The proof shows if $F$ is bounded in $\{x \mid \|x - x_0\| \leq r\} = B_{x_0}^r$, then

$$|F(x) - F(x_0)| \leq 2\|x - x_0\| r^{-1} \sup_{x \in B_{x_0}^r} |F(x)| \tag{1.41}$$

which yields the various uniform continuity results claimed below.

We immediately have, by Theorem 1.17 and Proposition 1.18,

**Theorem 1.19** *Let $U \subset \mathbb{R}^\nu$ be an open convex subset of $\mathbb{R}^\nu$ and $F\colon U \to \mathbb{R}$ a convex function. Then $F$ is continuous on $U$; indeed, for any compact subset $K \subset U$, there exists $C_K$ so if $x, y \in K$, then*

$$|F(x) - F(y)| \leq C_K |x - y| \tag{1.42}$$

Given the Arzelà–Ascoli theorem (see [303, Thm. I.28]), this immediately implies

**Theorem 1.20** *Let $U \subset \mathbb{R}^\nu$ be an open convex set and $K_1 \subset K_2 \subset \cdots \subset U$ compact subsets of $U$ so $\cup_{j=1}^\infty K_j = U$. Then for any sequence $c_k > 0$, $\mathcal{F} = \{f \mid f \text{ convex on } U, \sup_{x \in K_j} |f(x)| \leq c_k\}$ is compact in the topology generated by uniform convergence on compacts.*

*Proof* By (1.41), $\mathcal{F}$ is an equicontinuous family, so it has compact closure in $\|\cdot\|_\infty$. But convexity and the bounds are preserved under pointwise limit, so $\mathcal{F}$ is closed. □

For the next two results, we need another theorem on bounds for convex functions on $\mathbb{R}^\nu$.

**Proposition 1.21** *Let $f$ be convex on $U \subset \mathbb{R}^\nu$ and let $K \subset U$ be compact. Then*

$$\sup_{x \in K} |f(x)| \leq C \int_U |f(y)|\, d^\nu y \tag{1.43}$$

*where $C$ depends only on $K$ and $U$ and not on $f$.*

*Remark* In fact, $C$ depends only on $\operatorname{dist}(K, \mathbb{R}^\nu \backslash U)$.

*Proof* Pick $\delta$ so $\cup_{x \in K} B_x^\delta \subset U$. Since

$$f(x) \leq \tfrac{1}{2}[f(x+y) + f(x-y)]$$

if $|y| < \delta$, we have

$$f(x) \le |B_0^\delta|^{-1} \int_{B_x^\delta} |f(y)|\, d^\nu y \le |B_0^\delta|^{-1} \int_U |f(y)|\, d^\nu y \equiv \alpha(f)$$

On the other hand, if $|y| < \delta/2$,

$$\begin{aligned} f(x) &\ge 2f(x+y) - f(x+2y) \\ &\ge 2f(x+y) - \alpha(f) \end{aligned}$$

so

$$\begin{aligned} f(x) &\ge -2|B_0^{\delta/2}|^{-1} \int_{B_x^{\frac{\delta}{2}}} |f(y)|\, d^\nu y - \alpha(f) \\ &\ge -[2^{\nu+1}+1]|B_0^\delta|^{-1} \int_U |f(y)|\, d^\nu y \end{aligned}$$ □

The following result, which seems specialized, is useful in statistical mechanics (see [351]).

**Theorem 1.22** *Let $f_n, f$ be monotone functions on $(a,b) \subset \mathbb{R}$ so that $f_n \to f$ pointwise for a.e. $x \in (a,b)$. Suppose $f_n$ is $C^3$ with $f_n''' > 0$. Then $f$ is $C^1$, $f'$ is convex, $f_n \to f$ uniformly on each $[c,d] \subset (a,b)$, and $f_n' \to f'$ uniformly on each $[c,d] \subset (a,b)$.*

*Remark* One application will involve approximations $f_n = f * j_n$ where $j_n$ is an approximate identity.

*Proof* For any $c, d \subset (a,b)$, for which the limit exists $\int_c^d |f_n'(x)|\, dx = \int_c^d f_n'(x)\, dx = f_n(d) - f_n(c) \to f(d) - f(c)$ so

$$\sup_n \int_c^d |f_n'(x)|\, dx < \infty \tag{1.44}$$

It follows by Proposition 1.21 and Theorem 1.20 that there exists a convex function $g$ so $f_n' \to g$ uniformly on each $[c,d]$ with $[c,d] \subset (a,b)$. Thus, for any $x, y \in (a,b)$, $f_n(x) - f_n(y) = \int_y^x f_n'(z)\, dx \to \int_y^x g(z)\, dz$, so given that $f_n(x_0) \to f(x_0)$ at a single $x_0$, we conclude $f(x_0) + \int_{x_0}^x g(z)\, dz$ is the uniform limit of $f_n(x)$ and $g$ the uniform limit of $f_n'$.

By the presumed a.e. convergence, $f$ is equal a.e. to a $C^1$ function. Since $f$ is monotone, $f$ must be $g$ at all points (since $g(x) = \lim_{x_n \uparrow x} g(x_n) = \lim_{x_n \uparrow x} f(x_n) \le f(x) \le \lim_{x_n \downarrow x} f(x_n) = \lim_{x_n \downarrow x} g(x_n) = g(x)$ for points $x_n$ where $f = g$). Thus, $f$ is $C^1$, $g = f'$ is convex, and we have the claimed uniform convergence. □

**Theorem 1.23** *Let $f_n$ be a sequence of convex functions on U, a convex subset of $\mathbb{R}^\nu$. Let $f_\infty \in L^1(U, d^\nu x)$ and suppose either*

(i) $\int_U |f_n(x) - f_\infty(x)|\, d^\nu x \to 0$ *or*
(ii) $f_n(x) \to f_\infty(x)$ *for each fixed* $x \in U$.

*Then* $f_\infty$ *is convex (in case* (i) *after a possible change on a set of measure zero) and* $f_n \to f_\infty$ *uniformly on compact subsets of* $U$.

*Proof* If (ii) holds, the arguments in Lemma 1.16 show that $|f_n(x)|$ is uniformly bounded in $x$ and $n$ and $x$ runs through any $K \subset U$ which is compact. If (i) holds, (1.43) implies the same uniform boundedness. By (1.42), we have

$$|f_n(x) - f_n(y)| \le C_K |x - y|$$

where $C_K$ is independent of $n$. A simple equicontinuity argument (see [303]) shows the convergence is uniform on $K$. □

The following elementary fact related to limits is often useful.

**Theorem 1.24** *Let* $\{f_\alpha\}_{\alpha \in A}$ *be a family of convex functions on* $U$*, a convex subset of vector space* $V$*. Suppose*

$$f(x) = \sup_\alpha f_\alpha(x)$$

*is finite for every* $x \in U$*. Then* $f$ *is convex.*

*Remarks* 1. Similarly, infs of concave functions are concave.

2. This is often used when the $f_\alpha$'s are linear.

*Proof* If $x, y \in U$ and $\theta \in [0, 1]$, then

$$\begin{aligned} f_\alpha(\theta x + (1-\theta)y) &\le \theta f_\alpha(x) + (1-\theta) f_\alpha(y) \\ &\le \theta f(x) + (1-\theta) f(y) \end{aligned}$$

Taking a sup over $\alpha$ yields convexity of $f$. □

Now, we turn to differentiability where we consider first the one-dimensional case.

**Proposition 1.25** *Let* $F$ *be a convex function on an interval* $I \subset \mathbb{R}$*. Let* $x, y, z, w$ *be four points in* $I$ *with* $x < y \le z < w$*. Then*

$$\frac{F(y) - F(x)}{y - x} \le \frac{F(w) - F(z)}{w - z} \tag{1.45}$$

*Moreover, for* $x < y < w$*,*

$$\frac{F(y) - F(x)}{y - x} \le \frac{F(w) - F(x)}{w - x} \tag{1.46}$$

*and*

$$\frac{F(w) - F(y)}{w - y} \ge \frac{F(w) - F(x)}{w - x} \tag{1.47}$$

*Proof* See Figure 1.2 for the geometry.

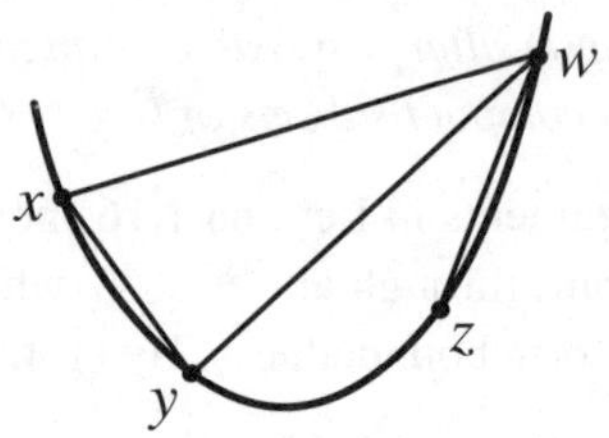


Figure 1.2 Comparing slopes

Suppose first that $y = z$. Write $y$ as a convex combination of $x$ and $w$; explicitly,

$$y = \left(\frac{w-y}{w-x}\right)x + \left(\frac{y-x}{w-x}\right)w$$

This implies

$$\begin{aligned} F(y) &\le \left(\frac{w-y}{w-x}\right)F(x) + \frac{(y-x)}{(w-x)}F(w) \\ &= F(x) + \frac{y-x}{w-x}(F(w)-F(x)) \end{aligned}$$

This implies (1.46). Moreover, it can be rewritten

$$F(y) \le F(x) + \left(\frac{y-x}{w-x}\right)(F(w) - F(y) + F(y) - F(x))$$

Thus,

$$(F(y)-F(x))\left(1 - \frac{y-x}{w-x}\right) \le \left(\frac{y-x}{w-x}\right)(F(w)-F(y))$$

or

$$(F(y)-F(x))\left(\frac{w-y}{w-x}\right) \le \left(\frac{y-x}{w-x}\right)(F(w)-F(y))$$

which implies (1.45) in this case. (1.47) follows from (1.46) if we note that

$$\frac{F(w)-F(x)}{w-x} = \left(\frac{F(w)-F(y)}{w-y}\right)\left(\frac{w-y}{w-x}\right) + \left(\frac{F(y)-F(x)}{x-y}\right)\left(\frac{x-y}{w-x}\right)$$

For the general case of (1.45), use the special case twice to note that

$$\frac{F(y)-F(x)}{y-x} \le \frac{F(z)-F(y)}{z-y} \le \frac{F(w)-F(z)}{w-z}$$

□

**Theorem 1.26** *Let $F$ be a convex function on an interval $I$. Then for any $x$ in the interior of $I$,*

$$(D^{\pm}F)(x) = \lim_{\varepsilon\downarrow 0} \frac{F(x\pm\varepsilon) - F(x)}{\pm\varepsilon} \tag{1.48}$$

*exists. $D^{\pm}F$ are monotone increasing in $x$, that is, $y < x$ implies*

$$D^{-}F(y) \le D^{+}F(y) \le D^{-}F(x) \le D^{+}F(x) \tag{1.49}$$

*with equality of $D^{+}F(y)$ and $D^{-}F(x)$ only if $F$ is affine on $[y, x]$ for $y < x$. Moreover, $D^{\pm}F$ are continuous from the right/left, that is,*

$$\lim_{\varepsilon\downarrow 0} (D^{\pm}F)(x-\varepsilon) = (D^{-}F)(x) \tag{1.50}$$

$$\lim_{\varepsilon\downarrow 0} (D^{\pm}F)(x+\varepsilon) = (D^{+}F)(x) \tag{1.51}$$

*In addition, for any $x$,*

$$(DF^{+})(x) \ge (D^{-}F)(x) \tag{1.52}$$

*with equality at all but a countable set of $x$'s. At points $x$ where equality in (1.52) holds, $F$ is differentiable.*

*Remark* The set at which equality fails in (1.52) is countable. We emphasize this allows it to be empty as it often is (e.g., $f(x) = x^2$ on $\mathbb{R}$).

*Proof* Define for $\varepsilon > 0$ with $x, x\pm\varepsilon \subset I$,

$$(D^{\pm}_{\varepsilon}F)(x) = \frac{F(x\pm\varepsilon) - F(x)}{\pm\varepsilon} \tag{1.53}$$

By Proposition 1.25, we have that for $\varepsilon_1 < \varepsilon_2$,

$$\pm(D^{\pm}_{\varepsilon_1}F)(x) \le \pm(D^{\pm}_{\varepsilon_2}F)(x) \tag{1.54}$$

and for any $\varepsilon, \tilde{\varepsilon} > 0$,

$$(D^{-}_{\varepsilon}F)(x) \le (D^{+}_{\tilde{\varepsilon}}F)(x) \tag{1.55}$$

and if $x < y$ and $\varepsilon < y - x$,

$$(D^{+}_{\varepsilon}F)(x) \le (D^{-}_{\varepsilon}F)(y) \tag{1.56}$$

(1.54) implies the limit in (1.48) is monotone and (1.55) implies that the terms are bounded from below (in plus case) and above (in minus case). Hence the limit (1.48) exists. (1.55)/(1.56) imply (1.49) and (1.52).

To see the $D^{-}$ part of (1.50), note that $(D^{-}F)(x-\varepsilon)$ increases as $\varepsilon \downarrow 0$ (by (1.49)) so $\alpha = \lim_{\varepsilon\downarrow 0}(D^{-}F)(x-\varepsilon)$ exists. By (1.49) again, $\alpha \le D^{-}F(x)$.

Let $y < x$. For $\varepsilon$ small, $y < x - \varepsilon < x$ and

$$(D^{-}F)(x-\varepsilon) \ge \frac{F(x-\varepsilon) - F(y)}{x - y - \varepsilon}$$

Taking $\varepsilon$ to zero, we see

$$\alpha \geq \frac{F(x) - F(y)}{x - y}$$

Now take $y \uparrow x$ and see that $\alpha \geq (D^- F)(x)$, that is, (1.50) holds for $D^-$. To get the $D^+$ result, use (1.49) to see $\lim_{\varepsilon \downarrow 0} (D^+ F)(x - \varepsilon) = \lim_{\varepsilon \downarrow 0} (D^- F)(x - \varepsilon)$. The proof of (1.51) is the same.

Given an interval $[a, b]$ and the bounds (1.49), if $DF^+(x_j) - DF^-(x_j) \geq n^{-1}$ with $x_j \in [a, b]$ and $j = 1, \dots, m$, then $(D^- F)(b) - (D^+ F)(a) \geq mn^{-1}$, so the number $m$ of such $x$'s is bounded. It follows the number of positive jumps is countable so $(D^+ F)(x) = (D^- F)(x)$ except at countably many $x$'s. It is obvious that if equality holds, then $F$ is differentiable at $x$. □

Under some circumstances, convergence of convex functions implies convergence of the derivatives.

**Theorem 1.27** *Let $F_n$ be a sequence of convex functions on some interval $I \subset \mathbb{R}$ and suppose $\lim_{n \to \infty} F_n(x) = F(x)$ exists for each $x \in I$. Then $F$ is convex and for any $x \in I$,*

$$(D^- F)(x) \leq \liminf_n (D^- F_n)(x) \leq \limsup_n (D^+ F_n)(x) \leq (D^+ F)(x) \tag{1.57}$$

*In particular, if $F$ is differentiable at $x$, then*

$$\lim_{n \to \infty} (D^- F_n)(x) = \lim_{n \to \infty} (D^+ F_n)(x) = (DF)(x) \tag{1.58}$$

*Proof* It is immediate that $F$ is convex. (1.57) implies (1.58). We will prove the first inequality in (1.57). The last is similar, and the middle is (1.52).

Fix $\varepsilon > 0$. Then

$$\frac{F_n(x - \varepsilon) - F_n(x)}{(-\varepsilon)} \leq (D^- F_n)(x)$$

Taking $n \to \infty$, we have

$$\frac{F(x - \varepsilon) - F(x)}{(-\varepsilon)} \leq \liminf (D^- F_n)(x)$$

Now take $\varepsilon$ to zero and get the first inequality in (1.57). □

*Remark* This result is especially important in statistical mechanics. It implies convergence of finite volume expectations to an infinite volume limit; see [351].

**Theorem 1.28** *Let $F$ be convex on an open interval $I \subset \mathbb{R}$. Suppose $[a, b] \subset I$. Then*

$$F(b) - F(a) = \int_a^b (D^- F)(x)\, dx = \int_a^b (D^+ F)(x)\, dx \tag{1.59}$$

*Remark* $D^{\pm}F$ are monotone functions and so Riemann integrable.

*Proof* Let $j_\varepsilon$ be an approximate identity. Let $F_\varepsilon = j_\varepsilon * F$. Then $F_\varepsilon$ is convex on $I_\varepsilon = \{x \mid (x-\varepsilon, x+\varepsilon) \subset J\}$. $F_\varepsilon$ is $C^\infty$ and so differentiable. Moreover,

$$\frac{F_\varepsilon(x+\delta) - F_\varepsilon(x)}{\delta} = \int j_\varepsilon(y)\, \frac{F(x-y+\delta) - F(x-y)}{\delta}\, dy$$
$$\to \int j_\varepsilon(y)(D^-F)(x-y)\, dy$$

as $\delta \downarrow 0$ by the monotone convergence theorem. Thus, $DF_\varepsilon = j_\varepsilon * D^{\pm}F$ and so

$$F_\varepsilon(b) - F_\varepsilon(a) = \int_a^b (j_\varepsilon * D^-F)(x)\, dx$$

(1.59) follows by taking $\varepsilon$ to zero. Since $(D^-F)(x)$ is continuous at a.e. $x$, $(j_\varepsilon * D^-F)(x) \to (D^-F)(x)$ for a.e. $x$, and the integral converges by the dominated convergence theorem. □

$D^-F$ is a monotone increasing function, continuous from below. At any point of continuity of $D^-F$, we have that $D^+F(x) = DF^-(x)$. At points, $x_0$, of discontinuity, $D^-F(x_0) = \lim_{\varepsilon\downarrow 0} DF^-(x_0-\varepsilon)$ and $(D^+F)(x_0) = \lim_{\varepsilon\downarrow 0} DF^-(x_0+\varepsilon)$. We can construct a Stieltjes measure, $\mu$, from $DF^-$ in the usual way (see Carothers [68]). Thus, $\mu$ is a measure on $\mathbb{R}$ so

$$\mu([a,b)) = D^-F(b) - D^-F(a) \tag{1.60}$$

and

$$\mu(\{a\}) = D^+F(a) - D^-F(a) \tag{1.61}$$

Combining (1.59) and (1.60), we see that for $x < y$,

$$F(y) - F(x) = (D^+F)(x)(y-x) + \int_x^y (y-z)\, d\mu(z) \tag{1.62}$$

where the integral in (1.62) is interpreted as

$$\int \chi_{(x,y)}(z)(y-z)\, d\mu(z)$$

It is important we take $\chi_{(x,y)}(z)$, not $\chi_{[x,y)}(z)$. If we took $\chi_{[x,y)}$, we would need to use $(D^-F)(x)$, not $(D^+F)(x)$. Since $(y-z)$ vanishes at $z=y$, there is no difference between $\chi_{(x,y)}$ and $\chi_{(x,y]}$.

It follows from (1.60) and (1.59) that $j_\varepsilon * d\mu$ is $d^2(j_\varepsilon * F)/dx^2$, and thus, we have the first half of

**Theorem 1.29** *Let $F$ be a convex function on an open interval $(a,b) \subset \mathbb{R}$. Then the second distributional derivative of $F$ is a (positive) measure. Conversely, if $F$ is*

*a distribution whose second distributional derivative is a (positive) measure, then $F$ is equal (as a distribution) to a convex function.*

*Proof* As noted, we have already proven the first half. To prove the converse, let $F$ be a distribution whose second derivative is a measure $d\mu$. Motivated by (1.62), pick $x_0 \in I$ not a pure point of $\mu$ and define

$$G(x) = \int_{x_0}^{x} (x - z)\, d\mu(z)$$

Then $G$ is a continuous function and $G * j_\varepsilon$ has a positive second derivative, so it is convex, and thus, taking $\varepsilon \downarrow 0$, $G$ is convex.

By construction as a distribution, $G'' = d\mu$ so $(F - G)'' = 0$. Any such distribution is an affine function, $H$, so $F = G + H$ is convex. □

We next want to define tangents and use them to prove the important Jensen's inequality.

**Definition** Let $F$ be a convex function on an open interval $I \subset \mathbb{R}$. Let $x_0 \in I$. A *tangent* to $F$ at $x_0$ is an affine function $G$ so

$$F(x) \geq G(x), \qquad \text{all } x \in I \tag{1.63}$$
$$F(x_0) = G(x_0)$$

Since affine functions have the form $G(x) = G(x_0) + \alpha(x - x_0)$ for some $\alpha$, (1.63) is equivalent to

$$F(x) - F(x_0) \geq \alpha(x - x_0) \tag{1.64}$$

We somewhat abuse the definition and call a linear function, $\alpha$, tangent to $F$ at $x_0$ if (1.64) holds.

**Theorem 1.30** (1.64) *holds if and only if*

$$D^- F(x_0) \leq \alpha \leq D^+ F(x_0) \tag{1.65}$$

*In particular, tangents exist at any point $x_0 \in I$.*

*Proof* If (1.64) holds,

$$\pm D_\varepsilon^\pm F(x_0) \geq \pm\alpha$$

so taking $\varepsilon \downarrow 0$, we obtain (1.65).

Conversely, by (1.54) which implies $\pm D_\varepsilon^\pm F(x) \geq \pm D^\pm F(x)$, we have

$$\begin{aligned} F(x) - F(x_0) &\geq (D^+ F)(x_0)(x - x_0) \qquad \text{if } x_0 < x \\ &\geq (D^- F)(x_0)(x - x_0) \qquad \text{if } x < x_0 \end{aligned}$$

from which (1.64) holds so long as $\alpha$ obeys (1.65). □

There is a converse to this result.

**Theorem 1.31** *Let $F$ be a function defined on an open interval $I \subset \mathbb{R}$. Suppose for each $x_0 \in I$, there exists $\alpha_0$ so that* (1.64) *holds. Then $F$ is convex.*

*Remark* Geometrically, (1.64) says $\{(x, \lambda) \mid \lambda \geq F(x)\}$ is an intersection of half spaces, and so a convex set. Equivalently, $F$ is a sup of linear functions.

*Proof* Let $x, y \in I$ and $\theta \in [0, 1]$. Let $x_0 = \theta x + (1 - \theta)y$. By hypothesis,

$$f(x) - f(x_0) \geq \alpha(x - x_0) \tag{1.66}$$

$$f(y) - f(x_0) \geq \alpha(y - x_0) \tag{1.67}$$

$x - x_0 = (1-\theta)(x-y)$ while $y - x_0 = -\theta(x-y)$. Thus, $\theta(x-x_0)+(1-\theta)(y-x) = 0$. It follows from (1.66)/(1.67) that

$$\theta(f(x) - f(x_0)) + (1 - \theta)(f(y) - f(x_0)) \geq 0$$

which is the convexity statement. □

If $F$ is a convex function on an open interval $(a, b) \subset \mathbb{R}$ ($a$ may be $-\infty$ and/or $b$ may be $\infty$), define

$$A^+(F) = \lim_{x \uparrow b} (D^- F)(x) \tag{1.68}$$

$$A^-(F) = \lim_{x \downarrow a} (D^- F)(x) \tag{1.69}$$

The limits exist, but may be $+\infty$ for $A^+$ and $-\infty$ for $A^-$, since $D^- F$ is monotone. Then:

**Proposition 1.32** *Let $F$ be a convex function on an open interval $I \subset \mathbb{R}$. For any $\alpha \in (A^-(F), A^+(F))$, there exists some $x_0 \in I$ so that* (1.64) *holds. The set of $x_0$ for which this is true is a closed interval $[x_0^-(\alpha), x_0^+(\alpha)]$. For all but a countable set of $\alpha$'s, $x_0$ is unique, that is, $x_0^-(\alpha) = x_0^+(\alpha)$, $x_0^-(\alpha) < x_0^+(\alpha)$ if and only if*

$$F(y) = F(x_0^-(\alpha)) + \alpha(y - x_0^-(\alpha)) \tag{1.70}$$

*for all $y \in [x_0^-(\alpha), x_0^+(\alpha)]$.*

*As $\alpha$ increases, $x_0(\alpha)$ increases, that is, if $\alpha_1 > \alpha_0$ and $x_1, x_0$ are points where* (1.64) *holds for $\alpha_1$ and $\alpha_0$, then $x_1 \geq x_0$. If equality holds, that is, $x_1 = x_0$ for $\alpha_1 > \alpha_0$, then $D^- F(x_0) < D^+ F(x_0)$. Explicitly,*

$$\{\alpha \mid x(\alpha) = x_0\} = [D^- F(x_0), D^+ F(x_0)] \tag{1.71}$$

*Proof* $(D^- F)(x)$ is a monotone function that runs from $A^-(F)$ up to $A^+(F)$ with some possible discontinuities. At any point, by (1.50)/(1.51), $\lim_{\varepsilon \downarrow 0} D^- F(x \pm \varepsilon) = D^\pm F(x)$. It follows that for any $\alpha \in (A^-(F), A^+(F))$, either $\alpha = DF^-(x_0)$

for some $x_0$ where $F$ is differentiable or $\alpha \in [DF^-(x_0), DF^+(x_0)]$ for some $x_0$. In either event, (1.64) holds.

If there are two $x$'s, say $x_0$ and $x_1$, for which (1.64) holds, for a given $\alpha$, let

$$G(x) = F(x) - F(x_0) - \alpha(x - x_0)$$

Then, by assumption, $G(x) \geq 0$. Moreover, $G(x) - G(x_1) \geq 0$, so $0 = G(x_0) \geq G(x_1)$ so $G(x_1) = 0$. By convexity, $G(x) \leq 0$ on $[x_0, x_1]$, so by the positivity, $G(x) = 0$ on $[x_0, x_1]$, that is, $F(x)$ is affine on $[x_0, x_1]$ and (1.64) holds for any $x_2 \in [x_0, x_1]$. Thus, the set of $x$'s is an interval as claimed. By continuity, it is a closed interval.

Since $F$ is linear on $[x_0, x_1]$, $\alpha$ is the unique tangent value for $x \in (x_0, x_1)$. It follows if $\alpha_1, \alpha_2$ are two different values of $\alpha$ for which $x$ is nonunique, $[x_0^-(\alpha_1), x_0^+(\alpha_2)]$ and $[x_0^-(\alpha_2), x_0^+(\alpha_2)]$ overlap at most in an endpoint. Thus, if $\alpha_1, \dots, \alpha_n, \dots$ are points with nonunique $x$'s,

$$\sum_{i=1}^{n} |x_0^+(\alpha_1) - x_0^-(\alpha_1)| \leq A^+(F) - A^-(F)$$

and thus, there are at most countable such $\alpha$'s.

Monotonicity of $x_0(\alpha)$ follows from monotonicity of $D^-F$. (1.71) is a restatement of Theorem 1.30 and it implies the claim $D^-F(x_0) < D^+F(x_0)$ at points with $x_0(\alpha_1) = x_0(\alpha_2) = x_0$ for some $\alpha_1 \neq \alpha_2$. $\square$

Thus, $\alpha \to x_0^-(\alpha)$ is an inverse to $x \to (D^-F)(x)$ with similar monotonicity and continuity properties to $D^-F$. This suggests there is a convex function $G(\alpha)$ with $(D^-G)(\alpha) = x_0^-(\alpha)$. We will see there is, and study its properties later in this chapter and in the next chapter when we discuss conjugate convex functions.

Theorem 1.30 lets us prove the following very useful result:

**Theorem 1.33** (Jensen's Inequality) *Let $F$ be a convex function on an open interval $I \subset \mathbb{R}$. Let $f\colon M \to I$ be a real-valued function, and let $\mu$ be a probability measure on $M$ (i.e., $\mu(M) = 1$). Suppose that*

$$\int |f(x)|\, d\mu(x) < \infty \tag{1.72}$$

*Then*

$$F\biggl(\int f(x)\, d\mu(x)\biggr) \leq \int F(f(x))\, d\mu(x) \tag{1.73}$$

*Remarks* 1. It is (1.73) that is known as Jensen's inequality. The special case $F(y) = e^y$, that is,

$$\int f(x)\, d\mu(x) \leq \log \int e^{f(x)}\, d\mu(x) \tag{1.74}$$

is also sometimes called Jensen's inequality.

2. (1.73) is intended in the sense that either $\int |F(f(x))|\, d\mu(x) < \infty$ or the interval diverges to $+\infty$.

3. If $d\mu$ is the point measure $\mu = \sum_{j=1}^{n} \theta_j \delta_{x_j}$ and $f(x) = x$, (1.73) is just (1.3). Jensen's inequality can be viewed as a limit of (1.3).

*Proof* Let $\lambda_0 = \int f(x)\, d\mu(x)$. Since $\mu$ is a probability measure and $f[M] \subset I$, $\lambda_0 \in I$, so by Theorem 1.30, $F(\lambda) - F(\lambda_0) \geq \alpha(\lambda - \lambda_0)$ for some $\alpha$. In particular, for each $x$,

$$F(f(x)) \geq F(\lambda_0) + \alpha(f(x) - \lambda_0) \tag{1.75}$$

so $x \mapsto F(f(x))$ is bounded below by an $L^1$ function. It follows that either $F(f(x)) \in L^1$ or $\int F(f(\lambda))\, d\mu = \infty$. In the former case, integrate (1.75) with respect to $d\mu(x)$ and use $\int (f(x) - \lambda_0)\, d\mu(x) = 0$ to obtain (1.73). □

**Corollary 1.34** *Let $d\mu$ be a finite measure. Then $F \mapsto \log \int \exp(F)\, d\mu = \mathcal{F}(F)$ (for $F \in L^\infty$) is a convex function.*

*Remark* Given Proposition 1.14, this is the third proof of Hölder's inequality.

*Proof* Let $G(t) = \mathcal{F}(F_0 + tF_1)$. We need to show that $G$ is convex. Note that

$$G(t) - G(t_0) = \log\left[\int \exp((t - t_0)F_1)\, d\nu\right] \tag{1.76}$$

where $d\nu = \exp(F_0 + t_0 F_1)\, d\mu / \int \exp(F_0 + tF_1)\, d\nu$ is a probability measure. Thus, by Jensen's inequality in the form (1.74),

$$\begin{aligned} G(t) - G(t_0) &\geq \log\left[\exp\left(\int (t - t_0)F_1\, d\nu\right)\right] \\ &= (t - t_0) \int F_1\, d\nu \end{aligned}$$

Thus, $G(t)$ has a tangent at every point, so $G(t)$ is convex by Theorem 1.31. □

Next, we turn to the existence of tangents in $\mathbb{R}^\nu$ (or more general spaces).

**Definition** A convex set $K \subset V$, a vector space is called *pseudo-open* if and only if for each $x \in K$ and $v \in V$, $\{\lambda \mid x + \lambda v \in K\}$ contains an open interval about $\lambda = 0$. If $\dim(V) < \infty$, it is easy to see that a convex set is pseudo-open if and only if it is open.

Just as one-dimensional convex functions have one-sided derivatives, we have

**Theorem 1.35** *Let $F$ be a convex function on a pseudo-open convex set $K \subset V$, a vector space. Then for each $x \in K$ and $\mathrm{e} \in V$, the directional derivative*

$$(D^{\mathrm{e}} F)(x) = \lim_{\lambda \downarrow 0} \frac{(F(x + \lambda \mathrm{e}) - F(x))}{\lambda} \tag{1.77}$$

*exists.*

*Proof* Let $g(\lambda) = F(x + \lambda e)$ for $\lambda \in \{\lambda \mid x + \lambda e \in K\}$. Then $g$ is a convex function on an open interval containing zero, so $(D^+ g)(\lambda)$ exists. But this is the limit in (1.77). □

The existence of tangents we are aiming towards depends on an extension result that is the heart of the Hahn–Banach theorem:

**Lemma 1.36** *Let $F$ be a convex function defined on a pseudo-open convex subset, $K$, of a vector space $V$. Suppose $0 \in K$. Let $W \subset V$ be a subspace and $v_0 \in V \backslash W$. Let $\ell$ be a linear function on $W$ that obeys*

$$F(w) - F(0) \geq \ell(w) \tag{1.78}$$

*for all $w \in W \cap K$. Then there exists a linear function $L$ on $W + [v_0] = \{w + \lambda v_0 \mid w \in W, \lambda \in \mathbb{R}\}$ so that*

$$F(x) - F(0) \geq L(x) \tag{1.79}$$

*for all $x \in (W + [v_0]) \cap K$ and so*

$$L(w) = \ell(w) \tag{1.80}$$

*for all $w \in W$.*

*Proof* We claim that for any $w, \tilde{w} \in W$ and $\lambda, \tilde{\lambda} > 0$ with $w + \lambda v_0 \in K$, $\tilde{w} - \tilde{\lambda} v_0 \in K$, we have

$$\frac{F(w + \lambda v_0) - F(0) - \ell(w)}{\lambda} \geq -\frac{F(\tilde{w} - \tilde{\lambda} v_0) - F(0) - \ell(\tilde{w})}{\tilde{\lambda}} \tag{1.81}$$

for (1.81) is equivalent to

$$\frac{\tilde{\lambda}}{\lambda + \tilde{\lambda}} F(w + \lambda v_0) + \frac{\lambda}{\lambda + \tilde{\lambda}} F(w - \tilde{\lambda} v_0) - F(0) - \ell(\eta) \geq 0 \tag{1.82}$$

where

$$\eta = \frac{\tilde{\lambda}}{\lambda + \tilde{\lambda}} w + \frac{\lambda}{\lambda + \tilde{\lambda}} \tilde{w} \tag{1.83}$$

Recognizing that

$$\frac{\tilde{\lambda}}{\lambda + \tilde{\lambda}} + \frac{\lambda}{\lambda + \tilde{\lambda}} = 1$$

and

$$\frac{\tilde{\lambda}}{\lambda + \tilde{\lambda}} (w + \lambda v_0) + \frac{\lambda}{\lambda + \tilde{\lambda}} (w - \tilde{\lambda} v_0) = \eta$$

we see, by the concavity of $F$, that

$$\text{LHS of (1.82)} \geq F(\eta) - F(0) - \ell(\eta)$$

which is nonnegative by hypothesis (1.78). Thus, (1.82) and so (1.81) is true.

(1.81) implies that as $(\lambda, w)$ runs through $\{(\lambda, w) \in (0, \infty) \times W \mid \lambda v_0 + w \in K\}$, the left side is bounded from below, and similarly, the right side is bounded from above. Thus, we can pick $\alpha$ so

$$\inf_{\substack{w\in W,\, \lambda>0 \\ w+\lambda v_0 \in K}} \frac{F(w+\lambda v_0) - F(0) - \ell(w)}{\lambda} \geq \alpha \geq \sup_{\substack{w\in W,\, \lambda>0 \\ w-\lambda v_0 \in K}} -\frac{F(w-\lambda v_0) - F(0) - \ell(w)}{\lambda} \tag{1.84}$$

Define $L$ on $W + [v_0]$ by

$$L(w + \lambda v_0) = \ell(w) + \lambda\alpha \tag{1.85}$$

Clearly, (1.80) holds and (1.84) implies (1.79). □

**Theorem 1.37** *Let $F$ be a convex function on a pseudo-open convex subset $K$ of a vector space, $V$. Let $x_0 \in K$. Then there is a linear function $\ell$ on $V$ so that*

$$F(x) - F(x_0) \geq \ell(x) - \ell(x_0) \tag{1.86}$$

*for all $x \in K$.*

*Proof* If $\dim(V) < \infty$, this is a simple induction starting with $w = 0$, using Lemma 1.36. For the infinite-dimensional case, we need to use Zorn's lemma as follows.

Consider pairs $p = (W, \ell)$ of linear functionals defined on a subspace $W$ of $V$ obeying (1.86). Write $p \subset p' = (W', \ell')$ if $W \subset W'$ and $\ell' \restriction W = \ell$. By Zorn's lemma, we can find a maximal chain in this order $\{p_\alpha\}_{\alpha\in I}$. Take $W_\infty = \cup_\alpha W_\alpha$ and define $\ell_\infty$ on $W_\infty$ by $\ell_\infty \restriction W_\alpha = \ell_\alpha$. This is possible because the $p_\alpha$'s are linearly ordered which implies $\ell_\infty$ is well defined.

If $W_\infty$ is not all of $V$, by Lemma 1.36, we can extend $\ell_\infty$ to one dimension more and so find $\tilde{W} \supsetneqq W_\infty$ and $\tilde{p} \supsetneqq p_\infty$, violating the maximality of the chain. Thus, $W_\infty = V$. □

The exact same proof shows:

**Theorem 1.38** (The Hahn–Banach Theorem) *Let $F$ be a convex function on a pseudo-open convex subset $K$ of a vector space $V$ with $0 \in K$. Suppose $W$ is a subspace of $V$, $\ell$ a linear functional on $W$ so that* (1.78) *holds for all $x \in W \cap K$. Then there is a linear functional $L$ on $V$ so* (1.79) *holds for all $x \in K$.*

In the next result, we demand $V$ be a normed space to be able to define integrals without a lot of machinery.

Given our proof of Jensen's inequality, we immediately have:

**Theorem 1.39** *Let $(M, d\mu)$ be a probability measure space. Let $V$ be a normed vector space and $K \subset V$ an open convex set. Let $f\colon M \to K$ be measurable and $F\colon K \to \mathbb{R}$ be convex. Suppose $\int \|f(x)\|\, d\mu(x) < \infty$. Then*

$$F\Big(\int f(x)\, d\mu(x)\Big) \le \int F(f(x))\, d\mu(x) \tag{1.87}$$

Next, we look at the issue of uniqueness of tangents:

**Lemma 1.40** *Let $K$ be an open convex subset of $\mathbb{R}^\nu$ and $F\colon K \to \mathbb{R}$ a convex function. Suppose $x_0 \in K$ and for $i = 1, \dots, n$, $t \mapsto F(x_0 + t\delta_i)$ is differentiable at $t = 0$ where $\delta_i$ is the vector in $\mathbb{R}^\nu$ with $(\delta_i)_j = \delta_{ij}$. Then there is a unique $\ell \in \mathbb{R}^\nu$ so*

$$F(x) - F(x_0) \ge \ell \cdot (x - x_0) \tag{1.88}$$

*Proof* If (1.88) holds, then

$$\ell_i = \frac{d}{dt} F(x_0 + t\delta_i)\Big|_{t=0}$$

so $\ell$ is uniquely determined. Existence is Theorem 1.37. □

*Remarks* 1. One can prove using Lemma 1.36 that if for some $i$, $(D^+F)(x_0 + 0\delta_i) \ne (D^-F)(x_0 + 0\delta_i)$, then there are multiple $\ell$'s obeying (1.88).

2. If $\ell$ is unique, then $\ell$ is the gradient of $F$ in the classical sense that $F(x) = F(x_0) + \ell(x - x_0) + o(|x - x_0)|)$.

**Theorem 1.41** *Let $K$ be an open convex subset of $\mathbb{R}^\nu$ and $F\colon K \to \mathbb{R}$ a convex function. Then for almost every $x_0 \in K$, there is a unique $\ell_{x_0} \in \mathbb{R}^\nu$ so that* (1.88) *holds.*

*Proof* For each $x_0$, $t \mapsto F(x_0 + t\delta_i)$ is differentiable in $t$ for all $t$ except for a countable set, and so for almost every $t$. By Fubini's theorem, $t \mapsto F(x_0 + t\delta_i)$ is differentiable at $t = 0$ for a.e. $x_0 \in K$. Thus, this holds for all $i = 1, \dots, n$ and a.e. $x_0$. By the lemma, $\ell$ is unique at a.e. $x_0$. □

*Remark* In fact, the set of $x$ where $F$ has multiple derivatives has codimension at least 1. See the discussion in the Notes.

There is also a result in the infinite-dimensional case. This will freely use the theory of dense $G_\delta$'s discussed, for example, in Oxtoby [281].

**Theorem 1.42** *Let $K$ be an open subset of a separable Banach space, $V$. Let $F$ be a continuous, convex function on $K$. Then $\{x \in K \mid$ there is a unique $\ell \in V^*$ tangent to $F$ at $x\}$ is a dense $G_\delta$.*

*Proof* If $y \in V$ and $(D^y F)(x_0)$ is the directional derivative (1.77), then $\ell$ is tangent at $x_0$ implies

$$\ell(y) \le (D^y F)(x_0) \tag{1.89}$$

for all $y$. Thus, $\ell(y)$ is uniquely determined if

$$(D^y F)(x_0) \equiv -(D^{-y} F)(x_0) \tag{1.90}$$

and so $\ell \in V^*$ is uniquely determined if (1.90) holds for a dense set of $y$ in $V$.

For $y \in V$, define

$$(\delta_y^\varepsilon F)(x) = (F(x+\varepsilon y) + F(x-\varepsilon y) - 2F(x))\varepsilon^{-1}$$

Then by convexity, $(\delta_y^\varepsilon F)(x) \ge 0$ and by Theorem 1.26, $\delta_y^\varepsilon$ is monotone in $\varepsilon$, and by (1.77),

$$\lim_{\varepsilon \downarrow 0} (\delta_y^\varepsilon F)(x) = (D^y F)(x) + (D^{-y} F)(x)$$

Thus, (1.90) holds if and only if

$$\forall n \, \exists m \, (\delta_y^{1/m} F)(x_0) < n^{-1}$$

so for each fixed $y$,

$$\begin{aligned} U_y &\equiv \{x_0 \mid (D^y F)(x_0) = -(D^y f)(x_0)\} \\ &= \bigcap_n \bigcup_m \{x_0 \mid (\delta_y^{1/m} F)(x_0) < n^{-1}\} \end{aligned}$$

is a $G_\delta$.

By Theorem 1.31, for any $x_0$, $U_y \cap \{x_0 + \alpha y \in K\}$ is $\{x_0 + \alpha y \in K\}$ except for a countable set of $\alpha$'s, and, in particular, there is $\alpha_n \to 0$ so $x_0 + \alpha_n y \in K \cap U_y$. Thus, $U_y$ is a dense $G_\delta$.

Let $\tilde{Y}$ be a countable, dense set in $X$. Then

$$\{x_0 \in K \mid \ell \in V^* \text{ tangent to } F \text{ at } x_0 \text{ is unique}\} = \bigcap_{y \in X} U_y$$

is a dense $G_\delta$. □

**Example 1.43** A tangent plane to $B$, the unit ball of a Banach space, $X$, at $x_0$ with $\|x_0\| = 1$ is an $\ell \in X^*$ with $B \subset \{x \mid \ell(x) \le 1\}$ and $\ell(x_0) = 1$, that is, an $\ell \in X^*$ with

$$\|\ell\| = 1, \qquad \ell(x_0) = 1 \tag{1.91}$$

It is easy to see that if $x_0 \neq 0$, tangents to the $F(x) = \frac{1}{2}\|x\|^2$ at $x_0$ are precisely those $L \in X^Y$ of the form $L = \|x_0\|\ell$ with $\ell$ a tangent plane to $B$ at $x_0/\|x_0\|$. Thus, Theorem 1.42 implies

$$\{x \mid \|x\| = 1, \text{ there is a unique } \ell \text{ in (1.91)}\}$$

is a dense $G_\delta$ in $\{x \mid \|x\| = 1\}$. □

We generalize the language of this last example. Let $V$ be any vector space. Let $K$ be a convex, absorbing subset in $V$ and let $g_K$ be its gauge. Let $x_0$ be such that $g_K(x_0) = 1$ (so for all small $\varepsilon > 0$, $(1-\varepsilon)x_0 \in K$ but $(1+\varepsilon)x_0 \notin K$). A linear functional $\ell$ on $V$ is called *tangent* to $K$ at $x_0$ if and only if $\ell(x_0) = 1$ and $\ell(x) \le g_K(x)$ (so $K \subset \{x \mid \ell(x) \le 1\}$). Geometrically, the plane $\{x \mid \ell(x) = 1\}$ is tangent to $K$ in the sense that $K$ lies on one side of the plane. If $V = \mathbb{R}^\nu$ and $\partial K$ is a smooth manifold, such planes are tangent in the usual intuitive sense.

**Theorem 1.44** *Let $K$ be a convex, absorbing subset in $V$ and $x_0$ a point with $g_K(x_0) = 1$ where $g_K$ is the gauge of $g_K$. Then there is a tangent to $K$ at $x_0$.*

*Proof* Let $W = \{\lambda x_0 \mid \lambda \in \mathbb{R}\}$. Define $\ell$ on $W$ by $\ell(\lambda x_0) = \lambda$. Since for $\lambda \ge 0$, $g_K(\lambda x_0) = \lambda$, and for $\lambda < 0$, $g_K(\lambda x_0) \ge 0 > \lambda$, we have $\ell \le g_K$ on $W$. Thus, by the Hahn–Banach theorem (Theorem 1.38), there is $L$ on $V$ with $L \le g_K$ and $L(x_0) = \ell(x_0) = 1$. □

Legendre transforms enter often in physics, for example, in the shift of Lagrangians to Hamiltonians and the relation of free energy to entropy. The standard physicists' definition, given a real-valued function $F$ on $\mathbb{R}^\nu$ and $y \in \mathbb{R}^\nu$, find $x_0 \in \mathbb{R}^\nu$ so

$$y_i = \frac{\partial F}{\partial x_i}(x_0) \tag{1.92}$$

and then

$$F^*(y) = x_0 \cdot y - F(x_0) \tag{1.93}$$

Notice that (1.92) is equivalent to $\nabla_x(x \cdot y - F(x))|_{x=x_0} = 0$, and this $x_0$ is an extreme point of $x \cdot y - F(x)$. This motivates the definition below (see (1.95)).

**Definition** A convex function, $F$, on $\mathbb{R}^\nu$ is called *steep* if and only if

$$\lim_{|x|\to\infty} \frac{F(x)}{|x|} = \infty \tag{1.94}$$

*Remark* Sometimes $F$ is called *accretive* if (1.94) holds.

The steepness condition will imply the Legendre transform is everywhere finite. One can extend the theory to infinite dimensions and to cases where $F$ and/or $F^*$ are only defined on suitable convex subsets. Since our primary interest is the one-dimensional case (in the next chapter), we restrict here to the steep case; see the discussion in Chapter 5 for the general case.

**Definition** Let $F$ be a steep convex function on $\mathbb{R}^\nu$. We define the *Legendre transform*, $F^*$, on $\mathbb{R}^\nu$ by

$$F^*(y) = \sup_{x\in\mathbb{R}^\nu} [x \cdot y - F(x)] \tag{1.95}$$

In the one-dimensional case, if $G$ is a convex function obeying

$$G(0) = 0, \qquad G(x) = G(-x) \tag{1.96}$$

then, as we will show, $G^*$ also obeys (1.96) and is also called the *conjugate convex function* to $G$.

**Theorem 1.45** *Let $F$ be a steep convex function on $\mathbb{R}^\nu$. For each $y \in \mathbb{R}^\nu$, there is an $x_0(y) \in \mathbb{R}^\nu$ so that $y$ is a tangent for $F$ at $x_0$, that is,*

$$F(x) - F(x_0(y)) \geq y \cdot (x - x_0(y)) \tag{1.97}$$

*$F^*$ is given by*

$$F^*(y) = x_0(y) \cdot y - F(x_0(y)) \tag{1.98}$$

*and, in particular, $F^*(y) < \infty$. $F^*$ is a steep convex function and $(F^*)^*(x) = F(x)$.*

*Remark* By (1.95), $F^*$ and $F$ obey

$$x \cdot y \leq F(x) + F^*(y) \tag{1.99}$$

(sometimes called Young's inequality). By (1.98), for each $y$, there is an $x_0(y)$ where equality holds, and for each $x$, there is a $y_0(x)$ where equality holds.

*Proof* Given $y$, by the steepness hypothesis, $F(x) \geq (\|y\| + 1)\|x\| - C$ for some constant $C$. Thus,

$$x \cdot y - F(x) \leq -\|x\| + C$$

and so

$$x \cdot y - F(x) < -F(0)$$

if $\|x\| > C + F(0)$. Thus, $\sup(x \cdot y - F(y))$ occurs somewhere on the closed ball $\{x \mid \|x\| \leq C + F(0)\}$. Since $x \cdot y - F(x)$ is continuous, the sup actually occurs at some point $x_0(y)$, that is, for all $x$,

$$x \cdot y - F(x) \leq x_0 \cdot y - F(x_0) \equiv F^*(y)$$

so (1.97) holds.

Each function $y \mapsto x \cdot y - F(x)$ is affine and a sup of affine functions is convex, so $F^*$ is convex. Moreover, by (1.99), with $x = \lambda y/\|y\|$, for some $\lambda > 0$,

$$F^*(y) \geq \lambda\|y\| - F\left(\frac{\lambda y}{\|y\|}\right)$$

which shows that

$$\liminf_{\|y\|\to\infty}\left(\frac{F^*(y)}{\|y\|}\right)\geq\lambda$$

Since $\lambda$ is arbitrary, $F^*$ is steep.

Finally, we need to show that $(F^*)^* = F$. Since

$$x\cdot y\leq F(x)+F^*(y)$$

for all $x, y$,

$$F(x)\geq x\cdot y-F^*(y)$$

for all $y$, so

$$F(x)\geq (F^*)^*(x) \tag{1.100}$$

On the other hand, given $x_0$, let $y_0$ be a tangent to $F$ at $x_0$. Then

$$F(x)-F(x_0)\geq y_0\cdot(x-x_0)$$

so $x\cdot y_0 - F(x)$ is maximized at $x_0$, that is,

$$F^*(y_0)=x_0\cdot y_0-F(x_0)$$

Thus,

$$\begin{aligned}F(x_0)&=x_0\cdot y_0-F^*(y_0)\\&\leq (F^*)^*(x_0)\end{aligned}$$

so with (1.100), we conclude that $F = (F^*)^*$. □

In the next chapter, we discuss one-dimensional Legendre transforms further, and in Chapter 5, we will discuss infinite-dimensional Legendre transforms as well as Legendre transforms when $F$ is not steep.

# 2
# Orlicz spaces

In this chapter, we study the Orlicz space introduced in Example 1.12.

**Definition** A *weak Young function* is a convex function on $\mathbb{R}$ that obeys
(i) $F(x) = F(-x)$
(ii) $F(x) = 0$ if and only if $x = 0$.

Since $F(0) = 0 \leq \frac{1}{2}(F(x) + F(-x)) = F(x)$, $F$ is strictly positive on $\mathbb{R}\backslash\{0\}$. Moreover, $F$ is monotone on $[0, \infty)$; indeed, since $x = \frac{x}{y}\, y + (1 - \frac{x}{y})0$, we have

$$0 \leq x \leq y \Rightarrow F(x) \leq \frac{x}{y}\, F(y) \tag{2.1}$$

so $F(x)/x$ is monotone.

**Definition** A *Young function* is a weak Young function that also obeys
(iii)

$$\lim_{x\to\infty} \frac{F(x)}{x} = \infty \tag{2.2}$$

(iv)

$$\lim_{x\downarrow 0} \frac{F(x)}{x} = 0 \tag{2.3}$$

Much of the theory of Orlicz spaces works for weak Young functions, but the duality theory requires Young functions.

**Example 2.1** $F(x) = |x|^p$ is a weak Young function if $p = 1$ and a Young function if $1 < p < \infty$. The function

$$F(x) = \exp(|x|) - 1 - |x| = \sum_{n=2}^{\infty} \frac{|x|^n}{n!} \tag{2.4}$$

is a Young function, as is

$$F(x) = (|x| + 1)\log(|x| + 1) - |x| \tag{2.5}$$

Throughout this chapter, $(M, d\mu)$ will be a probability measure space, that is, $\mu(M) = 1$. Given any weak Young function and $F: M \to \mathbb{C}$, we define

$$Q_F(f) = \int F(|f(x)|)\, d\mu(x) \tag{2.6}$$

which may be $+\infty$.

We defined the Orlicz space $L^{(F)}(M, d\mu)$ to be (equivalence classes of a.e. equal) functions, $f$, with

$$Q_F(\alpha f) < \infty, \qquad \text{for some } \alpha > 0 \tag{2.7}$$

and (Luxemburg norm)

$$\|f\|_F = \inf\{\lambda \mid Q_F(\lambda^{-1} f) \le 1\} \tag{2.8}$$

**Proposition 2.2** *$L^{(F)}(M, d\mu)$ with $\|\cdot\|_F$ is a complete space and*

$$Q_F\left(\frac{f}{\|f\|_F}\right) \le 1 \tag{2.9}$$

*Moreover, $L^\infty(M, d\mu) \subset L^{(F)}(M, d\mu) \subset L^1(M, d\mu)$ and*

$$\alpha\|f\|_1 \le \|f\|_F \le \beta\|f\|_\infty \tag{2.10}$$

*where*

$$\beta = [\sup\{y \mid F(y) \le 1\}]^{-1} \tag{2.11}$$

*and*

$$\alpha = [\inf\{y(1 + F(y)^{-1}) \mid y > 0\}]^{-1} \tag{2.12}$$

*Remark* One might guess that equality always holds in (2.9). Remarkably, we will see that for certain $F$, this is false!

*Proof* Let $\alpha_n = (1 - 1/n)$. Then $\lambda_n \equiv \|f\|_F \alpha_n^{-1} > \|f\|_F$, so by (2.8), $Q_F(\lambda_n^{-1} f) \le 1$, that is, $Q_F(\alpha_n f/\|f\|_F) \le 1$. By the monotone convergence theorem and continuity of $F$, $\lim_{\alpha \to \infty} Q_F(\alpha_n f/\|f\|_F) = Q_F(f/\|f\|_F)$. This proves (2.9).

Now suppose $f \in L^\infty$ and $F(y) \le 1$, then $F(f(x)y/\|f\|_\infty) \le 1$ for all $x$ and so, since $\mu(M) = 1$, $Q_F(yf/\|f\|_\infty) \le 1$ so $\|f\|_F \le y^{-1}\|f\|_\infty$. Thus, $L^\infty \subset L^{(F)}$ and $\|f\|_F \le \beta\|f\|_\infty$. By (2.1), for any $y > 0$,

$$\begin{aligned} Q_F(f) &\ge F(y)y^{-1} \int_{\{m \mid f(m) \ge y\}} |f(m)|\, d\mu(m) \\ &\ge F(y)y^{-1}[\|f\|_1 - y] \end{aligned}$$

By (2.9), we see that for any $y$,

$$F(y)y^{-1}\left[\frac{\|f\|_1}{\|f\|_F} - y\right] \leq 1$$

or

$$\|f\|_1 \leq (F(y)^{-1}y + y)\|f\|_F$$

which shows $\alpha\|f\|_1 \leq \|f\|_F$ so (2.10) is proven.

Finally, we turn to completeness, an argument patterned after the original Riesz–Fischer proof of completeness of $L^2$. Suppose $f_n$ is Cauchy in $\|\cdot\|_F$. By passing to a subsequence and considering $f_n - f_1$, we can suppose $\|f_{n-1} - f_n\|_F \leq 2^{-n}$ and $f_1 = 0$. If we show this subsequence converges to some $f \in L^{(F)}$, then the original sequence converges also.

Let $g_n = \sum_{j=1}^{n} |f_{j+1} - f_1|$ so $\|g_n\|_F < 1$, and thus, $Q(g_n) \leq 1$. Let $g_\infty = \sum_{j=1}^{\infty} |f_{j+1} - f_j|$ so $g_n$ converges monotonically to $g_\infty$. By the monotone convergence theorem (and continuity and monotonicity of $F$), $Q(g_\infty) \leq 1$ so $g_\infty \in L^{(F)}$ and $\|g_\infty\|_F \leq 1$. By the same argument,

$$\|g_\infty - g_{n-1}\|_F \leq 2^{1-n} \tag{2.13}$$

For, if $m \geq n$, $\|g_m - g_{n-1}\|_F \leq 2^{1-n}$ so $Q_F((g_m - g_{n-1})/2^{1-n}) < 1$ and $Q_F((g_\infty - g_{n-1})/2^{1-n}) \leq 1$ by the monotone convergence theorem.

Since $g_\infty(m) < \infty$ a.e., the series $\sum_{j=1}^{n}(f_{j+1} - f_j) = f_{n+1}$ (since $f_1 = 0$) converges absolutely, and thus, converges to a function $f_\infty$. Moreover, $|f_\infty - f_n| \leq |g_\infty - g_{n-1}|$ pointwise so $Q_F(\lambda^{-1}|f_\infty - f_n|) \leq Q_F(\lambda^{-1}|g_\infty - g_{n-1}|)$ by monotonicity of $F$. Thus, by (2.13), $\|f_\infty - f_n\|_F \leq 2^{1-n}$ so $f_n \to f_\infty$ in $L^{(F)}$, proving completeness. □

It turns out that Orlicz spaces have many properties so long as $F$ obeys an additional condition.

**Definition** Let $F$ be a weak Young function. We say that $F$ obeys the $\Delta_2$ condition if and only if there exist $x_0$ and $C$ so that

$$F(2x) \leq C\,F(x), \qquad \text{all } x \geq x_0 \tag{2.14}$$

**Example 2.3** $F(x) = |x|^p$ obeys the $\Delta_2$ condition since $F(2x) = 2^p F(x)$. $F(x) = \exp(|x|) - |x| - 1 = \sum_{n=2}^{\infty} |x|^n/n!$ is a Young function for which the $\Delta_2$ condition is *not* obeyed. Indeed, if $F$ obeys the $\Delta_2$ condition, for large $n$, $F(2^n) \leq DC^n$ for suitable $D$. Thus, if $2^{n-1} \leq x \leq 2^n$, we have, with $\alpha = \log C/\log 2$,

$$\begin{aligned} F(x) \leq F(2^n) &\leq DC\,\exp((n-1)\log C) \\ &\leq DC\,\exp(\alpha(\log 2)(n-1)) \\ &= DC\,2^{\alpha(n-1)} \leq DC\,x^\alpha \end{aligned}$$

so that the $\Delta_2$ condition implies that $F$ is polynomially bounded. Thus, more generally, if $F(x) = \exp(x^\gamma)$ for any $\gamma > 0$ for large $x$, then $\Delta_2$ fails. $F(x) = (|x|+1)\log(|x|+1) - |x|$ and, more generally, functions equal to $x^p \log x$ for large $x$ all obey the $\Delta_2$ condition. For these last functions, use (v) of the proposition below. □

**Proposition 2.4** *The following are equivalent:*

(i) *$F$ obeys the $\Delta_2$ condition.*

(ii)

$$\limsup_{x\to\infty} \frac{F(2x)}{F(x)} < \infty \tag{2.15}$$

(iii) *For any $k > 1$ and $\varepsilon > 0$, there is a $B$ (depending on $k$ and $\varepsilon$) so that for all $x$*

$$F(kx) \le B\,F(x) + \varepsilon \tag{2.16}$$

(iv) *For some $k > 1$ and $\varepsilon > 0$, there is a $B$ so that* (2.16) *holds for all $x$.*

(v)

$$\sup_{x\ge 1} \frac{x(D^- F)(x)}{F(x)} < \infty$$

*Moreover, if $F$ is such that $(D^-F)(x)$ is concave on $(x_0,\infty)$ for some $x_0$, then $F$ obeys the $\Delta_2$ condition.*

*Proof* (i) ⇔ (ii) Obviously, (2.14) implies that $\limsup_{x\to\infty} \frac{F(2x)}{F(x)} \le C$. Conversely, if (2.15) holds, and $2C$ is the lim sup, then (2.14) holds for large $x$.

(i) ⇒ (iii) Given $k$, pick $\ell$ so $k \le 2^\ell$. Then for $x \ge x_0$,

$$F(kx) \le F(2^\ell x) \le C\,F(2^{\ell-1}x) \le \cdots \le C^\ell F(x) \tag{2.17}$$

Pick $x_1 > 0$, so $F(kx_1) \le \varepsilon$. Since $F(kx)/F(x)$ is continuous on $[x_1,\infty)$ and bounded by $C^\ell$ as $x\to\infty$, $\sup_{x\ge x_1} F(kx)/F(x) \equiv B$ and (2.16) holds.

(iii) ⇒ (iv) is trivial.

(iv) ⇒ (i) Pick $x_0$ so $F(x_0) \ge \varepsilon$. For $x \ge x_0$,

$$F(kx) \le B\,F(x) + \varepsilon \le (B+1)F(x)$$

Now pick $\ell$ so $k^\ell \ge 2$. Then for $x \ge x_0$,

$$F(2x) \le F(k^\ell x) \le (B+1)F(k^{\ell-1}x) \le \cdots \le (B+1)^\ell F(x)$$

(i) ⇒ (v) Since $(D^-F)(x)$ is monotone, if (i) holds and $x \ge x_0$,

$$x(D^-F)(x) \le \int_x^{2x} D^-F(y)\,dy \le F(2x) \le C\,F(x)$$

so

$$\sup_{x \ge x_0} \frac{x(D^-F)(x)}{F(x)} \le C$$

Since

$$\sup_{1 \le x \le x_0} \frac{x(D^-F)(x)}{F(x)} < \infty$$

(v) holds.

<u>(v) ⇒ (iv)</u> Suppose (v) holds. Let the sup be $A$ and let $x \ge 1$ and $k > 1$. Since $F$ is monotone on $[x, kx]$, for $y \in [x, kx]$,

$$(D^-F)(y) \le \frac{AF(y)}{y} \le \frac{AF(kx)}{y}$$

Thus, by (1.59),

$$F(kx) - F(x) \le A(\log k)F(kx) \tag{2.18}$$

Pick $k$ so $A(\log k) \le \frac{1}{2}$. Then (2.18) implies

$$F(kx) \le 2F(x)$$

so (iv) holds.

Moreover, if $D^-F$ is concave for $x \ge x_0$, then for $x \ge 2x_0$,

$$(D^-F)(x) \ge \tfrac{1}{3}\,(D^-F)(2x) + \tfrac{2}{3}\,(D^-F)(\tfrac{x}{2}) \ge \tfrac{1}{3}\,(D^-F)(2x) \tag{2.19}$$

since $x = \frac{1}{3}(2x) + \frac{2}{3}(\frac{1}{2}x)$. Thus, integrating from $2x_0$ to $x$,

$$F(2x) \le 6F(x) + (F(4x_0) - 6F(2x_0))$$

which implies $F(2x) \le 7F(x)$ for $x$ large. □

**Example 2.5** One might think that the $\Delta_2$ condition is always true if $F(x) \le C|x|^p$ for some $p$, but this is not so. For let $x_1 = 0 < x_2 < x_3 < \cdots$ and define $F$ by

$$(D^-F)(x) = 2^{n^2}, \qquad x_n < x < x_{n+1}$$

Then, for $n \ge 2$ and $0 < y < x_n$,

$$(D^-F)(x_n + y) \ge 2^{n^2} = 2^{(2n-1)+(n-1)^2} \ge 2^{(2n-1)}(D^-F)(y)$$

and thus,

$$F(2x_n) \ge F(2x_n) - F(x_n) \ge 2^{(2n-1)}F(x_n)$$

so

$$\limsup \frac{F(2x)}{F(x)} = \infty$$

and the $\Delta_2$ condition fails. Suppose $x_n = 2^{2^{n^2}}$. Then $(D^- F)(x) \le C \log(x+2)$ and $F(x) \le C|x| \log(|x|+2)$. Notice in this case, $\liminf \frac{F(2x)}{F(x)} = 1$. This is no coincidence. If $\liminf_{x\to\infty} \frac{F(2x)}{F(x)} = \infty$, then

$$\frac{1}{n} \log F(2^n) = \frac{1}{n} \sum_{j=1}^{n} \log\left(\frac{F(2^j)}{F(2j-1)}\right) + \frac{1}{n} \log(F(1)) \to \infty$$

and it cannot be true that $F(x) \le Cx^p$ for any $p$. Thus, polynomially boundedness and the failure of the $\Delta_2$ condition requires $\liminf_n \frac{F(2x)}{F(x)} < \infty$. □

In Theorem 2.9 below, our proof that the $\Delta_2$ condition implies (ii)–(vi) depends on no assumption on the space $(M, d\mu)$, but our proof that the $\Delta_2$ condition is equivalent (basically our proof that (vi) ⇒ (i)) requires the space to have nonatomic components where

**Definition** A measure space $(M, \mu)$ is called *nonatomic* if, given $A \subset M$ with $\mu(A) > 0$ and $0 < \alpha < \mu(A)$, there is a measurable set $B \subset A$ with $\mu(B) = \alpha$. We say $M(\mu)$ has a *nonatomic component* if there exists $N \subset M$ with $\mu(N) > 0$ so $\mu \restriction N$ is *nonatomic*.

If $\mu$ is a Baire measure on a separable locally compact metric space, it is not hard to show that nonatomic is equivalent to having no pure points and having a nonatomic component is equivalent to not being a purely pure point measure.

Nonatomic measures are discussed further in Halmos [143].

**Lemma 2.6** *Let $(M, \mu)$ be a nonatomic probability measure space. Let $\alpha_1, \dots, \alpha_n, \dots$ be a sequence of nonnegative numbers with $\sum_{j=1}^\infty \alpha_j \le 1$. Then there exist disjoint measurable subsets $A_1, A_2, \dots$ with $\mu(A_j) = \alpha_j$.*

*Proof* By the nonatomic condition, pick $A_1$ so $\mu(A_1) = \alpha_1$. Since $\alpha_2 < 1 - \alpha_1$, we can find $A_2 \subset M \backslash A_1$ so $\mu(A_2) = \alpha_2$. Inductively, since $\alpha_{n+1} < 1 - \alpha_1 - \cdots - \alpha_n$, we can find $A_{n+1} \subset M \backslash \cup_{j=1}^n A_j$ so $\mu(A_{n+1}) = \alpha_{n+1}$. □

**Definition** $E^{(F)}(M, d\mu)$ is the closure of $L^\infty(M, d\mu)$ in $L^{(F)}(M, d\mu)$. $Y_F(M, d\mu) \equiv \{f \mid Q_F(f) < \infty\}$.

*Remarks* 1. By (2.10), the simple functions (i.e., finite linear combinations of characteristic functions) are dense in $E^{(F)}$.

2. As we shall see, unlike $L^{(F)}$ and $E^{(F)}$, $Y_F$ may not be a vector space.

**Definition** Let $\{f_n\}_{n=1}^\infty$ and $f$ lie in $L^F(M, d\mu)$. We say $f_n$ *converges in mean* to $f$ if and only if for $n$ large, $Q_F(f - f_n) < \infty$ and $\lim_{n\to\infty} Q_F(f - f_n) = 0$.

**Proposition 2.7** *$\|\cdot\|_F$ convergence implies mean convergence.*

*Proof* By (2.1),

$$Q_F(2^m g) \geq 2^m Q_F(g) \tag{2.20}$$

Thus, if $\|f - f_n\| \leq 2^{-m}$, then $Q_F(2^m(f - f_n)) \leq 1$ (by (2.9)), and so by (2.20), $Q_F(f_n - f) \leq 2^{-m}$. □

**Example 2.8** To see all that can fail if the $\Delta_2$ condition fails, consider the canonical example where $\Delta_2$ fails, namely, $F(x) = e^{|x|} - 1 - |x|$. Let $(M, d\mu) = ([0,1], dy)$. Let $f(y) = \log(y^{-1})$. Then $Q_F(\lambda f) < \infty$ if and only if $\lambda < 1$ since $F(\lambda f(y))$ diverges as $y^{-\lambda}$ as $y \downarrow 0$. In particular, $Y_F$ is not closed under scalar multiplication, and so $Y_F$ is not a vector space. Moreover, if $g$ is bounded, it is still true that $Q_F(\lambda(f - g)) < \infty$ if and only if $\lambda < 1$ so $\|f - g\| \geq 1$. Thus, $f \notin E^{(F)}$ so $E^{(F)} \neq L^{(F)}$. If $g_n(y) = \min(n, f(y))$, then $Q_F(\frac{1}{2}(f - g_n)) \to 0$ by the dominated convergence theorem, so $\frac{1}{2}g_n \to \frac{1}{2}f$ in mean, but since $\|\frac{1}{2}g_n - \frac{1}{2}f\| \geq \frac{1}{2}$ not in norm, so norm convergence is not equivalent to mean convergence.

Finally, let $f(x) = [\log(x^{-1}) - 2\log(1 + |\log(x)|)]\chi_{(0,\alpha)}$, with $\chi$ the characteristic function and $\alpha$ will be chosen in a moment. Then $Q_F(\lambda f) = \infty$ if $\lambda > 1$ and $Q_F(\lambda f) < \infty$ if $\lambda \leq 1$. Because $\exp(f(x)) = x^{-1}(1 + |\log(x)|)^{-2}$ is integrable, $Q_F(f) < \infty$ and, by taking $\alpha$ small, we can arrange that $Q_F(f) \leq \frac{1}{2}$. Thus, $\|f\|_F = 1$ and $Q_F(f/\|f\|) \leq \frac{1}{2} < 1$. □

We now turn to the main result on consequences of the $\Delta_2$ condition.

**Theorem 2.9** *Let $(M, \mu)$ be a measure space with a nonatomic component and $F$ a weak Young function. Then the following are equivalent:*

(i) *$F$ obeys the $\Delta_2$ condition.*
(ii) *Mean convergence implies* (*and so, by Proposition 2.7, is equivalent to*) *norm convergence.*
(iii) *$E^{(F)}(M, d\mu) = L^{(F)}(M, d\mu)$ (i.e., $L^\infty$ is dense in $L^{(F)}$).*
(iv) *$Y_F = L^{(F)}$*
(v) *$Y_F$ is a vector space.*
(vi) *For all nonzero $f \in L^{(F)}$,*

$$Q_F\left(\frac{f}{\|f\|}\right) = 1 \tag{2.21}$$

*Proof* We will show (i) ⇒ (ii) ⇒ (iii) ⇒ (iv) ⇒ (v) ⇒ (vi) ⇒ (i).

(i) ⇒ (ii) By Proposition 2.4 (i.e., (2.16)), we can find constant $C_k$ so that

$$Q_F(2^k g) \leq C_k Q_F(g) + \tfrac{1}{2} \tag{2.22}$$

(we use $\mu(M) = 1$ here). Thus, if $Q_F(g) \leq \frac{1}{2}C_k^{-1}$, then, by (2.22), $Q_F(2^k g) \leq 1$ so $\|g\|_F \leq 2^{-k}$. This shows if $Q_F(g_n) \to 0$, then $\|g_n\|_F \to 0$.

(ii) ⇒ (iii) Suppose (ii) holds and let $f \in L^{(F)}(M, d\mu)$. Pick $\lambda > 0$ so that $Q_F(\lambda f) < \infty$. Let

$$f_n(x) = \begin{cases} f(x) & \text{if } |f(x)| \le n \\ 0 & \text{if } |f(x)| > n \end{cases} \tag{2.23}$$

Then $f_n \in L^\infty$, and by the dominated convergence theorem, $Q_F(\lambda(f - f_n)) \to 0$. By hypothesis (ii), $\lambda\|f - f_n\| \to 0$, so since $\lambda > 0$, $f_n \to f$ in $\|\cdot\|_F$ and $f \in E^{(F)}$.

(iii) ⇒ (iv) We claim that for any $F$,

$$E^{(F)} \subset Y_F \subset L^{(F)}$$

so that (iii) implies (iv). $Y_F \subset L^{(F)}$ is evident from the definition of $L^{(F)}$. To see that $E^{(F)} \subset Y_F$, let $f \in E^{(F)}$. Pick $g \in L^\infty$ so $\|f - g\|_F \le \frac{1}{2}$. Thus, by (2.9), $Q_F(2(f - g)) \le 1$. By convexity of $Q_F$ and $f = \frac{1}{2}(2(f - g)) + \frac{1}{2}(2g)$,

$$Q_F(f) \le \tfrac{1}{2} + \tfrac{1}{2}\, Q_F(2g) < \infty$$

so $f \in Y_F$.

(iv) ⇒ (v) is trivial, since $L^{(F)}$ is a vector space.

(v) ⇒ (vi) Let $f \in L^{(F)}$ with $f$ nonzero. Then $\lambda_0 f \in Y_F$ for some $\lambda_0$ and so, since $Y_F$ is a vector space, $\lambda f \in Y_F$ for all $\lambda$. By the monotone convergence theorem, $\lambda \mapsto H(\lambda) \equiv Q_F(\lambda f)$ is a continuous function with $H(\lambda) = 0$ and $\lim_{\lambda\to\infty} H(\lambda) = \infty$ since $f$ is nonzero (i.e., strictly nonzero on a set of positive measure). By (2.1), if $\lambda > \mu$, $H(\lambda) \ge \lambda\mu^{-1} H(\mu) > H(\mu)$, so $H$ is strictly monotone, so there is $\lambda_0$ with $H(\lambda_0) = 1$ and $H(\lambda) > 1$ if $\lambda > \lambda_0$. It follows that $\lambda_0^{-1} = \|f\|$ and $Q_F(f/\|f\|) = H(\lambda_0) = 1$.

(vi) ⇒ (i) We will show that ∼(i) implies ∼(vi). So suppose that $F$ does not obey the $\Delta_2$ condition. Let $\beta_n = 2^{1/n}$. If $F(\beta_n x) \le C\, F(x)$ for all $x > x_0$, $F$ obeys the $\Delta_2$ condition by Proposition 2.4. It follows that $\limsup F(\beta_n x)/F(x) = \infty$. Thus, by induction, we can find $x_1 \le x_2 \le \cdots \le x_n \le \cdots$ so that $F(x_1) \ge 1$ and

$$F(\beta_n x_n) \ge 2^n F(x_n) \tag{2.24}$$

Let

$$\alpha_n = 2^{-n-1} F(x_n)^{-1} \tag{2.25}$$

Suppose that $(M, d\mu)$ is nonatomic. Since $F(x_n) \ge F(x_1) \ge 1$, $\sum_{n=1}^\infty \alpha_n \le \sum_{n=1}^\infty 2^{-n-1} = \frac{1}{2}$. So, by Lemma 2.6, we can find disjoint sets $A_1, A_2, \dots$ so $\mu(A_n) = \alpha_n$. Define

$$f(y) = \begin{cases} x_n, & y \in A_n \text{ for some } n \\ 0, & \text{otherwise} \end{cases}$$

Then

$$Q_F(f) = \sum_1^\infty F(x_n)\mu(A_n) = \sum_1^\infty 2^{-n-1} = \tfrac{1}{2} \tag{2.26}$$

while

$$\begin{aligned} Q_F(\beta_n f) &\geq \sum_{m=1}^\infty F(\beta_n x_m)\mu(A_m) \\ &\geq \sum_{m=n}^\infty F(\beta_n x_m)\mu(A_m) \\ &\geq \sum_{m=n}^\infty F(\beta_m x_m)\mu(A_m) \quad \text{(since } \beta_n \geq \beta_m \text{ if } n \leq m) \\ &\geq \sum_{m=n}^\infty 2^m F(x_m)\mu(A_m) \quad \text{(by (2.24))} \\ &= \sum_{m=n}^\infty 2^{-1} = \infty \end{aligned}$$

Since $\beta_n \downarrow 1$ as $n \to \infty$, we conclude $\|f\|_F = 1$. So, by (2.26), $Q_F(f/\|f\|) = \frac{1}{2}$ and (vi) is false.

If $M$ only has a nonatomic component, pick $A_\infty$ so $\mu_{A_\infty}$ is nonatomic and $\mu(A_\infty) > 0$. Pick $N_0$ so $\sum_{n=N_0}^\infty \alpha_n < \mu(A_\infty)$ and the disjoint $A_n, A_{n+1}, \dots$ and still find $\|f\| = 1$ while $Q_F(f/\|f\|) = 2^{-N_0}$. □

While monotone convergence upwards may not imply convergence *in* $\|\cdot\|_F$ norm, it does imply convergence of the $\|\cdot\|_F$ norms.

**Proposition 2.10** *Let $f \in L^{(F)}(M, d\mu)$ and suppose $|f_n|$ is monotone increasing with* $\lim|f_n(m)| = |f(m)|$. *Then*

(i) $\|f_1\|_F \leq \cdots \leq \|f_n\|_F \leq \cdots \leq \|f\|_F$
(ii) $\|f_n\|_F \to \|f\|_F$
(iii) *For some $\lambda > 0$, $\lambda f_n \to \lambda f$ in mean.*
(iv) *If $F$ obeys the $\Delta_2$ condition, $\| |f_n| - |f| \|_F \to 0$.*

*Proof* (i) If $0 \leq |g| \leq |h|$, then $Q_F(\lambda^{-1}|g|) \leq Q_F(\lambda^{-1}|h|)$ so $\{\lambda \mid Q_F(\lambda^{-1}|g|) \leq 1\} \supseteq \{\lambda \mid Q_F(\lambda^{-1}|h| \leq 1\}$ so $\|g\|_F \leq \|h\|_F$.

(ii) By (i), $\|f_n\|_F \leq \|f\|_F$. If $\alpha < \|f\|$, then $Q_F(\alpha^{-1}f) > 1$. By the monotone convergence theorem, $Q_F(\alpha^{-1}f_n) \to Q_F(\alpha^{-1}f)$, so $Q(\alpha^{-1}f_n) > 1$ for $n$ large and so, for $n$ large, $\|f_n\| > \alpha$. It follows that $\liminf \|f_n\| \geq \|f\|$.

(iii) Pick $\lambda$ so that $Q_F(\lambda f) < \infty$. Then $F(\lambda(f - f_n)) \leq F(\lambda f)$, so by the dominated convergence theorem, $Q_F(\lambda(f - f_n)) \to 0$.

(iv) follows from (iii) and (i) $\Rightarrow$ (ii) of Theorem 2.9. □

We are heading towards our next result that $E^{(F)}$ is often separable. We begin with a calculation of $\|\chi_A\|$ for $A$ a characteristic function.

**Lemma 2.11** *Let $A \subset M$ and $\chi_A$ its characteristic function. Then*

$$\|\chi_S\|_F = \left[F^{-1}\left(\frac{1}{\mu(S)}\right)\right]^{-1} \tag{2.27}$$

*In particular, if $S_j$ is a sequence of sets with $\mu(S_j) \to 0$, then $\|\chi_{S_j}\|_F \to 0$.*

*Remark* In (2.27), the $^{-1}$ in $F^{-1}$ is function inverse, while the final $^{-1}$ means $1/\cdot$.

*Proof* $Q_F(\lambda^{-1}\chi_S) = F(\lambda^{-1})\mu(S)$ so $Q_F(\lambda^{-1}\chi_S) \leq 1$ if and only if $\lambda^{-1} \leq F^{-1}(\mu(S)^{-1})$, that is, if and only if $\lambda \geq (F^{-1}(\mu(S)^{-1}))^{-1}$. Thus, (2.27) holds. If $\mu(S_j) \to 0$, then $\mu(S_j)^{-1} \to \infty$ so $F^{-1}(\mu(S_j)^{-1}) \to \infty$ so $\|\chi_{S_j}\|_F \to 0$. □

**Theorem 2.12** *Let $M$ be a locally compact separable metric space and $\mu$ a Baire measure with $\mu(M) = 1$. Then $C(M)$, the bounded continuous functions, are dense in $E^{(F)}(M, d\mu)$. Moreover, $E^{(F)}(M, d\mu)$ is separable. If $F$ obeys the $\Delta_2$ condition, $L^{(F)}(M, d\mu)$ is separable.*

*Proof* Let $A$ be an arbitrary Baire measurable set in $M$. By regularity of $\mu$, given $n \in \mathbb{Z}_+$, we can find $K_n \subset A \subset \mathcal{O}_n$ so $K_n$ is compact, $\mathcal{O}_n$ is open, and $\mu(\mathcal{O}_n \backslash K_n) < 1/n$, and then, by Urysohn's lemma, a continuous function $f_n$ with $0 \leq f_n \leq 1$, $f_n = 1$ on $K_n$, $f_n = 0$ on $M \backslash \mathcal{O}_n$. Then $|\chi_A - f_n| \leq \chi_{\mathcal{O}_n \backslash K_n}$ so, by (2.27), $\|\chi_A - f_n\|_F \leq [F^{-1}(n)]^{-1} \to 0$ as $n \to \infty$. Thus, $\chi_A$ is in $Q = \overline{C(M)}^{\|\cdot\|_F}$. Since $C(M)$ is a vector space, any simple function is in $Q$. Thus, $C(M)$ is dense in $E^{(F)}(M, d\mu)$.

Let $\{A_n\}_{n=1}^\infty$ be a sequence of compact sets with $A_n \subset A_{n+1}^{\text{int}}$ and $\cup A_n = M$. Let $g_n \in C(M)$ with $0 \leq g_n \leq 1$ and $g_n = 1$ on $A_n$ and $g_n = 0$ on $M \backslash A_{n+1}^{\text{int}}$. Since $\mu(M) = 1$, $\lim_{n\to\infty} \mu(M \backslash A_{n+1}) \to 0$ so, if $f \in C(M)$, $f(1 - g_n) \to 0$ in $\|\cdot\|_F$ by the dominated convergence theorem. It follows that $\cup_n \{f \in C(M) \mid \operatorname{supp} f \subset A_n\}$ is dense in $E^{(F)}$. But $C(A_n)$ is separable in $\|\cdot\|_\infty$ and so in $\|\ \|_F$. Thus, $E^{(F)}$ is separable. □

*Remark* Consider the degenerate function,

$$\begin{aligned} F(x) &= 0, \qquad |x| \leq 1 \\ &= \infty, \qquad |x| > 1 \end{aligned}$$

Then

$$Q_F(f) = \begin{cases} 0, & \|f\|_\infty \leq 1 \\ \infty, & \|f\|_\infty > 1 \end{cases}$$

and $\|\cdot\|_F$ is the $L^\infty$ norm. Of course, (2.27) fails, and it is not true that $\mu(S_j) \to 0$ implies $\|\chi_{S_j}\| \to 0$. $L^{(F)}$ is not separable. Despite this, it is a useful intuition that

$L^\infty$ is $L^{(F)}$ for this $F$. It even fits in with the duality theory for Orlicz spaces that we will discuss later. For formally, $F(x) = \sup_y (xy - |y|)$ so this $F$ can be viewed as the Legendre transform for $G(y) = |y|$, the generator of the Orlicz space in $L^1$.

**Example 2.13** It is generally true that if $F$ does not obey the $\Delta_2$ condition, then $L^{(F)}$ is not separable (see the reference in the Notes). We will settle for proving the result for $F(x) = e^x - 1 - x$ and $(M, d\mu) = ([0,1], dx)$. For $y_0 \in [0,1]$, let $f^{(y_0)}(y) = \log(|y - y_0|^{-1})$. Then $Q_F(\lambda(f^{(y_0)} - f^{(y_1)})) = \infty$ if $\lambda \geq 1$ and $y_0 \neq y_1$. Thus, $\|f^{(y_1)} - f^{(y_0)}\| \geq 1$ for all $y_0 \neq y_1$ in $[0,1]$ and so we have an uncountable set with pairwise distance at least 1. Thus, $L^{(F)}$ is not separable. □

We next turn to the duality theory for Orlicz spaces. The Legendre transform discussed at the end of Chapter 1 will play a critical role. We begin by discussing Legendre transforms of even functions in one dimension.

**Theorem 2.14** *Let $\tilde{F}$ be a convex function on $[0,\infty)$ with $\tilde{F}(0) = 0$, $\tilde{F}(x) \geq 0$, and $A^+(\tilde{F}) = \lim_{x\to\infty}(D^-\tilde{F})(x) = \infty$. Let $F(x) = \tilde{F}(|x|)$, which is convex. Define $\tilde{G}$ on $[0,\infty)$ as follows. For each $\alpha \in [0,\infty)$, let $[x_0^-(\alpha), x_0^+(\alpha)]$ be the set of $x \in [0,\infty)$ for which $\alpha$ is tangent to $\tilde{F}$ at $x$. (If $0 < \alpha < \lim_{x\downarrow 0}(D^-F)(x)$, we set $x_0^-(\alpha) = x_0^+(\alpha) = 0$.) Let $\tilde{G}(\alpha) = \int_0^\alpha x_0^-(\beta)\,d\beta$ and $G(\alpha) = \tilde{G}(|\alpha|)$. Then $F$ is steep, $G = F^*$, and $G(0) = 0$.*

*Remark* As discussed in Proposition 1.32, $x_0^-$ is an inverse to $D^-\tilde{F}$ which is continuous from below. By construction, $D^-\tilde{G}$ is $x_0^-$. In cases where $\tilde{F}$ is $C^1$ on $[0,\infty)$ with $\tilde{F}'(0) = 0$ and $\tilde{F}$ has no affine piece (so $D\tilde{F}$ is strictly monotone), $\tilde{G}$ is constructed as follows. Let $x_0$ be the inverse function to the strictly monotone function $\tilde{F}'$. Then $G(0) = 0$ and $G' = x_0$.

*Proof* That $F$ is steep is obvious. Let

$$\Gamma = \{(x,\alpha) \mid x \geq 0,\ \alpha \geq 0,\ \alpha \text{ is tangent to } F \text{ at } x\}$$

By construction of $F^*$, we have that if $(x,\alpha) \in \Gamma$, then

$$x\alpha = F(x) + F^*(\alpha) \tag{2.28}$$

Let $(x_0, \alpha_0) \in \Gamma$ and consider the rectangle, $R$, with vertices $(0,0)$, $(0,\alpha_0)$, $(x_0,\alpha_0)$, and $(x_0, 0)$. $\Gamma \cap R$ is the graph of $(D^-F)$ on $[0,x]$ with vertical line segments $\{(x,0) \mid (D^-F)(x) \leq \alpha \leq (D^+F)(\alpha)\}$ added at points where $F$ is not differentiable. Since $D^-$ is monotone, $R\backslash\Gamma \cap R$ is a union of two connected pieces. One is the area under the curve $\Gamma$ viewed as the graph of $D^-$. By Theorem 1.28, this area is $\int_0^{x_0}(D^-F)(x)\,dx = F(x_0)$. But $\Gamma$ is also the graph of $D^-G$ with coordinates interchanged. Thus, the area of the second component is the area under the curve $D^-G$, that is, $\int_0^{\alpha_0}(D^-G)(\alpha)\,d\alpha = G(\alpha_0)$. Thus, the area of $R$ is

$F(x_0) + G(\alpha_0)$, that is,

$$F(x_0) + G(\alpha_0) = \alpha_0 x_0 \tag{2.29}$$

Using (2.28), (2.29), and the fact that $\Gamma$ includes an $(x, \alpha)$ for each $\alpha > 0$, we conclude that

$$G(\alpha) = F^*(\alpha)$$ □

*Remarks* 1. Since $F^*(\alpha) = \sup_x x\alpha - F(x)$, (2.28) is complemented by the fact that for any $(x, \alpha) \in \mathbb{R}^2_+$, we have

$$\alpha x \leq F(x) + G(\alpha) \tag{2.30}$$

This is called *Young's inequality*. It has no relation to the convolution inequality (see Theorem 12.6), also called Young's inequality.

2. $G = F^*$ is called the *conjugate convex function* to $F$. If $F$ is a Young function, $G$ is called the *Young conjugate* or *conjugate Young* function.

**Example 2.15** This will contain our fourth and last proof of Hölder's inequality. Fix $1 < p < \infty$. Let $F(x) = x^p/p$. Then $F'(x) = x^{p-1}$. Since $1/p + 1/q = 1$ is equivalent to $(p-1)(q-1) = 1$, the inverse function to $F'$ is $G'(\alpha) = \alpha^{q-1}$, so $G(\alpha) = \alpha^q/q$ is the dual conjugate function. Thus, by Young's inequality, (2.30), for $x, y \geq 0$,

$$xy \leq \frac{x^p}{p} + \frac{y^q}{q} \tag{2.31}$$

Now let $f, g$ be two functions on $(M, d\mu)$ so that $\|f\|_p = \|g\|_q = 1$. Then, by

$$\begin{aligned}\int |f(m)|\,|g(m)|\,d\mu(m) &\leq \frac{1}{p}\int |f(m)|^p\,d\mu(m) + \frac{1}{q}\int |g(m)|^q\,d\mu \\ &= \frac{1}{p} + \frac{1}{q} = 1\end{aligned}$$

Thus, we see that $\|f\|_p = \|g\|_q = 1$ implies $\|fg\|_1 \leq 1$. For general $f, g \neq 0$ in $L^p$ and $L^q$, we have $\|f/\|f\|_p\|_p = \|g/\|g\|_q\| = 1$ so $\|fg\|_1/\|f\|_p\|g\|_q \leq 1$, which is Hölder's inequality. □

**Proposition 2.16** *Let $F$ be a weak Young function which is steep and let $G$ be the conjugate convex function. Then $G$ is a weak Young function if and only if*

$$\lim_{x \downarrow 0} \frac{F(x)}{x} = 0 \tag{2.32}$$

*Proof* If $y_0 > 0$ and $G(y_0) = 0$, then $y_0 x$ is a tangent to $F$ at 0, that is, $F(x) \geq xy$ for all $x$ so $\lim_{x\downarrow 0} F(x)/x \geq y_0$. Conversely, if $\lim_{x\downarrow 0} F(x)/x = y_0 > 0$, then $F(x) \geq xy_0$ so $G(y_0) = 0$. □

It is now clear why (2.32) is part of the definition of Young functions – it is needed for the conjugate function to be a weak Young function. Indeed,

**Proposition 2.17** *Let $F$ be a Young function and $G$ its convex conjugate. Then $G$ is a Young function.*

*Proof* By Theorem 1.45, $G$ is a convex function which is steep and nonnegative. By Proposition 2.16, $G(x) = 0$ if and only if $x = 0$. Since $F$ is even, $G$ is even. Finally, since $G^* = F$, Proposition 2.16 implies that $\lim_{x\downarrow 0} G(x)/x = 0$. □

**Example 2.18** If $F(x) = x^p, 1 < p < \infty$, the conjugate function is not $y^q$ (where $p^{-1} + q^{-1} = 1$) but $G(y) = p^{-q/p}q^{-1}y^q$ by a simple calculation. Thus,

$$\|g\|_G = \frac{1}{p^{1/p}q^{1/q}}\|g\|_q \equiv \beta_p\|g\|_q \tag{2.33}$$

For later purposes, notice that $\beta_p \to 1$ as $p \to 1$ or $\infty$, $\beta_{1/2} = \frac{1}{2}$, and $\beta$ is monotone decreasing on $[1, 2]$ and increasing on $[2, \infty]$. The easiest way to see this is to note that $\theta \to \theta\log\theta$ is convex on $(0,\infty)$, so $\theta \mapsto \theta\log\theta + (1-\theta)\log(1-\theta)$ is convex on $(0, 1)$ and invariant under $\theta \mapsto \frac{1}{2} - \theta$. Thus, it takes its maximum as $\theta \to 1$ and minimum at $\theta = \frac{1}{2}$. So $\exp(\frac{1}{p}\log(\frac{1}{p}) + \frac{1}{q}\log(\frac{1}{q}))$ has the same properties: maximum at $\frac{1}{p} \to 1$ and minimum at $\frac{1}{p} = \frac{1}{2}$. Thus,

$$\sup_p \beta_p = 1, \qquad \inf_p \beta_p = \beta_{1/2} = \tfrac{1}{2} \tag{2.34}$$

If $F(x) = e^{|x|} - 1 - |x|$, then $F'(x) = e^{|x|} - 1$ and $(F')^{-1}(y) = \log(y + 1)$ so $G(y) = \int_0^{|y|}\log(w+1)\,dw = (|y|+1)\log(|y|+1) - |y|$. The natural norm on $L^1\log_+ L$ is thus

$$\|f\|_{L^1\log_+ L} = \inf\left\{\lambda \,\middle|\, \int(\lambda|f|+1)(\log(\lambda|f|+1)-1) - 1\,d\mu \le 1\right\} \tag{2.35}$$

□

The following two preliminaries are needed for the duality theory:

**Lemma 2.19** *Let $f \in L^{(F)}(M, d\mu)$. Then*

$$\|f\|_F \le \max(1, Q_F(f)) \le 1 + Q_F(f) \tag{2.36}$$

*Proof* The second inequality is trivial, so we only need the first. Suppose first $f \in L^\infty$. Then if $\|f\|_F \ge 1$,

$$\begin{aligned} Q_F(f) &\ge Q_F\left(\frac{f}{\|f\|}\right)\|f\| \qquad \text{(by (2.1))} \\ &= \|f\| \end{aligned}$$

by the proof of (v) $\Rightarrow$ (vi) in Theorem 2.9 (which shows $Q_F(f/\|f\|) = 1$ if $f \in Y_F \supset L^\infty$). Thus, we have proven

$$\|f\|_F \leq \max(1, Q_F(f)) \tag{2.37}$$

for $f \in L^\infty$. For general $f$, let $f_n$ be given by (2.23) and use Proposition 2.10 to see that $\|f_n\|_F \to \|f\|_F$ and the monotone convergence theorem to see $Q_F(f_n) \to Q_F(f)$ to obtain (2.37) for $f$. □

**Lemma 2.20** *Let $F$ be a Young function. Let $f \in L^{(F)}(M, d\mu)$. Then*

$$\lim_{\alpha \downarrow 0} \alpha^{-1} Q_F(\alpha f) = 0$$

*Proof* By (2.1), for any $m$, $\alpha^{-1}F(\alpha f(m))$ is monotone decreasing, and by (2.3), the limit is zero. Since $f \in L^{(F)}$, $\int \alpha^{-1} F(\alpha f(m))\, d\mu(m) < \infty$ for some $\alpha$ and so, by the monotone convergence theorem, the limit is zero. □

The first major result in the duality theorem identifies many linear functionals on $L^{(F)}$.

**Theorem 2.21** *Let $F$ be a Young function and $G$ its conjugate Young function. Suppose that $g$ is a measurable function on $(M, d\mu)$ so that for some $c < \infty$ and all $f \in L^\infty(M, d\mu)$,*

$$\int |g(m)f(m)|\, d\mu(m) \leq c\|f\|_F \tag{2.38}$$

*Then $g \in L^{(G)}$ and*

$$\|g\|_G \leq c \tag{2.39}$$

*Conversely, if $g \in L^{(G)}$, then for all $f \in L^{(F)}$,*

$$\int |g(m)f(m)|\, d\mu(m) \leq 2\|f\|_F \|g\|_G \tag{2.40}$$

*In particular,*

$$L_g(f) = \int g(m)f(m)\, d\mu(m) \tag{2.41}$$

*defines a bounded linear functional on $L^{(F)}$ and*

$$\|g\|_G \leq \|L_g\|_{(L^{(F)})^*} \leq 2\|g\|_G \tag{2.42}$$

*In addition, $f \mapsto L_g(f)$ is continuous in mean, that is, $Q_F(f - f_n) \to 0 \Rightarrow L_g(f_n) \to L_g(f)$.*

*Proof* By replacing $g$ by $g/c$, we can suppose $c = 1$ in (2.38). Let $A_n = \{m \mid |g(m)| \leq n\}$ and $g_n = \chi_{A_n} g$. For any $x$, there exists $y(x)$ (e.g., $y = (DG^-)(x)$) so that

$$xy(x) = G(x) + F(y(x)) \tag{2.43}$$

Let

$$f_n(m) = \begin{cases} 0, & \text{if } g_n(m) = 0 \\ y(|g_n(m)|)\overline{g_n(m)}/|g_n(m)|, & \text{if } g_n(m) \neq 0 \end{cases}$$

Then, by (2.43), for any $m$,

$$f_n(m)g_n(m) = F(|f_n(m)|) + G(|g_n(m)|) \tag{2.44}$$

Thus, since $f_n \in L^\infty$ and we are assuming $c = 1$ in (2.38),

$$\begin{aligned} Q_F(f_n) + Q_G(g_n) &= \int f_n(m) g_n(m)\, d\mu(m) \\ &= \int g(m) f_n(m)\, d\mu(m) \\ &\leq \|f_n\|_F \leq 1 + Q_F(f_n) \end{aligned}$$

by Lemma 2.19. Since $f_n \in L^\infty$, $Q_F(f_n) < \infty$, and thus,

$$Q_G(g_n) \leq 1$$

By the monotone convergence theorem, $Q_G(g_n) \uparrow Q_G(g)$ so

$$Q_G(g) \leq 1$$

which implies $g \in L^{(G)}$ and $\|g\|_G \leq 1$ so (2.39) is proven.

Now let $f \in L^{(F)}$ and $g \in L^{(G)}$. Since

$$\begin{gathered} xy \leq G(x) + F(y) \\ \frac{|g(m)|}{\|g\|_G} \frac{|f(m)|}{\|f\|_F} \leq G\left(\frac{g}{\|g\|_G}\right) + F\left(\frac{f}{\|f\|_F}\right) \end{gathered}$$

so

$$\begin{aligned} (\|g\|_G \|f\|_F)^{-1} \int |g(m) f(m)|\, d\mu(m) &\leq Q_G\left(\frac{g}{\|g\|_G}\right) + Q_F\left(\frac{f}{\|f\|_F}\right) \\ &\leq 2 \end{aligned} \tag{2.45}$$

by (2.9) proving (2.40). This in turn implies $\|L_g\| \leq 2\|g\|_G$ while (2.38) $\Rightarrow$ (2.39) shows that $\|g\|_G \leq \|L_g\|$, so (2.42) is proven.

Finally, to prove that $L_g(\cdot)$ is continuous in mean, note that by the proof of (2.45), for any $f \in L^{(F)}$ and $g \in L^{(G)}$ and any $\alpha > 0$,

$$|L_g(f)| \leq \alpha^{-1}[Q_G(\alpha g) + Q_F(f)] \tag{2.46}$$

If $f_n \to f$ in mean, given $\varepsilon$, find $\alpha_0$ by Lemma 2.20 so $\alpha_0^{-1} Q_G(\alpha_0 g) < \varepsilon/2$ and then $N$ so $Q_F(f - f_n) < \varepsilon\alpha_0/2$ for $n \geq N$. By (2.46), if $n \geq N$,

$$|L_g(f) - L_g(f_n)| \leq \varepsilon$$

so $L_g(f_n) \to L_g(f)$ as claimed. □

*Remark* By (2.33), if $F(x) = |x|^p$, then $\|L_g\|_{(L^p)^*} = \|g\|_q = \beta_p^{-1}\|g\|_G$. By (2.34), $\beta_p^{-1}$ runs from 1 to 2 so the constants in (2.42) cannot be improved.

Here is the main duality theorem:

**Theorem 2.22** *Let $F$ be a Young function and $G$ its conjugate Young function. Then $(E^{(F)})^* = L^{(G)}$ in the sense that*

(i) *Every norm-continuous linear functional $L$ on $E^{(F)}$ is of the form $L_g$ (given by* (2.41)*) for some $g \in L^{(G)}$.*
(ii) *$g$ is unique.*
(iii) *For every $g \in L^{(G)}$, $L_g$ defines a bounded linear functional on $E^{(F)}$.*

*Remark* By (2.42), the isomorphism of $L^{(G)}$ and $(E^{(F)})^*$ via $g \mapsto L_g$ is norm bounded with norm-bounded inverse.

*Proof* (i) Let $L$ be a norm-bounded linear functional on $E^{(F)}$. For any measurable set $A \subset M$, define $\nu(A) = L(\chi_A) < \infty$ since $\chi_A \in L^\infty \subset E^{(F)}$. Since $\chi_{A\cup B} = \chi_A + \chi_B$, if $A \cap B = \emptyset$, $\nu(A \cup B) = \nu(A) + \nu(B)$, so $\nu$ is a finitely additive set of functions with $\nu(M) = L(\chi_M) \leq \|L\|\,\|\chi_M\|_F < \infty$ by (2.27).

Suppose $\{A_j\}_{j=1}^\infty$ are mutually disjoint and $A = \cup_{j=1}^\infty A_j$. Since

$$\sum_{j=1}^\infty \mu(A_j) = \mu(A) \leq \mu(M) < \infty$$

we have $\mu(A \backslash \cup_{j=1}^J A_j) \to 0$ so, by Lemma 2.11, $\|\chi_{A\backslash \cup_{j=1}^J A_j}\|_F \to 0$, and thus, $L(\chi_A) - \sum_{j=1}^J L(\chi_{A_j}) \to 0$, which means $\nu(A) = \sum_{j=1}^\infty \nu(A_j)$. It follows that $\nu$ is a bounded complex measure, clearly absolutely continuous with respect to $\mu$. By the Hahn decomposition theorem (see Halmos [143]) and the Radon–Nikodym theorem (see Halmos [143]), there is a function $g \in L^1(M, d\mu)$ so that

$$L(\chi_A) = \int_A g(m)\,d\mu(m)$$

Since characteristic functions are total in $L^\infty$, we have that

$$L(f) = \int f(m)g(m)\,d\mu(m)$$

for all $f \in L^\infty$. Moreover, since $|fg| = \tilde{f}g$ for suitable $\tilde{f}$ with $|\tilde{f}| = f$, we have

$$\int |g(m)f(m)|\, d\mu(m) \leq \|L\|\, \|f\|_F$$

so by Theorem 2.21, $g \in L^{(G)}$ and $L = L_g$ extends to a norm continuous functional on $L^{(F)}$. Since $L^\infty$ is dense in $E^{(F)}$ and by construction of $g$, $L = L_g$ on $L^\infty$, we have that $L = L_g$ on $E^{(F)}$.

(ii) If $L_g = 0$ on $E^{(F)}$, then $L_{|g|} = 0$ on $E^{(F)}$. Thus, $L_{|g|}(\chi_A) \equiv 0$ so $g = 0$ a.e. Since $g \mapsto L_g$ is linear, this proves uniqueness.

(iii) This is part of Theorem 2.21. □

**Corollary 2.23** *$F$ obeys the $\Delta_2$ condition if and only if $(L^{(F)})^* = L^{(G)}$.*

*Proof* If $L^\infty$ is not dense in $L^{(F)}$, the Hahn–Banach theorem says there exist nonzero elements $L \in (L^{(F)})^*$ so $L \restriction E^{(F)} = 0$. It follows $L \notin L^{(G)}$. Thus, $(L^{(F)})^* = L^{(G)}$ is equivalent to $E^{(F)} = L^{(F)}$. By Theorem 2.9, this is equivalent to the $\Delta_2$ property. □

**Corollary 2.24** *Let $F$ be a Young function and $G$ its conjugate Young function. Then $L^{(F)}$ is reflexive if and only if both $F$ and $G$ obey the $\Delta_2$ condition.*

*Proof* If both $F$ and $G$ obey the $\Delta_2$ condition, then by Corollary 2.23, $(L^{(F)})^* = L^{(G)}$ and $(L^{(G)})^* = L^{(F)}$, so $L^{(F)}$ is reflexive.

If $F$ does not obey the $\Delta_2$ condition, $L^{(G)}$ is a closed proper subspace of $(L^{(F)})^*$ by Corollary 2.23 (closed because it is complete in $\|\cdot\|_G$ and so in $\|\ \|_{(L^{(F)})^*}$). By the Hahn–Banach theorem, there exists a nonzero $L \in (L^{(F)})^{**}$ vanishing on $L^{(G)}$. Since nonzero functionals on $(L^{(F)})^*$ induced by $f \in L^{(F)}$ do not vanish identically on $L^{(G)}$, $(L^{(F)})^{**}$ is strictly bigger than $L^{(F)}$.

If $G$ does not obey the $\Delta_2$ condition but $F$ does, there exist nonzero $L$'s in $(L^{(G)})^*$ which vanish on $E^{(G)}$, but any nonzero linear functional induced by any $f \in L^{(F)}$ does not vanish on $E^{(G)}$. Thus, $L^{(F)}$ is not reflexive. □

$L^1 \log_+ L^*$ is $L^{(G)}$ for $G(x) = \exp(|x|) - 1 - |x|$, but $(L^{(G)})^*$ is strictly bigger than $L^1 \log_+ L$ and $L^1 \log_+ L$ is not reflexive.

The following sheds light on the last few results:

**Theorem 2.25** *Let $L$ be a linear functional on $L^{(F)}$ which is continuous in the sense of mean convergence, that is, $Q_F(f_n - f) \to 0 \Rightarrow L(f_n) \to L(f)$. Then $L = L_g$ for a unique $g \in L^{(G)}$.*

*Proof* By Proposition 2.7, continuity in mean implies continuity in norm, so by Theorem 2.22, there is a unique $g \in L^{(G)}$ so that $L \restriction E^{(F)} = L_g$. By Theorem 2.21, $L_g$ extends to a mean-continuous function on all of $L^{(F)}$. We have to show $L = L_g$ on $L^{(F)}\backslash E^{(F)}$. Let $f \in L^{(F)}$ and suppose $Q_F(\lambda_0 f) < \infty$. Let $f_n$ be defined by (2.23). Then, by the dominated convergence theorem, $Q_F(\lambda_0(f - f_n)) \to 0$ so $L(\lambda_0(f - f_n)) \to 0$ and $L_g(\lambda_0(f - f_n)) \to 0$. It follows that $L(f_n) \to L(f)$ and $L_g(f_n) \to L_g(f)$ so $L(f) = L_g(f)$. □

We once again see that Corollary 2.23 holds since under the $\Delta_2$ condition, mean continuity is equivalent to norm continuity.

# 3

# Gauges and locally convex spaces

A *topological vector space* is a vector space $X$ (over $\mathbb{R}$ or $\mathbb{C}$) with a Hausdorff topology so that the maps $(x, y) \mapsto x + y$ of $X \times X \to X$ and $(\lambda, x) \mapsto \lambda x$ of $\mathbb{R} \times X$ or $\mathbb{C} \times X \to X$ are continuous. In this chapter, we'll study a class of topological vector spaces which, because of convexity considerations, have a large number of linear functionals. Virtually all spaces that arise in applications are in this class.

For now, the two main examples to bear in mind are Banach spaces in the norm topology and Banach spaces in the weak or weak-$*$ topology. Later we will discuss $L^p$ and $H^p$ for $0 < p < 1$ and the spaces $\mathcal{S}(\mathbb{R}^\nu)$, $\mathcal{D}(\Omega)$ for $\Omega \subset \mathbb{R}^\nu$ and their duals, the spaces of distributions.

To avoid having to say "$\mathbb{R}$ or $\mathbb{C}$" repeatedly, we will use the symbol "$\mathbb{K}$" to stand for one or the other.

If $X$ is a real vector space, we defined $K \subset X$ to be *balanced* if and only if $x \in K$ implies $-x \in K$. If $V$ is a complex vector space, $K \subset V$ is called *balanced* if and only if $x \in K$ and $\lambda \in \partial\mathbb{D} = \{\lambda \in \mathbb{C} \mid |\lambda| = 1\}$ implies $\lambda x \in K$. Sometimes the phrase "circled" is used in the complex case, but we settle for a single term. Also "absolutely convex" is sometimes used for "balanced and convex." We call $K$ *pseudo-open* (resp. *pseudo-closed*) if for all $x \in K$, $y \in X$, $\{\lambda \in \mathbb{K} \mid x + \lambda y \in K\}$ is open (resp. closed) in $\mathbb{K}$. Notice that if $0 \in K$ and $K$ is pseudo-open, then $K$ is absorbing.

**Proposition 3.1** *Let $X$ be a topological vector space.*

(i) *Every open* (*resp. closed*) *set is pseudo-open* (*resp. pseudo-closed*).

(ii) *Let $U$ be a neighborhood of* $0$. *Then there exists an open neighborhood, $V$, of* $0$ *with $V \subset U$ and $V$ balanced.*

(iii) *Let $U$ be a neighborhood of* $0$. *Then there exists an open neighborhood, $V$, of* $0$ *with $V + V \subset U$.*

*Proof* (i) For $x \in K$, $y \in X$, $f \colon \lambda \mapsto (x + \lambda y)$ is continuous so $f^{-1}$ of an open (resp. closed) set is open (resp. closed).

(ii) For the real case, take $V = U \cap (-U)$. For the complex case, let $F\colon \mathbb{C}\times X \to X$ by $f(\lambda,x) = \lambda x$. $F^{-1}[U]$ is a neighborhood of $(\lambda,0)$ for each $\lambda \in \partial\mathbb{D}$. Thus, for each $\lambda \in \partial\mathbb{D}$, we can find $N_\lambda$ an open neighborhood of $\lambda$ in $\partial\mathbb{D}$ and $W_\lambda$ an open neighborhood of 0 in $X$ so that $\mu \in N_\lambda$ and $x \in W_\lambda$ implies $\mu x \in U$. By compactness of $\partial\mathbb{D}$, pick $\lambda_1,\dots,\lambda_n$ so $\cup_{i=1}^n N_{\lambda_i} = \partial\mathbb{D}$. Let $V_1 = \cap_{i=1}^n W_{\lambda_i}$. Then $V_1$ is open and for any $\mu \in \mathbb{D}$, $\mu V_1 \subset U$ since $\mu \in N_{\lambda_i}$ for some $i$ and then $\mu V_1 \subset \mu W_{\lambda_i} \subset U$. Let $V = \cup_{\mu\in\partial D} \mu V_1$. Then $V \subset U$, $V$ is open, $0 \in V$, and $V$ is circled.

(iii) Since $P\colon X \times X \to X$ by $(x,y) \mapsto x+y$ is continuous, we can find open $V_1, V_2$ neighborhoods of 0 so $V_1 + V_2 \subset U$. Pick $V = V_1 \cap V_2$. □

**Definition** Let $X$ be a topological vector space. $B \subset X$ is called *bounded* if and only if for any open neighborhood $U$ of 0, there is $\lambda$ in $\mathbb{K}$ so that $B \subset \lambda U$.

Note that since an open set is pseudo-open, $\cup_\lambda \lambda U = X$.

If $X$ is a Banach space with the norm topology picking $U = \{x \mid \|x\| \le 1\}$, $B \subset \lambda U$ if and only if $\sup_{x\in B}\|x\| \le |\lambda|$ so this extends the natural notion of bounded from the Banach space context. To treat the case of the weak topology on a Banach space, we need

**Proposition 3.2** *Let $X$ be a Banach space and let $A$ be a set of elements of $X^*$, so for each $x \in X$, $\{\ell(x) \mid \ell \in A\}$ is a bounded subset of $\mathbb{K}$. Then*

$$\sup_{\ell\in A}\|\ell\| < \infty \tag{3.1}$$

*Remarks* 1. This is just the Banach–Steinhaus principle. The direct proof is so short, we give it. The usual proof appeals to the Baire category theorem whose proof is essentially included in the proof below.

2. If $A$ is a subset of $X$ with $\{\ell(x) \mid x \in A\}$ bounded for each $\ell \in X^*$, then we can view $A \subset (X^*)^*$ and apply the proposition to see that $\sup_{x\in X}\|x\| < \infty$.

*Proof* Suppose we find a ball $B_{x_0}^r \subset X$ so

$$s = \sup\{|\ell(x)| \mid \ell \in A, x \in B_{x_0}^r\} < \infty \tag{3.2}$$

Then, since

$$\|\ell\| = \sup_{x\in B_0^1}|\ell(x)|$$

and

$$x_0 + rB_0^1 = B_{x_0}^r$$

we see that

$$\sup_{\ell \in A} \|\ell\| \leq r^{-1}\left[s + \sup_{\ell \in A} |\ell(x_0)|\right] < \infty$$

We see that if (3.1) fails, then (3.2) fails for all $x_0$ and $r > 0$.

Suppose (3.1) fails. Pick $x_1, x_2, \dots \in X$, $r_1 > r_2 > \cdots$ and $\ell_1, \ell_2, \dots \in A$ as follows: Pick $\ell_1 \in A$ so $\|\ell_1\| \geq 1$, then $x_1$ so $\ell_1(x_1) \geq 1$, and set $r_1 = (2\|\ell\|)^{-1} \leq \frac{1}{2}$. Thus, if $x \in \overline{B_{x_1}^{r_1}}$, $\ell_1(x) \geq 1 - \frac{1}{2} \geq \frac{1}{2}$. Assuming we have picked $\ell_1, \dots, \ell_{n-1}$, $x_1, \dots, x_{n-1}$, and $r_1, \dots, r_{n-1}$, pick $\ell_n, x_n, r_n$ as follows. Since (3.2) must fail for $B_{x_n-1}^{r_n-1}$, pick $x_n \in B_{x_n-1}^{r_n-1}$ and $\ell_n \in A$ so $|\ell_n(x_n)| \geq n$. Pick $r_n$ so $B_{x_n}^{r_n} \subset B_{x_n-1}^{r_n-1}$, $r_n < 2^{-n}$, and $r_n \leq \frac{n}{2}\|\ell_n\|^{-1}$. Thus, if $x \in \overline{B_{x_n}^{r_n}}$, then $|\ell_n(x)| \geq |\ell_n(x_n)| - \|x_n - x\|\,\|\ell_n\| \geq n/2$.

Thus, $x_j, x_{j+1}, \dots \subset B_{x_j}^{r_j}$ so $\|x_m - x_j\| \leq 2^{-j}$ for $m \geq j$. Since $x_j$ is Cauchy, it has a limit $x_\infty \subset \overline{B_{x_j}^{r_j}}$. Thus, $|\ell_j(x_\infty)| \geq j/2$ and so $\sup_A |\ell(x_\infty)| = \infty$, contradicting the hypothesis. It follows that (3.1) must hold. □

**Corollary 3.3** *Let $X$ be a Banach space viewed as a topological vector space with the weak topology (or weak-$*$ topology if $X$ is a dual space). Then any set $A$ is bounded if and only if* $\sup\{\|y\| \mid y \in A\} < \infty$.

*Proof* A base of neighborhoods of 0 are the $U_{\ell_1,\dots,\ell_n} = \{x \mid |\ell_1(x)| < 1, \dots, |\ell_n(x)| < 1\}$ for arbitrary $\ell_1, \ell_2, \dots, \ell_n$ in $X^*$. If $r_A = \sup_{y \in A} \|y\| < \infty$, then $A \subset [r_A \sup_{j=1,\dots,n} \|\ell_j\|] U_{\ell_i,\dots,\ell_n}$ so $A$ is weakly bounded.

Conversely, if $A$ is weakly bounded, for each $\ell \in X^*$, $A \subset r_\ell\{x \mid |\ell(x)| < 1\}$, that is, $\sup_{y \in A} |\ell(y)| \leq r_\ell$. By Proposition 3.2, $r_A$ is bounded. □

Completeness of a metric space is not a function of the topology alone: $(0,1)$ and $\mathbb{R}$ with their usual metrics and topologies are homeomorphic, but only the latter is complete as a metric space. What one needs for completeness is a way of comparing nearby points to $x_n$ and nearby points to another $x$. This can be done in general with the notion of uniform structure (see the Notes), but can be done naturally for vector spaces using the additive structure to compare points.

**Definition** Let $X$ be a topological vector space. A net $\{x_\alpha\}_{\alpha \in I}$ is called *Cauchy* if and only if for any neighborhood, $U$, of 0, there exists $\gamma \in I$ so if $\alpha, \beta > \gamma$, then $x_\alpha - x_\beta \in U$. $X$ is called *complete* if every Cauchy net converges, *boundedly complete* if every bounded Cauchy net converges, and *sequentially complete* if every Cauchy sequence converges. Some authors use *quasi-complete* where we use boundedly complete.

Of course, if the topology is given by a metric, sequentially complete is the same as complete. If 0 has a bounded neighborhood (we will prove below that in the locally convex case, such a space is a normed space!), every Cauchy sequence is eventually bounded, and so bounded completeness is equivalent to completeness. In particular, for topologies given by a norm, all three notions are equivalent.

Two Cauchy nets, $\{x_\alpha\}_{\alpha\in I}$ and $\{y_a\}_{a\in J}$, are called *equivalent* if and only if for any neighborhood, $U$, of 0, there exist $\gamma \in I$ and $c \in J$ so that $\alpha \geq \gamma$ and $a \geq c$ implies $x_\alpha - y_c \in U$. As with metric spaces, there is a natural way to make the set of equivalence classes of Cauchy nets into a complete topological vector space in which $X$ is dense. It is called the completion of $X$.

While completeness is the right notion for topologies given by lots of norms, it is not a useful notion for weak topologies, since we are about to show that spaces with weak topologies are essentially never complete. To state this in great generality, it pays to consider a general notion of weak topologies that will play a major role in Chapter 5.

**Definition** Let $X$ and $Y$ be two vector spaces, both over the same $\mathbb{K}$. A *duality* between $X$ and $Y$ is a nondegenerate pairing, that is, a bilinear map $x, y \mapsto \langle x, y\rangle$ of $X \times Y$ to $\mathbb{K}$ so that for every $x \neq 0$ in $X$, there is $y$ with $\langle x, y\rangle \neq 0$, and for every $y \neq 0$ in $Y$, there is $x$ with $\langle x, y\rangle \neq 0$. A *dual pair* is a pair of two vector spaces with a duality between them.

*Remarks* 1. In other words, $Y$ is a family of linear functionals on $X$ that separates points.

2. In some references, duality does not include nondegeneracy, and if nondegeneracy holds, one speaks of a "strict duality."

The following will be useful here and later in Chapter 5:

**Proposition 3.4** *Let $X, Y$ be a dual pair. Let $y_1, \dots, y_\nu$ be linearly independent elements of $Y$. Let $\alpha_1, \dots, \alpha_\nu$ be arbitrary elements of $\mathbb{K}$. Then there is an $x \in X$ with $\langle x, y_j\rangle = \alpha_j$ for $j = 1, \dots, \nu$.*

*Proof* Let $\Phi\colon X \to \mathbb{K}^\nu$ by $\Phi(x)_j = \langle x, y_j\rangle$. If $\operatorname{Ran}\Phi$ is not all of $\mathbb{K}^\nu$, there exists $\gamma \in \operatorname{Ran}\Phi^\perp$ with $^\perp$ computed in the Euclidean inner product. Thus, for all $x$, $\langle x, \sum \gamma_j y_j\rangle = 0$. By the nondegeneracy, $\sum \gamma_j y_j = 0$, violating independence. □

**Definition** Let $X, Y$ be a dual pair. The *$Y$-weak topology*, denoted $\sigma(X, Y)$ on $X$, is the weakest topology on $X$, making each $\langle\,\cdot\,, y\rangle$ into a continuous linear map of $X$ to $\mathbb{K}$. The family $\{x \mid |\langle x, y_1\rangle| < 1, \dots, |\langle x, y_\ell\rangle| < 1\}$ of sets for all $y_1, \dots, y_\ell \in Y$ is a base of neighborhoods of 0 in the topology.

Dual topologies will be studied extensively in Chapter 5. Since a duality separates points, $\sigma(X, Y)$ makes $X$ into a topological vector space.

**Proposition 3.5** *Let $X, Y$ be a dual pair. $X$ is complete in the $\sigma(X, Y)$-topology if and only if $X$ is the algebraic dual of $Y$, that is, the space of all linear functionals on $Y$. In general, the completion of $X$ in the $\sigma(X, Y)$-topology is $Y^*_{\text{alg}}$, the algebraic dual of $Y$.*

*Proof* We will show $Y^*_{\text{alg}}$ is complete in the $\sigma(Y^*_{\text{alg}}, Y)$-topology and any $X \subset Y^*_{\text{alg}}$ that separates points of $Y$ is dense in $Y^*_{\text{alg}}$ in the $\sigma(Y^*_{\text{alg}}, Y)$-topology. The proposition then follows. If $\ell_\alpha \in Y^*_{\text{alg}}$ is Cauchy, then $\ell_\alpha(y)$ is Cauchy in $\mathbb{K}$ for each $y$, and so converges to some $\ell(y)$. Since each $\ell_\alpha$ in $\mathbb{K}$ is linear, so is $\ell$, that is, $\ell \in Y^*_{\text{alg}}$, so $\ell_\alpha \to \ell$, proving $Y^*_{\text{alg}}$ is complete.

If $X$ separates points in $Y$, given $\ell \in Y^*_{\text{alg}}$ and $y_1, \dots, y_n \in Y$, pick, by Proposition 3.4, $x_{y_1,\dots,y_n} \in X$ so $\langle x, y_j \rangle = \ell(y_j)$ for $j = 1, \dots, n$. Order finite subsets of $Y$ by inclusion. $x_{y_1,\dots,y_n}$ is a net and it clearly converges to $\ell$ in the $\sigma(Y^*_{\text{alg}}, Y)$-topology. □

Thus, completeness is much too strong a notion for weak topologies. Boundedly completeness is not, in some cases.

**Theorem 3.6** *Let $X$ be a Banach space. Let $X^*$ be given the $\sigma(X^*, X)$- (i.e., weak-$*$) topology. Then $X^*$ is boundedly complete and also sequentially complete.*

*Proof* Since balls in $X^*$ are $\sigma(X^*, X)$-compact (see Theorem 5.12), any bounded net has a convergent subnet. Hence, any bounded Cauchy net converges. If $\{\ell_n\}$ is a $\sigma(X^*, X)$ Cauchy sequence in $X^*$, thus for any $x \in X$, $\ell_n(x)$ is a Cauchy sequence in $\mathbb{K}$, so bounded. Thus, for any $x$, $\sup_n |\ell_n(x)| < \infty$. By Proposition 3.2, $\sup_n \|\ell_n\|$ is bounded so, by compactness, $\ell_n$ converges. □

**Theorem 3.7** *If $X$ is a finite-dimensional topological vector space, $X$ is homeomorphic to $\mathbb{K}^\nu$ by a linear map.*

*Proof* Let $e_1, \dots, e_\nu$ be a basis of $X$. Let $f\colon \mathbb{K}^\nu \to X$ by $f(\beta_1, \dots, \beta_n) = \sum_{i=1}^n \beta_i e_i$ and let $\lambda\colon X \to \mathbb{K}^\nu$ be its inverse. Because the vector space operators are continuous, $f$ is continuous. So we need only show $\lambda$ is continuous, that is, if $\{x_\alpha\}_{\alpha\in I}$ is a net and $x_\alpha \to x_\infty$, then $\lambda(x_\alpha) \to \lambda(x_\infty)$. By adding $-x_\infty + e_1$ to $x_\alpha$, we can suppose $x_\infty = e_1$. Notice if $\lambda_\infty$ is a limit point of $\lambda(x_\alpha)$, then $x(\lambda_\infty)$ is a limit point of $x_\alpha$ and so $\lambda_\infty = (1, 0, \dots, 0)$.

Suppose first $\|\lambda(x_\alpha)\|$ is eventually bounded where $\|\cdot\|$ is the Euclidean norm on $\mathbb{K}^\nu$. Then, since closed balls in $\mathbb{K}^\nu$ are compact and $(1, 0, \dots, 0)$ is the only limit point of $\{\lambda(x_\alpha)\}$, $\lambda(x_\alpha) \to (1, 0, \dots, 0) = \lambda(x_\infty)$.

If $\|\lambda(x_\alpha)\|$ is not eventually bounded, pick $\{y_\alpha\}_{\alpha\in I}$ as follows. $y_\alpha = x_{\beta(\alpha)}$ with $\beta(\alpha)$ is chosen so $\beta(\alpha) > \alpha$ and so $\|\lambda(x_{\beta(\alpha)})\| \geq 2$. Since $\beta(\alpha) > \alpha$, $y_\alpha \to x_\infty$ also. Since $\|\lambda(y_\alpha)\|^{-1} \leq \frac{1}{2}$, we can find a subnet with $\|\lambda(y_\alpha)\|^{-1} \to \mu_\infty \leq \frac{1}{2}$, and a further subnet so $\lambda(y_\alpha)/\|\lambda(y_\alpha)\| \to \beta_\infty \in \mathbb{K}^\nu$ with $\|\beta_\infty\| = 1$ since $\{\beta \in \mathbb{K}^\nu \mid \|\beta\| = 1\}$ is compact. Thus, since $x$ is continuous, $y_\alpha/\|\lambda(y_\alpha)\| \to x(\beta_\infty)$. On the other hand, $y_\alpha/\|\lambda(y_\alpha)\| \to \mu_\infty e_1$ so $x(\beta_\infty) = \mu_\infty e_1$ or $\beta_\infty = (\mu_\infty, 0, \dots, 0)$, violating $\mu_\infty \leq \frac{1}{2}$. This contradiction shows $\|\lambda(x_\alpha)\|$ is eventually bounded. Thus, $\lambda$ is continuous. □

**Corollary 3.8** *All norms on $\mathbb{K}^\nu$ are equivalent, that is, for $\|\cdot\|_1$ and $\|\cdot\|_2$ on norms, there exist c and d $\in (0,\infty)$ so*

$$c\|x\|_1 \le \|x\|_2 \le d\|x\|_1$$

*Proof* By the theorem, the identity map from $(\mathbb{K}^\nu, \|\cdot\|_1)$ to $(\mathbb{K}^\nu, \|\cdot\|_2)$ is continuous, that is, $\|x\|_2 \le d\|x\|_1$. By symmetry, the other relation holds. □

**Corollary 3.9** *If $X$ is a topological vector space and $M \subset X$ is a finite-dimensional subspace, then $M$ is closed.*

*Proof* Let $\{m_\alpha\}_{\alpha\in I}$ be a net in $M$ with $m_\alpha \to x \in X$. Then $m_\alpha$ is Cauchy in the topology $X$ induces on $M$ (where open sets are $M \cap U$ with $U$ open in $X$). By the theorem, $M$ with this induced topology is $\mathbb{K}^\nu$ which is complete, so $m_\alpha \to m$ in $M$ and so $x = m \in M$, that is, $m$ is closed. □

**Theorem 3.10** *Let $X$ be a topological vector space and let $K \subset X$ be compact. If $K^{\text{int}}$ is nonempty, then $X$ is finite-dimensional.*

*Remark* In other words, the only locally compact topological vector spaces are $\mathbb{K}^\nu$.

*Proof* If $x_0 \in K^{\text{int}}$, then $K - x_0$ is compact and $0 \in (K - \{x_0\})^{\text{int}}$. So, without loss, suppose $0 \in U \equiv K^{\text{int}}$. Since $\frac{1}{2}U$ is open and $K$ is covered by $\cup_{x\in K}(x+\frac{1}{2}U)$, we can find $x_1,\dots,x_\ell$ so $K \subset \cup_{i=1}^{\ell}(x_i + \frac{1}{2}U)$. Let $M$ be the subspace generated by $\{x_i\}$ which has dimension at most $\ell$. Then

$$U \subset M + \tfrac{1}{2}U$$

so

$$U \subset M + \tfrac{1}{2}(M + \tfrac{1}{2}U) = M + \tfrac{1}{4}U$$

By induction,

$$U \subset M + \frac{1}{2^n}U$$

Thus, if $u \in U$, there exist $m_n \in M$ and $u_n \in U$ so

$$u = m_n + \frac{1}{2^n}u_n \tag{3.3}$$

Since $\{u_n\} \subset U \subset K$, $u_n$ has a limit point $v$ so $\frac{1}{2^n}u_n$ has a limit point 0 so, by (3.3), $u$ is a limit point of $m_n$. Since $M$ is closed, $U \subset M$.

But $U$ is an open set, so pseudo-open, so absorbing, that is, $\cup_{a=1}^{\infty} aU = X$. Thus, $X = M$ is finite-dimensional. □

In the remainder of this chapter, convexity, which has not appeared yet, will play a critical role. We begin with

**Proposition 3.11** *Let V be a topological vector space and K convex. Then*
(i) $\bar{K}$ *is convex.*
(ii) $K^{\text{int}}$ *is convex.*
(iii) *If K is balanced and convex and* $K^{\text{int}} \neq \emptyset$*, then* $0 \in K^{\text{int}}$.

*Proof* (i) Let $x \in \bar{K}$ and $y \in K$. There is a net $\{x_\alpha\}$ in $K$ so $x_\alpha \to x$. Then, by continuity of scalar multiplication and addition, $\theta x_\alpha + (1-\theta)y \to \theta x + (1-\theta)y$, so the later points are in $\bar{K}$. Now let $x \in \bar{K}$, $y \in \bar{K}$. Pick a net $\{y_\beta\}$ in $K$ with $y_\beta \to y$. Then $\theta x + (1-\theta)y_\beta \in \bar{K}$ and it converges to $\theta x + (1-\theta)y$ which is also in $\bar{K}$. It follows that $K$ is convex.

(ii) Let $x, y \in K^{\text{int}}$. Let $\theta \in (0,1)$. Let $U$ be an open neighborhood of 0 with $x + U \subset K$. Then $\theta(x+U) + (1-\theta)y = \theta x + (1-\theta)y + \theta U \subset K$ and $\theta U$ is open since multiplication is continuous and $\theta U$ is the inverse image of $U$ under $x \mapsto \theta^{-1}x$.

(iii) Let $x \in K^{\text{int}}$ and let $U$ be an open neighborhood of 0, with $x + U \subset K$. Then $-x \in K$ and so $\frac{1}{2}(-x) + \frac{1}{2}(x+U) = \frac{1}{2}U \subset K$. Since $\frac{1}{2}U$ is an open neighborhood of 0 (as above), $0 \in K^{\text{int}}$. □

**Proposition 3.12** *Let X be a topological vector space. Let U be a convex neighborhood of* 0*. Then there exists* $V \subset U$ *with* $0 \in V$ *so that V is convex, open, and balanced.*

*Proof* By Proposition 3.1, we can find $V_1$ balanced and open with $0 \in V_1 \subset U$. Then

$$\begin{aligned} V_2 &= \{\theta x + (1-\theta)y \mid 0 \le \theta \le 1,\, x, y \in V_1\} \\ &= V_1 \cup \bigcup_{0<\theta<1} \bigcup_{x \in V_1} [\theta x + (1-\theta)V_1] \end{aligned}$$

is open, balanced, and in $U$ since $U$ is convex and $0 \in V_1 \subset V_2 \subset U$. If $V_{n+1} = \{\theta x + (1-\theta y) \mid 0 \le \theta \le 1,\, x, y \in V_n\}$, then $V_n$ is balanced, open, and $0 \in V \subset V_1 \subset \cdots \subset V_n \subset U$. $V = \cup V_n$ is balanced, open, convex, and in $U$. □

**Proposition 3.13** *Let X be a topological vector space. Let U be an open, balanced, convex neighborhood of* 0*. Then U is absorbing and its gauge,* $\rho_U$*, given by Corollary 1.10 is a continuous function on X with* $U = \{x \mid \rho_U(x) < 1\}$ *and is a seminorm. If K is a closed, balanced, convex set and* $U \equiv K^{\text{int}}$ *is nonempty, then* $K = \overline{K^{\text{int}}}$ *and* $K = \{x \mid \rho_U(x) \le 1\}$.

*Proof* Let $x \neq 0$ in $X$ and let $I_x = \{\lambda \in \mathbb{R} \mid \lambda x \in U\}$. $I_x$ is open (by Proposition 3.1) and contains $\lambda \ge 0$ so for some $\varepsilon > 0$, $\varepsilon x \in U$, that is, $U$ is absorbing. Thus, the gauge is convex and a seminorm. Moreover, by Remark 5 after Corollary 1.10, $U = \{x \mid \rho_U(x) < 1\}$.

By scaling, for any $\lambda > 0$, $U_\lambda \equiv \{x \mid \rho_U(x) < \lambda\} = \lambda^{-1}U$ is open. If $\rho_U(x) > 1$, then since $\rho_U$ is a seminorm,

$$A(x) \equiv \{y \mid \rho_U(x-y) < \rho_U(x) - 1\} \subset \{y \mid \rho_U(y) > 1\}$$

(since $\rho_U(y) \geq \rho_U(x) - \rho_U(x-y)$). $A(x) = x + U_{\rho_U(x)-1}$ is open, so $W = \{x \mid \rho_U(x) > 1\}$ is open and so $W_\lambda = \{x \mid \rho_U(x) > \lambda\} = \lambda^{-1}W$ is open. It follows for $0 < \mu < \lambda$, $\{x \mid \mu < \rho_U(x) < \lambda\} = U_\lambda \cap W_\mu$ is open. Thus, $\rho_U$ is continuous.

Suppose $K^{\text{int}} \neq \emptyset$. By Proposition 3.11(i), $K^{\text{int}}$ is convex and since $K$ is balanced, $K^{\text{int}}$ is balanced. Thus, $U \equiv K^{\text{int}}$ is an open, balanced, convex neighborhood of 0.

Let $x \in K$ and let $0 \leq \theta < 1$. Then $\theta x + (1-\theta)U \subset K$ and so $\theta x \in K^{\text{int}} = U$. It follows that for any $x \in K$, $\{\lambda > 0 \mid \lambda x \in U\}$ and $\{\lambda > 0 \mid \lambda x \in K\}$ differ by at most a single point, and thus, $\rho_U = \rho_K$. By Remark 5 after Corollary 1.10, since $K$ is closed, $K = \{x \mid \rho_K(x) \leq 1\}$. It follows that $\bar{U} = K$. □

*Remark* The proof of continuity of $\rho_U$ did not use the fact that $U$ is balanced. If $U$ is an arbitrary open convex set with $0 \in U$, then $\rho_U$ obeys $\rho_U(x) \leq \rho_U(y) + \rho_U(x-y)$, and that implies continuity of $\rho_U$.

**Theorem 3.14** (Kolmogorov's Theorem) *Let $X$ be a topological vector space. Then the topology of $X$ is given by a norm* (i.e., *$X$ is a normed linear space*) *if and only if* 0 *has a bounded convex neighborhood.*

*Proof* If the topology of $X$ comes from a norm, $\{x \mid \|x\| < 1\}$ is a bounded convex neighborhood of 0.

Conversely, suppose 0 has a bounded convex neighborhood. By Proposition 3.12, we can find $U$, bounded, open, balanced, convex, and a neighborhood of 0. If $x \neq 0$, let $W$ be a neighborhood of 0 with $x \notin W$. Since $U$ is bounded, $U \subset \lambda W$ for some $\lambda$. But $\lambda x \notin \lambda W$ so $\lambda x \notin U$ and $\rho_U(x) \geq \lambda^{-1}$, so $\rho_U(x) \neq 0$. It follows that $\rho_U$ is a norm.

If $Y$ is any neighborhood of 0, $U \subset \lambda Y$ for some $\lambda$, so $Y \supset \lambda^{-1}U$. Thus, $\{n^{-1}U \mid n = 1, 2, \dots\}$ is a base for the neighborhood of 0, so the topology is given by the norm $\rho_U$. □

**Example 3.15** ($L^p$, $0 < p < 1$) Let $0 < p < 1$. Define

$$L^p(0,1) = \left\{ f \,\middle|\, \int_0^1 |f(x)|^p \, dx \equiv \rho_p(f) < \infty \right\}$$

We claim that

$$\rho_p(f+g) \leq \rho_p(f) + \rho_p(g) \tag{3.4}$$

For Minkowski's inequality in $\mathbb{R}^2$ on $(a,0) + (0,b)$ shows for $a, b \geq 0$, ($p < 1$ means $\frac{1}{p} > 1$)

$$(a^{1/p} + b^{1/p})^p \leq a + b$$

so with $\alpha = a^{1/p}$, $\beta = b^{1/p}$,

$$(\alpha + \beta)^p \le \alpha^p + \beta^p$$

which means

$$|f(x) + g(x)|^p \le (|f(x)| + |g(x)|)^p \le |f(x)|^p + |g(x)|^p$$

yielding (3.4).

Thus, $\rho_p$ has one of the two properties of a norm, but instead of homogeneity of degree one, we have

$$\rho_p(\alpha f) = |\alpha|^p p(f) \tag{3.5}$$

for $\alpha$ in $\mathbb{K}$. Place a metric on $L^p$ by

$$m(f, g) = \rho_p(f - g) \tag{3.6}$$

Then (3.4) implies the triangle inequality and continuity of addition. (3.5) implies continuity of scalar multiplication; indeed, all one needs is $\rho_p(\alpha f) \le G(\alpha)\rho_p(f)$ where $\lim_{\alpha \downarrow 0} G(\alpha) = 0$. Finally, given $f \neq g$, if

$$U = \{h \mid \rho_p(f - h) < \tfrac{1}{2}\, \rho_p(f - g)\}$$
$$V = \{h \mid \rho_p(g - h) < \tfrac{1}{2}\, \rho_p(f - g)\}$$

then $U \cap V = \emptyset$ and both are open, so the topology is Hausdorff (as is any metric topology) so $L^p$ is a topological vector space. It is not hard to see that $L^p(0, 1)$ is complete.

The hypothesis of Theorem 3.14 fails in a very strong way for we claim that the only nonempty, convex open set is all of $L^p(0, 1)$! For let $U$ be a convex open neighborhood of 0. Then $U$ must contain some set $\{g \mid \rho_p(g) \le \varepsilon\}$ for some $\varepsilon > 0$. By scaling, we can suppose $\varepsilon = 1$ (for if $\varepsilon^{-1/p} U = L^p$, then $U$ also equals $L^p$).

Given $f \in L^p(0, 1)$ with $\rho_p(f) = 1$, note that $H(y) = \int_0^y |f(x)|^p \, dx$ is continuous, running from 0 to 1. Define $y_j^N = \inf_y \{y \mid H(y) = \frac{j}{N}\}$ and let $f_j^N = N^{1/p} f \chi_{[y_{j-1}^N, y_j^N]}$. Then $\rho_p(f_j^N) = [N^{1/p}]^p [H(y_j^N) - H(y_{j-1}^N)] = NN^{-1} = 1$ so $f_j^N \in U$. Therefore, since $U$ is convex,

$$\sum_{j=1}^{N} \frac{1}{N} f_j^N = N^{1/p-1} f \in U$$

Since $N^{1/p-1} \to \infty$ and $0 \in U$, $\lambda f \in U$ for all $\lambda > 0$, so $U = L^p$.

This fact not only shows that the topology of $L^p$ is not given by a norm. It implies that $L^p$, $0 < p < 1$, has no nonzero continuous linear functionals! For let $\ell$ be such a functional. Then $\{x \mid |\ell(x)| < 1\}$ is a convex open neighborhood of 0. Thus, for any $f$ and all $\lambda$, $|\ell(\lambda f)| \le 1$ so $|\ell(f)| \le |\lambda|^{-1}$, which means $\ell(f) = 0$. Thus, $\ell \equiv 0$.

This motivates our focusing below on locally convex spaces: to have lots of linear functionals, we will want lots of open convex sets. □

**Example 3.16** ($H^p, 0 < p < 1$) Let $f$ be analytic on $D$ and let

$$N_p(f)(r) = \frac{1}{2\pi} \int_0^{2\pi} |f(re^{i\theta})|^p \, d\theta \tag{3.7}$$

For any $p \in (0, \infty)$, $N_p$ is monotone in $r$ (see, e.g., Duren [106]) and $H^p(D)$ is defined as those analytic $f$ with

$$\rho_p(f) = \sup_{0<r<1} N_p(f)(r) = \lim_{r \uparrow 1} N_p(f)(r) < \infty \tag{3.8}$$

As with $L^p$, if $0 < p < 1$, $\rho_p(H)$ obeys (3.4) so (3.6) makes $H^p$ into a metric topological vector space.

Unlike $L^p$, $H^p$ has many continuous linear functionals. For monotonicity of $N_p$ in $r$ shows

$$\begin{aligned} |f(0)|^p &= \lim_{r \downarrow 0} N_p(f)(r) \\ &\leq \int_0^\rho \frac{N_p(f)(r) \, r \, dr}{\frac{1}{2}\rho^2} \\ &= \frac{1}{2\pi} \int_0^\rho \frac{|f(re^{i\theta})|^p \, dr d\theta}{\frac{1}{2}\rho^2} \\ &= \frac{1}{\pi\rho^2} \int_{|z|<\rho} |f(z)|^p \, d^2 z \end{aligned}$$

Thus, for any $z_0 \in \mathbb{D}$,

$$\begin{aligned} |f(z_0)|^p &\leq \frac{1}{\pi(1 - |z_0|)^2} \int_{|z - z_0| < 1 - |z_0|} |f(z)|^p \, d^2 z \\ &\leq \frac{1}{\pi(1 - |z_0|^2)} \int_{|z|<1} |f(z)|^p \, d^2 z \\ &\leq \frac{1}{1 - |z_0|^2} \rho_p(f) \end{aligned}$$

This means if

$$\ell_{z_0}(f) = f(z_0)$$

then

$$|\ell_{z_0}(f) - \ell_{z_0}(g)| \leq \left[ \frac{1}{1 - |z_0|^2} \rho_p(f - g) \right]^{1/p}$$

so $\ell_{z_0}$ is a continuous linear functional on $H^p(0, 1)$. Thus, there are enough continuous linear functionals to distinguish points of $H^p$, and so enough convex open sets to separate points.

It can be shown, however, that there exists a closed proper $V \subset H^p$ so that any continuous linear functional that vanishes on $V$ vanishes on all of $H^p$; see the references in the Notes. That means if $w \notin V$, there is no open convex neighborhood of $W$ disjoint from $V$. We will settle here for showing $\{f \mid \rho_p(f) < 1\}$ does not contain any open convex set. By Theorem 3.14, this implies that $H^p$ is not a normed linear space. It will also imply – once we describe locally convex spaces – that $H^p$ is not locally convex.

Define $b_1^N$ on $[0, 2\pi]$ by

$$b_1^N(\theta) = \begin{cases} \frac{N\theta}{\pi}, & 0 \le \theta \le \frac{\pi}{N} \\ 1 - \frac{N\theta}{\pi}, & \frac{\pi}{N} \le \theta \le \frac{2\pi}{N} \\ 0, & \frac{2\pi}{N} \le \theta \le 2\pi \end{cases}$$

$b_1^N$ is a continuous function so, since $\{e^{im\theta}\}_{m=-\infty}^{\infty}$ are total by the Weierstrass theorem, for any $\delta$, we can find $g_{1,\delta}^N(\theta) = \sum_{j=-K_{N,\delta}}^{K_{N,\delta}} e^{ij\theta}\eta_{j,\delta}^N$ so $\|g_{1,\delta}^N - b_1^N\| \le \delta$. Let

$$f_{1,\delta}^N(e^{i\theta}) = N^{1/p} e^{ik_{N,\delta}\theta} g_{1,\delta}^N(\theta)$$

so $f_{1,\delta}^N$ is a polynomial in $z = e^{i\theta}$ and so in $H^p$.

Define $b_j^N(\theta) = b_1^N(\theta - (j-1)\pi/N)$ and $f_{j,\delta}^N(z) = f_{1,\delta}^N(z\exp(-i(j-1)\pi/N))$. Then

$$\begin{aligned} \lim_{\delta\downarrow 0} \rho_p(f_{j,\delta}^N) &= \frac{N}{2\pi}\int |b_1^N(\theta)|^p\, d\theta \\ &= 2\int_0^{1/2} (2x)^p\, dx \\ &= \frac{1}{1+p} < 1 \end{aligned} \tag{3.9}$$

On the other hand,

$$\begin{aligned} \lim_{\delta\downarrow 0} \rho_p\left(\frac{1}{N}\sum_{j=1}^N f_{j,\delta}^N\right) &= N^{1-p} \lim_{\delta\downarrow 0} \rho_p\left(\sum_{j=1}^N N^{-1/p} f_{j,\delta}^N\right) \\ &= N^{1-p} \int_0^{2\pi} \sum_{j=1}^N |b_j^N(\theta)|^p \frac{d\theta}{2\pi} \\ &= \frac{N^{1-p}}{(1+p)} \end{aligned} \tag{3.10}$$

Suppose $\{f \in H^p \mid \rho_p(H) < 1\}$ contains an open convex set $U$ containing 0. Then $U \supset \{f \in H^p \mid \rho_p(f) < \varepsilon\}$. Letting $W = \varepsilon^{-1}U$, we see

$$\{f \in H^p \mid \rho_p(f) < 1\} \subset W \subset \{f \in H^p \mid \rho_p(f) < \varepsilon^{-1}\} \tag{3.11}$$

Pick $N$ so $N^{1-p}/(1+p) > \varepsilon^{-1}$. For $\delta$ small, by (3.9) and (3.11), each $f_{j,\delta} \in W$ so, since $W$ is convex, $\frac{1}{N}\sum_{j=1}^{N} f_{j,\delta} \in W$ for all small $\delta$. Thus, by (3.10) and (3.11), $N^{1-p}/(1+p) \le \varepsilon^{-1}$, violating our choice of $N$. It follows that $\{f \in H^p \mid \rho_p(f) < 1\}$ contains no open convex set containing 0. □

**Definition** A *locally convex space* is a topological vector space with the property that there is a family of convex neighborhoods of 0 which forms a base for the neighborhoods of 0.

The above analysis proves that for $0 < p < 1$, neither $L^p$ nor $H^p$ is locally convex.

By Proposition 3.12, if $X$ is locally convex, we can suppose 0 has a neighborhood base of open, balanced, convex sets.

**Theorem 3.17** *Let $X$ be a locally convex vector space. Then there exists a family of continuous seminorms $\{\rho_\alpha\}_{\alpha\in J}$ on $X$ so that*

(i) *The sets $\{x \in X \mid \rho_{\alpha_1}(x) < \lambda_1, \dots, \rho_{\alpha_n}(x) < \lambda_n\}$ for all $\alpha_1, \dots, \alpha_n \in J$ and $\lambda_1, \dots, \lambda_n > 0$ is a neighborhood base for* 0.

(ii) *A net $\{x_\beta\}_{\beta\in I}$ converges to $x \in X$ if and only if for each $\alpha \in J$, $\rho_\alpha(x - x_\beta) \to 0$ as $\beta \to \infty$.*

*Conversely, if $X$ is a topological vector space with a family of seminorms so that either* (i) *or* (ii) *holds, then $X$ is locally convex and both* (i) *and* (ii) *hold.*

*Proof* Suppose $X$ is locally convex. Let $J$ be a set of convex, balanced, open neighborhoods of 0 that form a neighborhood base. For $U \in J$, let $\rho_U$ be the gauge of $U$. Then $\rho_U$ is a seminorm and it is continuous by Proposition 3.13.

It follows that each set in (i) is open and those sets are a neighborhood base since they include $U = \{x \in X \mid \rho_U(x) < 1\}$. To see (ii), note that if $x - x_\beta \to \infty$, then by continuity of $\rho_U$, each $\rho_U(x - x_\beta) \to 0$. Conversely, if $\rho_U(x - x_\beta) \to 0$ for all $U$, then $x - x_\beta$ is eventually in each neighborhood in (i).

To see the converse, note that (ii) implies (i), and if (i) holds, the sets given are a convex neighborhood base. □

**Theorem 3.18** *Let $X$ be a locally convex vector space. The following are equivalent:*

(i) *The topology of $X$ is given by a metric.*

(ii) 0 *has a countable neighborhood base.*

(iii) *The topology of $X$ is generated by a countable family of seminorms.*

*In case these conditions hold, the metric can be chosen so that*

$$d(x, y) = d(x - y, 0) \tag{3.12}$$

*Proof* (i) ⇒ (ii) If $d$ is the metric, $\{x \mid d(x, 0) < \frac{1}{n}\}$ is a countable neighborhood base.

(ii) $\Rightarrow$ (iii) Since $X$ is locally convex, each set in the presumed countable neighborhood base for 0 contains an open, balanced, convex set, and so 0 has a countable base of such sets. By the last theorem, the countable set of gauges generates the topology.

(iii) $\Rightarrow$ (i) and (3.12) If $\{\rho_n\}_{n=1}^\infty$ is the set of seminorms,

$$d(x,y) = \sum_{n=1}^\infty \min(2^{-n}, \rho_n(x-y)) \tag{3.13}$$

form the required metric. For the sum is convergent and $d(x_m, x) \to 0$ if and only if $\rho_n(x_m - x) \to 0$. □

**Definition** A locally convex space whose topology is given by a metric and which is complete is called a *Fréchet space*.

**Definition** Two sets of seminorms, $\{\rho_\alpha\}_{\alpha\in I}$, $\{\eta_\beta\}_{\beta\in J}$, are called *equivalent* if and only if they generate the same topology. It is easy to see this is true if and only if for all $\alpha \in I$, there are $\beta_1, \dots, \beta_n \in J$ and $C$ so

$$\rho_\alpha(x) \le C\biggl(\sum_{i=1}^n \eta_{\beta_i}(x)\biggr) \tag{3.14}$$

and conversely with the roles of $\rho$ and $\eta$ reversed.

Similarly, it is easy to see if $\{\rho_\alpha\}_{\alpha\in I}$ is a generating family of seminorms, then a linear map $\ell$ from $X$ to $\mathbb{R}$ or $\mathbb{C}$ is continuous if and only if for some $\alpha_1, \dots, \alpha_n$ and $C$,

$$|\ell(x)| \le C\biggl(\sum_{i=1}^n \rho_{\alpha_i}(x)\biggr) \tag{3.15}$$

More generally, if $X$ and $Y$ are locally convex spaces with topologies generated by $\{\rho_\alpha\}_{\alpha\in I}$ and $\{\eta_\beta\}_{\beta\in J}$, a linear map $T\colon X \to Y$ is continuous if and only if

$$|\eta_\beta(Tx)| \le C_\beta\biggl(\sum_{i=1}^n \rho_{\alpha_i}(x)\biggr) \tag{3.16}$$

Here is another concept made easy by the use of seminorms:

**Proposition 3.19** *Let $X$ be a locally convex space and $\{\rho_\alpha\}_{\alpha\in I}$ a set of seminorms generating the topology on $X$. Let $A \subset X$. Then $A$ is bounded if and only if for each $\alpha \in I$,*

$$\sup_{x\in A} \rho_\alpha(x) < \infty \tag{3.17}$$

*Proof* Suppose $A$ is bounded. Since $B_\alpha = \{x \mid \rho_\alpha(x) < 1\}$ is an open set containing 0, $A \subset \lambda_\alpha B_\alpha$ for some $\lambda_\alpha$, that is, $x \in A$ implies $\rho_\alpha(\lambda_\alpha^{-1}x) < 1$ implies $\rho_\alpha(x) < \lambda_\alpha$ so $\sup_{x\in A} \rho_\alpha(x) \le \lambda_\alpha$ and (3.17) holds.

Conversely, if (3.17) holds and $U = \{x \mid \rho_{\alpha_1}(x) < \mu_1, \ldots, \rho_{\alpha_n}(x) < \mu_n\}$ and $\lambda_\alpha = \sup_{x \in A} \rho_\alpha(x)$, then $A \subset \alpha U$ where $\alpha = \max_{i=1,\ldots,n} \frac{\lambda_{\alpha_i}}{\mu_{\alpha_i}}$, so $A$ is bounded. □

**Example 3.20** (Test function spaces) For any multi-index $\alpha$, $D^\alpha \equiv \partial^{|\alpha|}/\partial x_1^{\alpha_1} \ldots \partial x_\alpha^{\alpha_\alpha}$. $\mathcal{S}(\mathbb{R}^\nu)$, the *Schwartz space* of test functions is defined to be the $C^\infty$ functions on $\mathbb{R}^\nu$ with $(1 + |x|^2)^\ell D^\alpha f$ bounded for each $\alpha$ and $\ell$. For each pair of multi-indices, $\alpha$, $\beta$, define the norm

$$\|f\|_{\alpha,\beta} = \|x^\alpha D^\beta f\|_\infty \tag{3.18}$$

(note that $\|f\|_{\alpha,\beta} = 0$ means $D^\beta f = 0$ so $f$ is a polynomial which must be zero if $f$ is bounded). $\mathcal{S}$ with this countable family of seminorms is a metrizable locally convex space. It is often useful to use instead the equivalent class of norms

$$\|f\|_{(n)} = \|(-\Delta + |x|^2 + 1)^n f\| \tag{3.19}$$

It is not hard to see that $\mathcal{S}(\mathbb{R}^\nu)$ is complete, and so a Fréchet space. □

**Example 3.21** (Distribution spaces) Let $\{\alpha_n\}_{n=0}^\infty$ be a family of multi-indices and $\{a_n\}_{n=1}^\infty$ an arbitrary sequence of nonnegative numbers. Let $\mathcal{D}(\mathbb{R}^\nu)$ be the $C^\infty$ functions of compact support. For $f \in \mathcal{D}$, define

$$\|f\|_{\{\alpha\},\{a\}} = \sum_{n=0}^\infty a_n \sup_{|x| \geq n} |D^{\alpha_n} f(x)| \tag{3.20}$$

Since $f$ has compact support, the sum is always finite These norms make $\mathcal{D}$ into a locally convex space whose topology is not given by a metric. It is not hard to see that $\mathcal{D}$ is complete.

More generally, if $\Omega \subset \mathbb{R}^\nu$ and $K_1 \subset K_2 \subset \cdots \subset \Omega$ with $K_j$ compact, $K_j \subset K_{j+1}^{\text{int}}$ and $\cup K_j = \Omega$, we define $\mathcal{D}(\Omega)$ to be the $C^\infty$ functions of compact support in each $\Omega$ and define norms like (3.20) with $\sup_{|x| \geq n}$ replaced by $\sup_{x \in K_n}$. □

In the next chapter, we will see locally convex spaces have lots and lots of continuous linear functions. The set of such linear functionals on $X$ is called the dual space, $X^*$. In Chapter 5, we will see there are several natural topologies. The topology that is the analog of the norm topology on the dual of a Banach space is the $\beta(X^*, X)$-topology of uniform convergence on bounded subsets of $X$.

The duals of $\mathcal{S}$ and $\mathcal{D}$ are called $\mathcal{S}'(\mathbb{R}^\nu)$, the space of tempered distributions, and $\mathcal{D}'$, the space of ordinary distributions. Tempered distributions involve a finite number of derivatives and do not grow faster than polynomially at infinity. It can be seen that in the $\beta$-topology, $\mathcal{S}'$ and $\mathcal{D}'$ are complete.

**Example 3.22** (Holomorphic spaces) Let $\Omega \subset \mathbb{C}^\nu$ be open and let $K_1 \subset K_2 \subset \cdots \subset K_n \subset \cdots$ be compact sets with $\cup K_i = \Omega$. Let $\mathcal{H}(\Omega)$ be the set of holomorphic functions on $\Omega$. $\mathcal{H}$ is topologized by letting

$$\rho_i(f) = \sup_{z \in K_i} |f(z)|$$

and taking the associated locally convex topology. This space is complete (since uniform limits of analytic functions are analytic) and is a Fréchet space. □

**Definition** A *barrel* in a locally convex space, $X$, is a closed, balanced, convex, absorbing set.

By Proposition 3.13, if $U$ is a balanced, convex, open set, $\bar{U} \subset 2U$, so any locally convex topology has the property that those barrels with nonempty interior form a neighborhood base; but in some spaces, not all barrels have nonempty interior. For example, if $X$ is a Banach space, the unit ball is a barrel in the $\sigma(X, X^*)$-topology but it has empty interior.

**Definition** A locally convex space is called *barreled* if every barrel has nonempty interior, that is, if and only if the set of *all* barrels is a neighborhood base for 0.

**Proposition 3.23** *Every Fréchet space is barreled.*

*Proof* Let $K$ be a barrel in $X$. Then, since $K$ is absorbing, $X = \cup_{n=1}^\infty nK$. Therefore, by the Baire category theorem, some $nK$ must have nonempty interior. But then $K = n^{-1}(nK)$ has nonempty interior. □

Since any locally convex space has a neighborhood base at 0 of barrels, spaces with every barrel a neighborhood of 0 have lots of open sets, and so their topology is especially strong. We will make this precise in Chapter 5; see Theorem 5.24.

**Definition** A *Montel space* is a barreled space with the property that any closed bounded set is compact.

By Theorem 3.10, a Banach space with the norm topology is Montel if and only if it is finite-dimensional. The dual $X^*$ of a Banach space has every closed bounded set compact in the $\sigma(X^*, X)$-topology, but it is not barreled in this topology, and so not Montel. $\mathcal{S}(\mathbb{R}^\nu)$, $\mathcal{D}(\mathbb{R}^\nu)$, and $\mathcal{H}(\Omega)$ are all Montel spaces. For the first two, use equicontinuity ideas, and for the last, the Vitali convergence theorem.

# 4
# Separation theorems

What makes locally convex spaces special is not only that they have lots of continuous linear functionals and so closed hyperplanes, but enough to slip between disjoint convex sets. In this brief chapter, we'll explain this interface of geometry and analysis. The hard work has already been done in Theorem 1.38 (the Hahn–Banach theorem). By a real linear functional, we mean a linear functional if the space is over $\mathbb{R}$ and a real-valued function linear on $X$ as a real vector space if the space is over $\mathbb{C}$.

*Remark* If $\ell$ is a real linear functional on a complex vector space, $L(x) = \ell(x) - i\ell(ix)$ is a complex linear functional since $L(ix) = \ell(ix) - i\ell(-x) = i[\ell(x) - i\ell(ix)]$. Conversely, if $L$ is complex linear, $\ell(x) = \operatorname{Re} L(x)$ is real linear. So we could talk about complex linear functionals and place Re[ ] in front of the functional to handle the complex case.

**Definition** Two sets $A$ and $B$ in $X$, a topological vector space, are said to be *separated* if and only if there exists a nonzero continuous real linear functional $\ell$ on $X$ and $\alpha \in \mathbb{R}$ so that

$$A \subset \{x \mid \ell(x) \leq \alpha\}, \qquad B \subset \{x \mid \ell(x) \geq \alpha\} \tag{4.1}$$

If the inequalities in (4.1) can be taken strict (i.e., = dropped), we say that $A$ and $B$ are *strictly separated*.

**Theorem 4.1** *Let $A$ and $B$ be disjoint convex subsets of a locally convex space $X$ with $A$ open. Then $A$ and $B$ can be separated.*

*Proof* $A-B = \cup_{x\in B}(A-x)$ is open. Pick $-x_0 \in A-B$ and let $C = x_0+A-B$ so $0 \in C$, $C$ is open, convex, $x_0 \notin C$ (since $x_0 \in C \Rightarrow 0 \in A-B \Rightarrow A\cap B \neq \emptyset$). Let $\rho_C$ be the gauge of $C$. Then, since $x_0 \notin C$, $\rho_C(x_0) \geq 1$. Let $W = \{\lambda x_0 \mid \lambda \in \mathbb{R}\}$ and $\ell\colon W \to \mathbb{R}$ by $\ell(\lambda x_0) = \lambda\rho_C(x_0)$. Then $\lambda > 0$ implies that

$$\ell(\lambda x_0) = \lambda\rho_C(x_0) = \rho_C(\lambda x_0)$$

while $\lambda < 0$ implies

$$\ell(\lambda x_0) = \lambda \rho_C(x_0) < 0 \leq \rho_C(\lambda x_0)$$

so (1.78) holds for $F(w) = \rho_C(w)$. Thus, by the Hahn–Banach theorem, there is $L\colon X \to \mathbb{R}$ so $L(x_0) = \rho_C(x_0) \geq 1$ and $L(x) \leq \rho_C(x)$ so for $x \in C$, $L(x) \leq 1$.

By the remark after the proof of Proposition 3.13, $\rho_C$ is a continuous convex function. Thus, $\tilde{\rho}_C(x) = \max(\rho_C(x), \rho_C(-x))$ is also continuous. Since $|L(x)| = \max(L(x), L(-x))$ obeys

$$|L(x)| \leq \tilde{\rho}_C(x)$$

we see that $L$ is a continuous linear functional.

It follows if $a \in A$, $b \in B$, then

$$L(x_0 + a - b) \leq L(x_0)$$

or

$$L(a) \leq L(b)$$

Since this holds for all pairs,

$$\sup_{a \in A} L(a) \equiv \alpha \leq \inf_{b \in B} L(b)$$

and $L$ separates $A$ and $B$. □

**Lemma 4.2** *Let $A$ be an open convex set and $L\colon X \to \mathbb{R}$ be a nonzero linear continuous functional. Then $L[A]$ is open.*

*Proof* Pick $x_0$ with $L(x_0) \neq 0$. For any $y \in A$, $\{t \mid y + tx_0 \in A\}$ is an open interval about 0, so $L[A]$ contains an open interval about $L(y)$. □

**Theorem 4.3** *Let $A$ and $B$ be disjoint open convex subsets of a locally convex space $X$. Then $A$ and $B$ can be strictly separated.*

*Proof* By Theorem 4.1, there is a nonzero linear functional $\ell$ and $\alpha \in \mathbb{R}$ so $\ell[A] \subset (-\infty, \alpha]$ and $\ell[B] \subset [\alpha, \infty)$. Since $A$ and $B$ are open, $\ell[A] \subset (-\infty, \alpha)$ and $\ell[B] \subset (\alpha, \infty)$ by the lemma. □

**Lemma 4.4** *Let $A$ and $B$ be disjoint closed convex sets with $B$ compact. Then there exist disjoint open convex sets $U$ and $V$ with $A \subset U$ and $B \subset V$.*

*Proof* Let $C = A - B$. If $x_\alpha = a_\alpha - b_\alpha$ is a net in $C$ so $x_\alpha \to x$, then by passing to a subnet, we can suppose $b_\alpha \to b$ in $B$ since $B$ is compact. Thus, $a_\alpha = x_\alpha + b_\alpha \to x + b = a \in A$ since $A$ is closed. Thus, $x = a - b \in C$, that is, $C$ is closed. Since $0 \notin C$, we can find $W$ open, balanced, and convex so $0 \in W$ and $W \cap C = \emptyset$. Let $U = A + \frac{1}{2}W$ and $V = B + \frac{1}{2}W$. Then $U \cap V$ is empty and $U, V$ are open and convex. □

**Theorem 4.5** *Let $A$ and $B$ be disjoint closed convex subsets of a locally convex space $X$ with $B$ compact. Then $A$ and $B$ can be strictly separated.*

*Proof* This follows immediately from Theorem 4.3 and Lemma 4.4. □

*Remark* In $\mathbb{R}^2$, let $A = \{(x, y) \mid x \leq 0\}$ and $B = \{(x, y) \mid x \geq y^{-1}\}$ (see Figure 4.1). Then $A$ and $B$ are disjoint closed convex sets. They cannot be strictly separated. This shows it is essential that $B$ be compact.

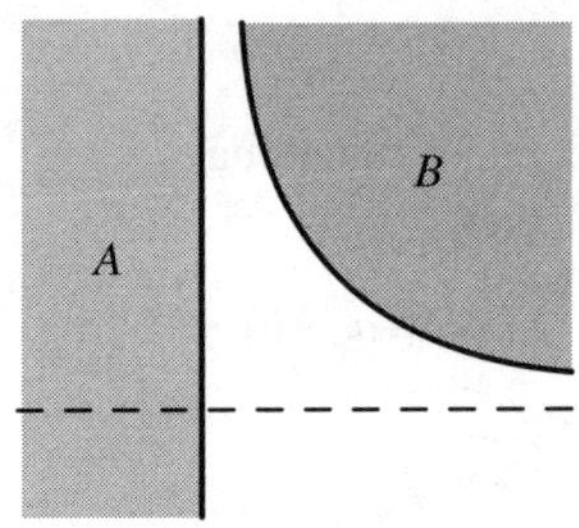


Figure 4.1 Closed convex sets which are not strictly separated

**Corollary 4.6** *Let $X$ be a locally convex vector space and $x, y \in X$ with $x \neq y$. Then there exists $\ell \in X$ with $\ell(x) \neq \ell(y)$.*

*Proof* Take $A = \{x\}$ and $B = \{y\}$. □

**Corollary 4.7** *Let $X$ be a locally convex vector space. Let $W \subset X$ be a closed subspace and $x \notin W$. Then there exists $\ell \in X^*$ so $\ell \restriction W = 0$ and $\ell(x) \neq 0$.*

*Proof* Let $A = W$ and $B = \{x\}$. Let $\ell$ separate $A$ and $B$. Since $\ell[W]$ is a subspace of $\mathbb{R}$ and it is semibounded, it must be 0, that is, $\ell[W] = \{0\}$. $\ell(x)$ is then nonzero. □

Here is the chapter's final application of the separation theorem.

**Definition** A *closed half-space* is a set of the form

$$\{x \mid \ell(x) \geq \alpha\} \tag{4.2}$$

for some continuous, nonzero linear function and some $\alpha \in \mathbb{R}$.

*Remark* One might also want to take sets of the form $\{x \mid \ell(x) \leq \alpha\}$, but since we can take $\ell \to -\ell$ and $\alpha \to -\alpha$, that is unnecessary!

**Theorem 4.8** *A set $A$ is a closed convex set if and only if it is an intersection of closed half-spaces. If $0 \in A$ and $A$ is a closed convex set, it is the intersection of a family of half-spaces of the form* (4.2) *with $\alpha = -1$.*

*Proof* An arbitrary intersection of closed sets is closed and an arbitrary intersection of convex sets is convex. Therefore, any intersection of closed half-spaces is a closed convex set.

Conversely, if $A$ is a closed convex set and $x \notin A$, by Theorem 4.5, we can find $\ell_x$ and $\alpha_x$ so $\ell_x(x) < \alpha_x$ and

$$A \subset \{y \mid \ell_x(y) > \alpha_x\} \tag{4.3}$$

We claim

$$A \equiv \bigcap_{x \notin A} \{y \mid \ell_x(y) \geq \alpha_x\} \tag{4.4}$$

By (4.3), $A \subset \cap_{x \notin A}\{y \mid \ell_x(y) > \alpha_x\} \subset \cap_{x \notin A}\{y \mid \ell_x(y) \geq \alpha_x\}$ and by construction, for any $x \notin A$, $x \notin \{y \mid \ell_x(y) \geq \alpha_x\}$ so $x \notin \cap_{x \in A}\{y \mid \ell_x(y) \geq \alpha_x\}$.

Finally, if $0 \in A$, the $\alpha_x$'s above are negative. Replace $\ell_x$ by $|\alpha_x|^{-1}\ell_x = \tilde{\ell}_x$. We have

$$A = \bigcap_{x \in A} \{y \mid \tilde{\ell}_x(y) \geq -1\} \tag{4.5}$$

□

**Corollary 4.9** *Let $X$ be a locally convex space. Let $A \subset X$ be a closed convex set. Then $A$ is closed in the $\sigma(X, X^*)$-topology. If it is compact in the original topology, it is also compact in the $\sigma(X, X^*)$-topology.*

*Proof* Each closed half-space is weakly closed and so $A$ is weakly closed by Theorem 4.8.

If $\mathcal{T}$ is the topology on $X$, the identity map $x\colon X_{\mathcal{T}} \to X_\sigma$ is continuous and a bijection on $A$. Therefore, $i[A]$ is $\sigma$-compact. □

# 5

# Duality: dual topologies, bipolar sets, and Legendre transforms

Throughout this chapter, we will have a dual pair, $X, Y$, of vector spaces with a pairing $(x, y) \mapsto \langle x, y\rangle$ of $X \times Y$ into $\mathbb{K}$. We will be interested in topologies $\mathcal{T}$ on $X$ that make it into a topological vector space in which $Y = X^*_\tau$, the set of $\mathcal{T}$-continuous linear functions on $X$. Such topologies are called *dual topologies* and this chapter classifies them all. Then we'll prove a duality theorem for Legendre transforms.

Recall that the $\sigma(X, Y)$-topology is the weakest topology on $X$ in which each $y \in Y$ defines a continuous functional on $X$ via $x \mapsto \langle x, y\rangle$. We begin by noting

**Proposition 5.1** *The $\sigma(X, Y)$-topology is an $X, Y$ dual topology.*

*Proof* By definition, each $x \mapsto \langle x, y\rangle$ defines a $\sigma(X, Y)$-continuous functional, so we need only show that every such continuous functional lies in $Y$. By the definition of the $\sigma(X, Y)$-topology, $\{|\langle\,\cdot\,, y\rangle|\}$ are a generating family of seminorms. Thus, by (3.15), if $\ell \in X^*_\sigma$, there exist $y_1, \dots, y_n$ on $Y$ and $C > 0$ so

$$|\ell(x)| \le C\sum_{j=1}^{n}|\langle x, y_j\rangle| \tag{5.1}$$

Without loss, we can suppose that the $\{y_j\}_{j=1}^n$ are linearly independent.

If $\ell$, as an element of $X^*_{\text{alg}}$, is independent of $\{y_i\}_{i=1}^n$, then by Proposition 3.4, we can find $x \in X$ so $\ell(x) = 1$, but $\langle x, y_j\rangle = 0$, $j = 1, \dots, n$. This contradicts (5.1), so $\ell$ is a linear combination of the $y_j$'s, and so in $Y$. □

Clearly, by definition, $\sigma(X, Y)$ is the weakest dual topology. We will later identify the strongest dual topology. The following notions will be critical also in Chapters 8–11.

**Definition** Let $A$ be a subset of a locally convex vector space, $X$, with an $\langle X, Y\rangle$ dual topology. $\text{ch}(A)$, the *convex hull* of $A$, is the smallest convex set containing $A$. $\text{cch}(A)$, the *closed convex hull* of $A$, is the smallest closed convex set containing $A$.

Since arbitrary intersections of closed and/or convex sets are closed and/or convex, such smallest sets exist.

**Theorem 5.2** *Let $A \subset X$, a locally convex space. Then*

(i) $\text{ch}(A) = \cup_{n=1}^{\infty}\{\theta_1 x_1 + \cdots + \theta_n x_n \mid \theta_i \in [0,1], \sum_{i=1}^{n} \theta_i = 1, x_i \in A\}$

(ii) $\overline{\text{ch}(A)} = \text{cch}(A)$

(iii) $\text{cch}(A)$ *is the intersection of all closed half-spaces containing $A$.*

(iv) $\text{cch}(A)$ *is independent of which dual topology the closure is computed in.*

(v) *If $A$ is bounded,* $\text{cch}(A)$ *is bounded.*

(vi) *If $A_1, \dots, A_n$ are convex sets, then*

$$\text{ch}(A_1 \cup \cdots \cup A_n) = \left\{\theta_1 x_1 + \cdots + \theta_n x_n \mid x_i \in A_i,\, \theta_i \geq 0,\, \sum_{i=1}^{n} \theta_i = 1\right\} \tag{5.2}$$

(vii) *If $A_1, \dots, A_n$ are compact convex sets, then* $\text{ch}(A_1 \cup \cdots \cup A_n)$ *is compact.*

*Proof* (i) It is easy to see that if $B_n = \{\theta_1 x_1 + \cdots + \theta_n x_n \mid \dots\}$, then $\theta B_n + (1-\theta)B_n \subset B_{2n}$ so $\cup_{n=1}^{\infty} B_n$ is convex. On the other hand, if $A \subset B$ and $B$ is convex, then $B_n \subset B$ so $\cup_{n=1}^{\infty} B_n \subset B$. Thus, $\cup_{n=1}^{\infty} B_n$ is the convex hull.

(ii) Since $\text{cch}(A)$ is convex, $\text{ch}(A) \subset \text{cch}(A)$, and thus, $\overline{\text{ch}(A)} \subset \text{cch}(A)$. On the other hand, $\overline{\text{ch}(A)}$ is closed and convex by Proposition 3.11.

(iii) This is essentially Theorem 4.8. Explicitly, since each half-space is convex and closed, their intersection contains $\text{cch}(A)$. Conversely, by Theorem 4.8 if $x \notin \text{cch}(A)$, there is a half-space containing $\text{cch}(A)$ and so $A$, and not containing $x$. Thus, the intersection of half-spaces is contained in $\text{cch}(A)$.

(iv) Continuous functionals are the same in all dual topologies, so closed half-spaces are the same, so by (iii), $\text{cch}(A)$ is the same.

(v) If $A$ is bounded and $U$ is an open convex neighborhood of 0, then $A \subset \lambda U$ for some $\lambda$, and thus, $\text{ch}(A) \subset \lambda U$ so $\text{ch}(A)$ is bounded. Moreover, the closure of a bounded set is bounded.

(vi) The right side of (5.2) is clearly contained in any convex set containing $A_1 \cup \cdots \cup A_n$. Since

$$\varphi\left(\sum_{i=1}^{n} \theta_i x_i\right) + (1-\varphi)\left(\sum_{i=1}^{n} \eta_i y_i\right) = \sum_{i=1}^{n} \gamma_i z_i$$

with $\gamma_i = \varphi\theta_i + (1-\varphi)\eta_i$, $z_i = \gamma_i^{-1}(\varphi\theta_i x_1 + (1-\varphi)\eta_i y_i)$, $z_i \in A_i$ (since $A_i$ is convex), and $\sum \gamma_i = \varphi \sum \theta_i + (1-\varphi)\sum \eta_i = \varphi + (1-\varphi) = 1$, we see the right side of (5.2) is convex.

(vii) Let $\Delta_{n-1}$ be the *simplex* in $\mathbb{R}^{\nu}$:

$$\Delta_{n-1} = \left\{(\theta_1, \dots, \theta_n) \mid \theta_i \geq 0,\, \sum_{i=1}^{n} \theta_i = 1\right\} \tag{5.3}$$

(called $\Delta_{n-1}$ since its dimension is $n-1$) and let $\Phi\colon X^n \times \Delta_{n-1} \to X$ by

$$\Phi(x_1, \dots, x_n, \theta) = \sum_{i=1}^{n} \theta_i x_i$$

Thus, $\Phi$ is continuous, so $\Phi[A_1 \times \cdots \times A_n \times \Delta_{n-1}]$ is compact. But by (5.2), this set is $\mathrm{ch}(A_1 \cup \cdots \cup A_n)$. □

The following theorem of Mazur illustrates the use of separation theorems and convex hulls.

**Theorem 5.3** *Let $X$ be a locally convex space and let $Y$ be its space of continuous functionals. Let $\{x_n\}$ be a sequence in $X$ with $x_n \to x_\infty$ in the $\sigma(X, Y)$-topology. Then $x_\infty$ is a limit in the $X$-topology of some convex combination of $\{x_n\}$, explicitly,*

$$x_\infty = \bigcap_n \mathrm{cch}(\{x_m\}_{m \geq n})$$

*Remark* Think of the example of an orthonormal basis $\{e_j\}_{j=1}^{\infty}$ in a Hilbert space, $e_j \to 0$ weakly, $\|e_j\| \not\to 0$ by $\|\sum_{j=1}^{n} \frac{1}{n} e_j\| = n^{-1/2} \to 0$.

*Proof* Let $C_n = \mathrm{cch}(\{x_m\}_{m \geq n})$. If $x_\infty \notin C_n$, there exists $y \in Y$ so $\langle y, x_\infty\rangle > \sup_{x \in C_n} \langle y, x\rangle \geq \sup_{m \geq n} \langle y, x_n\rangle$, which is incompatible with $\langle y, x_n\rangle \to \langle y, x_\infty\rangle$. □

**Definition** Let $A$ be a subset of $X$, a locally convex space, which is half of a dual pair $X, Y$, so that functionals in $Y$ are continuous on $X$. Then the *polar* of $A$, denoted $A^\circ$, is defined by

$$A^\circ = \{y \in Y \mid \langle x, y\rangle \geq -1 \text{ for all } x \in A\}$$

This definition is in the case of real vector spaces. In accordance with the remark before Theorem 4.1, in the complex vector space case, $\langle x, y\rangle \geq -1$ should be replaced by $\mathrm{Re}\langle x, y\rangle \geq -1$. For simplicity, we will use notation consistent with the real case.

**Proposition 5.4** *$A^\circ$ is a closed convex set containing* 0.

*Proof*

$$A^\circ = \bigcap_{x \in A} \{y \in Y \mid \langle x, y\rangle \geq -1\} \tag{5.4}$$

is clearly an intersection of closed convex sets and obviously $0 \in A^\circ$. □

**Theorem 5.5** (The Bipolar Theorem) $(A^\circ)^\circ = \mathrm{cch}(A \cup \{0\})$

*Remark* cch taken in any dual topology with respect to the dual pair $X, Y$ is used for computing the polars.

*Proof* If $x \in A$, then $\langle x, y\rangle \geq -1$ for all $y \in A^\circ$, so $x \in (A^\circ)^\circ$, that is, $A \subset (A^\circ)^\circ$. By Proposition 5.4, $0 \in (A^\circ)^\circ$ and $(A^\circ)^\circ$ is a closed convex set. Thus, $\text{cch}(A \cap \{0\}) \subset (A^\circ)^\circ$. On the other hand, by Theorem 4.8 and Theorem 5.2(iv),

$$\text{cch}(A \cup \{0\}) = \bigcap_{y \in S} \{x \mid \langle x, y\rangle \geq -1\} \tag{5.5}$$

where $S$ is the set of all $y$'s with $A \cup \{0\} \subset \{x \mid \langle x, y\rangle \geq -1\}$, that is, $S$ is $A^\circ$ and the intersection is $(A^\circ)^\circ$ by (5.4). □

*Remark* It is not hard to see that $\text{cch}(A \cup \{0\})$ is the closure of all sums of the form $\sum_{i=1}^n \alpha_i x_i$ with $x_i \in A$ and $\alpha_i \in [0,1]$ obeying $\sum_{i=1}^n \alpha_i \leq 1$. $(A^\circ)^\circ$ is sometimes written in those terms.

**Example 5.6** Let $A$ be a balanced set. Then

$$A^\circ = \{y \mid |\langle x, y\rangle| \leq 1 \text{ for all } x \in A\} \tag{5.6}$$

In the real case, $\pm\langle x, y\rangle \geq -1$ is equivalent to $-1 \leq \langle x, y\rangle \leq 1$ or $|\langle x, y\rangle| \leq 1$. In the complex case, $\text{Re}\langle e^{i\theta} x, y\rangle \geq -1$ is equivalent to $|\langle x, y\rangle| \leq 1$. (5.6) implies $A^\circ$ is then also balanced. □

This lets us also see a connection between polars and gauges.

**Theorem 5.7** *Let $A \subset X$ be a balanced, closed convex neighborhood of* 0 *in some topology on $X$ stronger than the $\sigma(X, Y)$-topology. Let $\rho_A$ be its gauge. Then*

$$\rho_A(x) = \sup_{y \in A^\circ} |\langle y, x\rangle| \tag{5.7}$$

*Proof* Since $A$ is closed in the given topology on $X$, $X \backslash A$ is open in the given topology, and so in the $\sigma(X, Y)$-topology since we are assuming the given topology is stronger (has more open sets). Thus, $A$ is closed in the $\sigma(X, Y)$-topology, and thus, $(A^\circ)^\circ = A$.

Let $\tilde{\rho}_A$ be the right side of (5.7). Then $\tilde{\rho}_A$ is obviously convex and homogeneous of degree 1. Since $A = \{x \mid \rho_A(x) \leq 1\}$, we need only show $A = \{x \mid \tilde{\rho}_A(x) \leq 1\}$. But $\tilde{\rho}_A(x) \leq 1$ if and only if $x \in (A^\circ)^\circ = A$. □

**Example 5.8** Let $A$ be a subspace of $X$. Then $\langle \lambda x, y\rangle \geq -1$ for all $\lambda \in \mathbb{R}$ (or $\mathbb{C}$) implies $\langle x, y\rangle = 0$, that is, $A^\circ = \{y \mid \langle y, \cdot\,\rangle$ vanishes on $A\}$. Since $A$ is convex and $0 \in A$, $(A^\circ)^\circ = \bar{A}$ and the bipolar theorem generalizes the fact that $A^{\perp\perp} = \bar{A}$ for subspaces of a Hilbert space. □

**Example 5.9** Let $A$ be a closed convex cone in $X$. Then $\langle \lambda x, y\rangle \geq -1$ for all $\lambda \geq 0$ implies $\langle x, y\rangle \geq 0$, that is, $A^\circ = \{y \in X \mid \langle y, \cdot\,\rangle$ is nonnegative on $A\}$, the *dual cone* of $A$. The canonical example is $X = C(U)$ with $U$ a compact Hausdorff space and $A = \{f \mid f \geq 0\}$. Then if $Y = \mathcal{M}(U)$, the measures on $U$,

$A^\circ = \mathcal{M}_+(U)$, the positive measures on $U$. Some authors define $A^\circ$ by demanding $\langle x, y\rangle \leq 1$ for all $x \in A$. Their $A^\circ$ is the negative of our $A^\circ$ (so their $(A^\circ)^\circ$ is our $(A^\circ)^\circ$). That our definition is the "right" one is seen by this example of dual cones.

Still other authors define $A^\circ$ by demanding $|\langle x, y\rangle| \leq 1$ for all $x \in A$. Their $A^\circ$ is $B \cap (-B)$ where $B$ is our $A^\circ$. This is fine for subspaces and some other sets but awful for cones where typically $B \cap (-B) = \{0\}$! With this definition, the bipolar is the closure of $\{\sum \alpha_i x_i \mid \sum |\alpha_i| \leq 1,\, x_i \in A\}$.

Still other authors define $A^\circ$ by demanding $\langle x, y\rangle \leq 0$ for all $x \in A$. $A^\circ$ is then a cone and the bipolar is the closed convex cone generated by $A$, that is, the closure of $\{x = \sum_{i=1}^n \lambda_i x_i \mid x_i \in A,\, \lambda_i \geq 0\}$. □

**Example 5.10** In $\mathbb{R}^\nu$ with the Euclidean inner product on any real Hilbert space, $X$ is in duality with itself, and one can ask for sets with $A = A^\circ$. There are many such sets. For example, in $\mathbb{R}^\nu$, the set $\mathbb{R}_n^+ = \{x \mid x_i \geq 0\}$ is its own polar, and that is true for the image of $\mathbb{R}_n^+$ under any orthogonal map.

However, there is a unique set with $A = -A^\circ$, namely, $\{x \mid \langle x, x\rangle \leq 1\} \equiv B$. $B^\circ = B = -B$ since $y \in B^\circ \Rightarrow \langle y, -y/\|y\|\rangle \geq -1 \Rightarrow \|y\| \leq 1$ and $\|y\| \leq 1 \Rightarrow |\langle x, y\rangle| \leq 1$ for all $x \in B \Rightarrow \langle x, y\rangle \geq -1 \Rightarrow y \in B^\circ$. On the other hand, if $A = -A^\circ$ and $x \in A$, then $\langle x, -x\rangle \geq -1$ so $\|x\| \leq 1$, that is, $A \subset B$. But then $B = -B^\circ \subset -A^\circ = A$ since, in general, $C \subset D$ implies $D^\circ \subset C^\circ$. Thus, $A = B$. This example is clearer if $A^\circ$ is defined by $\langle x, y\rangle \leq 1$, in which case $B$ is the unique self-polar. □

The following will be useful when we discuss the $\beta(X, Y)$-topology below. Recall a barrel is a closed, absorbing, balanced convex set.

**Proposition 5.11** *Let $X, Y$ be a dual pair. Let $A \subset X$ be a closed, balanced convex set in some topology stronger than the $\sigma(X, Y)$-topology. Then $A$ is a barrel (i.e., $A$ is absorbing) if and only if $A^\circ$ is bounded in the $\sigma(Y, X)$-topology.*

*Remarks* 1. Intuitively absorbing, that is, that $\cup_{\lambda>0} \lambda A = X$ and boundedness, that is, $A \subset \lambda U$ for certain $U$'s are dual notions. This makes that precise.

2. The topology on $X$ need not be a dual topology and will not be in an application below.

*Proof* Since $A$ is closed in a topology stronger than $\sigma(X, Y)$, it is closed in the $\sigma(X, Y)$-topology, and thus, $(A^\circ)^\circ = A$. As noted, since $A$, and thus, $A^\circ$, are balanced,

$$A^\circ = \{y \mid |\langle x, y\rangle| \leq 1 \text{ for all } x \in A\} \tag{5.8}$$

and

$$A = \{x \mid |\langle x, y\rangle| \leq 1 \text{ for all } y \in A^\circ\} \tag{5.9}$$

Since the $\sigma(Y,X)$-topology is generated by $\{y \mid |\langle x_1, y\rangle| \le 1, \dots, |\langle x_n, y\rangle| \le 1\}$ for arbitrary $x_1, \dots, x_n \in X$, $A^\circ$ is $\sigma(Y,X)$-bounded if and only if for all $x \in X$,

$$\sup_{y \in A^\circ} |\langle x, y\rangle| \equiv \alpha_x < \infty \tag{5.10}$$

But $x \in \lambda_x A$ if and only if $\lambda_x^{-1} x \in A$ which, by (5.9), holds if and only if

$$\sup_{y \in A^\circ} |\langle x, y\rangle| \le \lambda_x$$

By (5.10), this holds for all $x \in X$ (i.e., $A$ is absorbing) if and only if $A^\circ$ is bounded. □

Here is an interesting use of polars:

**Theorem 5.12** (The Bourbaki–Alaoglu Theorem) *Let $X, Y$ be a dual pair and let $A$ be a closed convex subset of $X$ in one and, hence all, dual topologies. Then*

(i) *If $A$ is compact in a dual topology, then it is $\sigma(X,Y)$-compact and the topologies restricted to $A$ are identical.*
(ii) *If $A^\circ$ is a neighborhood of $0 \in Y$ in any dual topology on $Y$, then $A$ is $\sigma(X,Y)$-compact.*

*Remarks* 1. If $X$ is a reflexive Banach space, its unit ball is $\sigma(X,X^*)$-compact, although it is not compact in the norm topology which is a dual topology. Thus, the converse of (i) is false.

2. The subtlety of this theorem is seen by considering the unit ball, $A$, on a nonreflexive Banach space under the $X, X^*$ duality. $A^\circ = \{\ell \in X^* \mid \|\ell\| \le 1\}$, the unit ball in $X^*$. $A^\circ$ is a neighborhood of 0 in the norm topology on $X^*$ but $A$ is not $\sigma(X,X^*)$-compact. This is compatible with (ii) because the norm topology on $X^*$ is not an $X, X^*$ dual topology since the dual of $X^*$ is not $X$ but the larger $X^{**}$.

*Proof* (i) The identity map from $A$ with the dual topology to $A$ with the $\sigma(X,Y)$-topology is continuous because each function $x \mapsto \langle x, y\rangle$ is continuous in the dual topology. Thus, $A$ is compact in the $\sigma(X,Y)$-topology as the image of a compact set and the topologies are identical since a continuous bijection between compact spaces is a homeomorphism.

(ii) Pick $U$, a balanced convex neighborhood of 0, with $U \subset A^\circ$. If $x \in A$, $u \in U$, and $|\lambda| \le 1$ in $\mathbb{K}$, then $\mathrm{Re}\langle x, \lambda u\rangle \ge -1$ or $|\langle x, u\rangle| \le 1$. It follows that if $y \in Y$ and $x \in A$, then

$$|\langle x, y\rangle| \le \rho_U(y) \tag{5.11}$$

with $\rho_U$ the gauge of $U$.

Let $\Omega$ be the compact space $\prod_{y \in Y} \{\lambda \mid |\lambda| \le \rho_U(y)\}$ with $\lambda$ in $\mathbb{K}$. Map $A$ into $\Omega$ by mapping $x$ into $\{\langle x, y\rangle\}_{y \in Y}$. This map is a bijection onto its range. Moreover, we claim the range is closed because if $\langle x_\alpha, y\rangle$ converge for each $y$, the limit $\ell(y)$ defines a linear functional with $|\ell(y)| \le \rho_U(y)$, hence a continuous functional in $Y$ in the dual topology, hence a point $x_\infty$ in $X$. Since $A$ is $\sigma(X,Y)$-closed (since it is convex and closed), $x_\infty \in A$, that is, the image of $A$ in $\Omega$ is closed, hence compact. The weak topology in $A$ is precisely the induced topology by $\Omega$, so $A$ is compact. □

*Remark* We will show later (see Theorem 5.22) that this theorem has a converse. If $A$ is a $\sigma(X,Y)$-compact convex subset of $X$, then $A^\circ$ is a neighborhood of 0 in a suitable dual topology on $Y$.

Next, we turn to Legendre transforms. We want to show that the bipolar theorem is not merely reminiscent of the theorem that the double Legendre transform recovers a convex function but that they are variants on virtually identical themes. In doing so, we will extend the theory not merely to infinite dimensions but to finite-dimensional situations where $f$ is not necessarily steep or even everywhere finite. To figure out the right class of functions to consider, it pays to think of the formula for the Legendre transform $f^*(x) = \sup_y[\langle x,y\rangle - f(y)]$. For each fixed $y$, the function is continuous in $x$ and so $f^*$ is a sup of continuous functions, and thus, $f^*$ is lsc. To add to the understanding that lsc is the right condition, a function $f$ is lsc if and only if

$$\Gamma^*(f) = \{(x,\lambda) \in X \times \mathbb{R} \mid \lambda \ge f(x)\} \tag{5.12}$$

is closed so lsc convex functions will have $\Gamma^*(f)$ a closed convex set, a good property for double duals, as we have seen.

Also, to allow nonsteep $f$, we want to allow $f^*$ to be infinite, so we will also allow $f$ to be infinite.

**Definition** Let $X, Y$ be a dual pair. A *regular convex function* on $X$ is a map $f\colon X \to \mathbb{R} \cup \{\infty\}$ (*Note*: infinity is allowed) not identically $+\infty$ so that $\Gamma^*(f)$ given by (5.12) is convex and is closed in the product topology of $\sigma(X,Y)$ on $X$ and the usual topology on $\mathbb{R}$.

*Remark* As usual, closed convex sets are the same in all dual topologies, so one can replace $\sigma(X,Y)$ by any dual topology.

**Proposition 5.13** *Let $f$ be a regular convex function and let*

$$D(f) = \{x \in X \mid f(x) < \infty\} \tag{5.13}$$

*Then,*

(i) *$D(f)$ is convex.*
(ii) *$f \restriction D(f)$ is convex in the ordinary sense.*
(iii) *$f \restriction D(f)$ is lsc in the usual sense.*
(iv) *If $x_0 \in \partial D(f)$ but $x_0 \notin D(f)$, then $\lim_{x\to x_0,\, x\in D(f)} f(x) = \infty$.*

*Conversely, if $f$ is a (finite) convex function on a convex set, $A \subset X$ which is lsc on $A$ and which obeys*

(a) *If $x_0 \in \partial D(f)$, but $x_0 \notin A$, then $\lim_{x \to x_0,\, x \in A} f(x) = \infty$,*

*then the extension of $f$ to $X$ obtained by setting $f = \infty$ on $X \backslash A$ is a regular convex function.*

*Proof* Elementary. □

**Example 5.14** The canonical example of a convex function on $\mathbb{R}$ which is not steep is $f(x) = |x|$. If we try to define its Legendre transform by

$$f^*(y) = \sup_{x \in \mathbb{R}} [xy - f(x)]$$

then

$$f^*(y) = \begin{cases} 0, & |y| \leq 1 \\ \infty, & |y| > 1 \end{cases}$$

This is a regular convex function. Notice that

$$\sup_{y \in D(f^*)} [xy - f^*(y)] = \sup \pm x = |x|$$

so in a suitable sense, $(f^*)^* = f$ despite the fact that $f^*$ is infinite much of the time! This example will be generalized below in Example 5.20. □

**Definition** Let $X, Y$ be a dual pair. If $f$ is a function from $X$ to $\mathbb{R} \cup \{\infty\}$ so that for some $y \in Y$ and $a \in \mathbb{R}$ and all $x \in X$,

$$f(x) \geq a + \langle x, y \rangle \tag{5.14}$$

we define the *convex envelope*, $f_*$, by

$$f_*(x) = \sup\{g(x) \mid g \text{ is regular convex and } g \leq f\} \tag{5.15}$$

Notice that since a sup of lsc functions is lsc and a sup of convex functions is convex, $f_*$ is a regular convex function.

**Theorem 5.15** *Let $X, Y$ be a dual pair and let $f$ be a regular convex function. Then*

$$f(x) = \sup\{\langle x, y \rangle + a \mid y \in Y \textit{ and } a \in \mathbb{R} \text{ so (5.14) } \textit{holds}\} \tag{5.16}$$

*More generally, if $f$ is any function so that* (5.14) *holds for some $y \in Y$ and $a \in \mathbb{R}$, then*

$$f_*(x) = \sup\{\langle x, y \rangle + a \mid y \in Y \textit{ and } a \in \mathbb{R} \text{ so (5.14) } \textit{holds}\} \tag{5.17}$$

*Proof* Given any function $f$ for which (5.14) holds for some $y \in Y$, $a \in \mathbb{R}$, let $\tilde{f}(x)$ be the right side of (5.17). Clearly,

$$f(x) \geq \tilde{f}(x) \tag{5.18}$$

We are going to use separation theorems on $X \times \mathbb{R}$ where linear functionals are of the form $(y, \alpha) \in Y \times \mathbb{R}$ with $\langle (x, \beta), (y, \alpha) \rangle = \langle x, y \rangle + \beta\alpha$. Since separating functionals can be multiplied by nonzero constants, we can limit ourselves to the case $\alpha = 0$ and $\alpha = -1$.

Suppose $f$ is regular convex and $x_0 \in D(f)$ and $b < f(x_0)$. Since

$$\langle (x_0, f(x_0)), (y, 0) \rangle = \langle (x_0, b), (y, 0) \rangle$$

we cannot use an $\alpha = 0$ functional to separate $(x_0, b)$ from $\Gamma^*(f)$. Thus, there has to be an $\alpha = -1$ functional that separates. Since $\lim_{\lambda \to +\infty} \langle (x_0, \lambda), (y, -1) \rangle = -\infty$, there must be $y \in Y$ so

$$\langle (x_0, b), (y, -1) \rangle > \sup_{(x,\lambda) \in \Gamma^*(f)} \langle (x, \lambda), (y, -1) \rangle \tag{5.19}$$

that is, for all $x \in D(f)$,

$$-a \equiv \langle x_0, y \rangle - b \geq \langle x, y \rangle - f(x)$$

that is,

$$f(x) \geq \langle x, y \rangle + a$$

By definition of $\tilde{f}$,

$$\tilde{f}(x_0) \geq \langle x_0, y \rangle + a = b$$

Since $b$ is an arbitrary number less than $f(x)$, we see $f(x_0) \leq \tilde{f}(x_0)$, so by (5.18), $f(x_0) = \tilde{f}(x_0)$.

Now suppose $x_0 \notin D(f)$. If for every $b \in \mathbb{R}$, there is a $y$ so (5.19) holds, then as above, $\tilde{f}(x_0) \geq b$ and so $\tilde{f}(x_0) = \infty$. If for some $b$, (5.19) fails, then since $\Gamma^*(f)$ is strictly separated from $(x_0, b)$, the separating functional must have $\alpha = 0$, that is, there is $y_0 \in Y$ so

$$\langle x_0, y_0 \rangle > \sup_{x \in D(f)} \langle x, y_0 \rangle$$

so let

$$-\beta = \sup_{x \in D(f)} \langle x - x_0, y_0 \rangle < 0$$

or for $\beta > 0$,

$$\langle x - x_0, y_0 \rangle + \beta \leq 0 \tag{5.20}$$

for all $x \in D(f)$. By the first part of this proof (taking a tangent functional at some $x_1 \in D(f)$), we can find $y_1$ and $a$ so

$$f(x) \geq \langle x, y_1 \rangle + a \tag{5.21}$$

Then, by (5.20) for any $\lambda > 0$ and all $x \in D(f)$,

$$f(x) \geq \langle x, y_1 \rangle + a + \lambda[\langle x - x_0, y_0 \rangle + \beta] \tag{5.22}$$

For any $\lambda > 0$, (5.22) trivially holds if $f(x) = \infty$. The right side of (5.22) is an affine functional so

$$\tilde{f}(x_0) \geq \langle x_0, y_1 \rangle + a + \lambda\beta \tag{5.23}$$

Since $\beta > 0$ and $\lambda$ can be taken to infinity, $\tilde{f}(x_0) = \infty$. We have therefore proven (5.16).

To prove (5.17), use $\tilde{g}$ to denote

$$\tilde{g}(x_0) = \sup\{\langle x_0, y \rangle + a \mid y \in Y,\, a \in \mathbb{R},\, g(x) \geq \langle x, y \rangle + a \text{ for all } x \in X\}$$

for any function $g$ where the set of such $(y, a)$ is nonempty. Clearly, if $g_1 \leq g_2$, $\tilde{g}_1 \leq \tilde{g}_2$. Thus, since $f \geq f_*$, $\tilde{f} \geq \tilde{f}_*$, but since $f_*$ is a regular convex function, we have just proven that $\tilde{f}_* = f_*$ so

$$f_* \leq \tilde{f}$$

On the other hand, by construction for any $g$, $\tilde{g} \leq g$ so $\tilde{f} \leq f$. Since $\tilde{f}$, as the sup of continuous affine functions, is regular convex, we have

$$\tilde{f} \leq f_*$$

Thus, $\tilde{f} = f_*$, that is, (5.19) holds. □

**Definition** Let $X, Y$ be a dual pair and let $f$ be a function from $X$ to $\mathbb{R} \cup \{\infty\}$ for which (5.14) holds for some $y \in Y$ and $a \in \mathbb{R}$ (in particular, let $f$ be a regular convex function). We define the *Legendre transform*, $f^*$, from $Y$ to $\mathbb{R} \cup \{\infty\}$ by

$$f^*(y) = \sup_{x \in X} \, [\langle x, y \rangle - f(x)]$$

**Proposition 5.16** *$f^*$ takes values in $\mathbb{R} \cup \{\infty\}$ and is not identically infinite. $f^*$ is a regular convex function.*

*Proof* Since there is an $x_0$ with $f(x_0) < \infty$,

$$f^*(y) \geq \langle x_0, y \rangle - f(x_0)$$

and so takes values in $\mathbb{R} \cup \{\infty\}$. Since (5.14) holds for some $y_0, a$,

$$\langle x, y_0 \rangle - f(x) \leq -a$$

so $f^*(y_0) \leq -a$ is finite. Finally, as a sup of continuous linear functions, $f^*$ is convex and lsc. □

Just as the bipolar theorem is essentially a rewriting of Theorem 4.8, the theorem on double Legendre transforms in essentially a rewriting of Theorem 5.15.

**Theorem 5.17** (Fenchel's Theorem) *Let $X, Y$ be a dual pair and let $f$ be a function from $X$ to $\mathbb{R} \cup \{\infty\}$ for which* (5.14) *holds for some $y \in Y$ and $a \in \mathbb{R}$. Then $(f^*)^* = f_*$. In particular, if $f$ is regular convex, then $(f^*)^* = f$.*

*Proof* $\langle x, y\rangle + a \leq f(x)$ for all $x$ if and only if $-a \geq \sup_x[\langle x, y\rangle - f(x)] = f^*(y)$. Thus, (5.17) says that

$$\begin{aligned} f_*(x) &= \sup\{\langle x, y\rangle + a \mid y \in Y \text{ so that } -a \geq f^*(y)\} \\ &= \sup\{\langle x, y\rangle - f^*(y)\} = (f^*)^*(x) \end{aligned}$$ □

**Example 5.18** We have already seen many Legendre transforms in the case $f(x)$ is an even convex function on $\mathbb{R}$. Here are some examples where $f$ is not even. In general, one finds $f^*$ by solving

$$\frac{df}{dx}(x_0) = y_0 \tag{5.24}$$

and setting $f^*(y_0) = x_0 y_0 - f(x_0)$. If $f$ is convex, this works so long as (5.24) has a solution.

(i) $f(x) = e^x$ on $\mathbb{R}$; then

$$f^*(y) = \begin{cases} y \log y - y, & y > 0 \\ 0, & y = 0 \\ \infty, & y < 0 \end{cases}$$

(ii) $f(x) = -x^p/p$ for $0 < p < 1$ and $x \geq 0$; $f(x) = \infty$ for $x < 0$. Then

$$f^*(y) = \begin{cases} \infty, & y \geq 0 \\ |y|^q/|q|, & y < 0 \end{cases}$$

where $\frac{1}{p} + \frac{1}{q} = 1$ and $q < 0$, since $p < 1$.

(iii) $f(x) = \alpha^{-1}x^{-\alpha}$, for $\alpha > 0$ and $x > 0$, $f(x) = \infty$ for $x < 0$. This is essentially the same as (ii) given Fenchel's theorem. Explicitly, if $p = \alpha/(\alpha + 1) \in (0, 1)$, then

$$f^*(y) = \begin{cases} \infty, & y > 0 \\ -|y|^p/p, & y \leq 0 \end{cases}$$

(iv) $f(x) = -\log x$ for $x > 0$, $= \infty$ for $x \leq 0$. Then

$$f^*(y) = \begin{cases} \infty, & y \geq 0 \\ -1 - \log(|y|), & y < 0 \end{cases}$$

This example will be central in Chapter 16 when we study entropy. □

**Example 5.19** This is closely related to Example 5.10. Part (iv) of the last example says if $f(x) = -\frac{1}{2} - \log x$ for $x > 0$ and $\infty$ for $x \le 0$, then $f^*(-x) = f(-x)$. It is one of many functions with the property. However, even in the context of a general Hilbert space where $X$ is in duality with itself, there is a unique function with $f^*(x) = f(x)$, naturally $f_0(x) = \frac{1}{2}x^2$. It is an easy computation (a special case of $x^p/p$ already discussed in Example 2.15) to see that $f_0^* = f_0$. On the other hand, suppose $f = f^*$. Then by Young's inequality,

$$\|x\|^2 \le f(x) + f^*(x) = 2f(x)$$

so $f \ge f_0$. But then $f_0^* \ge f^*$ (since, in general, $g \ge h$ implies $h^* \ge g^*$), so $f = f_0$. □

**Example 5.20** This example, which generalizes Example 5.14, will show that Fenchel's theorem and the bipolar theorem are not merely analogs that look similar, but that the bipolar theorem is a special case of Fenchel's theorem!

Given any set $S \subset X$, a locally convex space, define its indicator function, $I_S$, by

$$I_S(x) = \begin{cases} 0, & x \in A \\ \infty, & x \notin A \end{cases}$$

Then $I_S$ is convex if and only if $A$ is convex and $I_S$ is lsc if and only if $S$ is closed. We claim

$$(I_S)_* = I_{\text{cch}(S)} \tag{5.25}$$

For by the above, $I_{\text{cch}(S)}$ is a regular convex function, and if $h \le I_S$ and $h$ is regular and convex, by convexity, $h \le 0$ on $\text{ch}(S)$ and then by lsc, $h \le 0$ on $\text{cch}(S)$ so $h \le I_{\text{cch}(S)}$.

Next, we generalize the notion of gauge of a convex set $A$ containing 0 to not require that $A$ be absorbing by allowing $\rho_A$ to take the value $\infty$, that is,

$$\rho_A(x) = \begin{cases} \infty, & \text{if } \{\lambda \mid \lambda x \in A\} = \{0\} \\ \sup\{\lambda \mid \lambda^{-1}x \in A\}, & \text{if } \lambda^{-1}x \in A \text{ for some } \lambda > 0 \end{cases}$$

Notice with this definition, if $A$ is a subspace, then $\rho_A = I_A$!

Now let $X, Y$ be a dual pair and let $S$ be any subset of $X$. We claim

$$I^*_{\{0\}\cup S} = \rho_{-S^\circ} \tag{5.26}$$

where $S^\circ \subset Y$ is the polar of $S$. For, by definition of the Legendre transform,

$$I^*_{\{0\}\cup S}(\ell) = \sup_{x \in \{0\}\cup S} \ell(x) \tag{5.27}$$

This is clearly a nonnegative, homogeneous function of degree 1 and, by (5.27), $I^*_{\{0\}\cup S}(\ell) \le 1$ if and only if $\sup_S \ell(x) \le 1$ if and only if $\inf_{-S} \ell(x) \ge -1$ if and only if $\ell \in (-S)^\circ = -S^\circ = -(S \cup \{0\})^\circ$. This proves (5.26).

If $A$ is a convex set in $X$ with $0 \in A$, we claim that

$$\rho_A^* = I_{-A^\circ} \tag{5.28}$$

for

$$\begin{aligned} \rho_A^*(\ell) &= \sup_{\substack{x\in X \\ \rho_A(x)<\infty}} [\ell(x) - \rho_A(x)] \\ &= \begin{cases} 0, & \text{if } \ell(x) \le \rho_A(x) \text{ for all } x \\ \infty, & \text{otherwise} \end{cases} \end{aligned} \tag{5.29}$$

since, if $\ell(x) > \rho_A(x)$ for some $x$, $\sup_{x\in[0,\infty)}[\ell(\lambda x) - \rho_A(\lambda x)] = \sup_{\lambda\in(0,\infty)} \lambda[\ell(x) - \rho_A(x)] = \infty$. Since $\ell \in -A^\circ$ if and only if $\inf_{x\in -A} \ell(x) \ge -1$ if and only if $\sup_{x\in A} \ell(x) \le 1$ if and only if $\ell(x) \le \rho_A(x)$ for all $x \in A$, (5.29) implies (5.28).

With these calculations, we see for any set $S \subset X$,

$$\begin{aligned} I_{\text{cch}(S\cup\{0\})} &= (I_{S\cup\{0\}})_* && \text{(by (5.25))} \\ &= (I^*_{S\cup\{0\}})^* && \text{(by Fenchel's theorem)} \\ &= (\rho_{-S^\circ})^* && \text{(by (5.26))} \\ &= I_{-(-S^\circ)^\circ} && \text{(by (5.28))} \\ &= I_{S^{\circ\circ}} \end{aligned}$$

Since $I_A$ determines $A$, we see $S^{\circ\circ} = \text{cch}(S \cup \{0\})$, that is, the bipolar theorem is a special case of Fenchel's theorem. □

We end this chapter by using bipolar theory to return to the issue of dual topologies. In understanding the next definition, keep in mind Theorem 5.7 which says the gauge associated to $A$ is the uniform norm associated to functions on $A^\circ$. The weak topology is uniform convergence on finite subsets of $Y$, that is, pointwise convergence.

**Definition** Let $X, Y$ be a dual pair. The *Mackey topology*, $\tau(X,Y)$, on $X$ is the topology of uniform convergence on balanced, convex, $\sigma(Y,X)$-compact subsets of $Y$. The *strong topology*, $\beta(X,Y)$, on $X$ is the topology of uniform convergence on $\sigma(Y,X)$-bounded subsets of $Y$.

Thus, a net $x_\alpha$ in $X$ converges to $x$ in $X$ in $\tau(X,Y)$ (resp. $\beta(Y,X)$) if and only if for every compact (resp. bounded) $A \subset Y$,

$$\sup_{y\in A} |\langle x_\alpha - x, y\rangle| \to 0$$

We will see below that $\tau$ is a dual topology. $\beta$ may not be a dual topology (i.e., $\beta$ can have more open sets than $\tau$ and so more continuous functionals), but as we will see, it has an intrinsic description that makes no mention of $Y$.

**Example 5.21** Let $X$ be any Banach space. Then the unit ball in $X$ is $\sigma(X, X^*)$-bounded and the unit ball in $X^*$ is $\sigma(X^*, X)$-bounded and the norm is bounded on any bounded set. It follows that the $\beta(X, X^*)$-topology on $X$ is the norm topology, and similarly, the $\beta(X^*, X)$-topology is the norm topology on $X^*$. This is no coincidence: we will show below that if $X, Y$ is any dual pair, then any topology in which $X$ is a complete metric space and in which every $\langle \cdot\,, y\rangle$ is continuous is the $\beta(X, Y)$-topology. The Mackey topology $\tau(X, X^*)$ is also the norm topology since every bounded set in $X^*$ is $\sigma(X^*, X)$- (i.e., weak-$*$) compact. But if $X$ is not reflexive, then $\tau(X^*, X)$ is strictly weaker than $\beta(X^*, X)$. For we will see immediately following that the $\tau(X^*, X)$ dual of $X^*$ is $X$, but since $\beta(X^*, X)$ is the norm topology, the $\beta$-dual of $X^*$ is $X^{**}$, which is strictly bigger than $X$. Incidentally, this shows $\beta$ may not be a dual topology, and also proves that $X$ is reflexive if and only if every $\sigma(X, X^*)$-bounded, closed set is $\sigma(X, X^*)$-compact. □

**Theorem 5.22** *Let $X, Y$ be a dual pair. Then the Mackey topology, $\tau(X, Y)$, is a dual topology, that is, every $\tau$-continuous linear functional on $X$ is of the form $x \mapsto \langle x, y\rangle$ for some $y \in Y$.*

*Proof* Let $y \in Y$ and let $x_\alpha \to x$ in $\tau$. Then $A = \{y\}$ is a compact set, so $|\langle x_\alpha - x, y\rangle| \to 0$, that is, $\langle \cdot\,, y\rangle$ is a $\tau$-continuous linear functional. We thus need to show that any $\tau$-continuous linear functional is in $Y$.

Let $Z = X_\tau^*$ and suppose $Z$ is strictly bigger than $Y$. Pick $\ell \in Z\backslash Y$. Since $\ell$ is $\tau$-continuous, by (3.15), there is a $c$ and $\sigma(Y, X)$-compact, balanced convex sets $A_1, \dots, A_n$ in $Y$ so

$$|\ell(x)| \leq c \sum_{i=1}^{n} \sup_{y\in A_i} |\langle y, x\rangle| \tag{5.30}$$

Let

$$A = (cn)\text{ch}(A_1 \cup \cdots \cup A_n) \tag{5.31}$$

Then $A$ is compact, balanced, and convex by Theorem 5.2(vii). Moreover,

$$\begin{aligned} \sup_{y\in A_i} |\langle y, x\rangle| &\leq \sup_{y\in A_1\cup\cdots\cup A_n} |\langle y, x\rangle| \\ &= \sup_{y\in \text{ch}(A_1\cup\cdots\cup A_n)} |\langle y, x\rangle| \\ &= (cn)^{-1} \sup_{y\in A} |\langle y, x\rangle| \end{aligned} \tag{5.32}$$

since $\sup_{y\in\lambda A}|\langle x,y\rangle| = \lambda \sup_{y\in A}|\langle y,x\rangle|$ for $\lambda \geq 0$. (5.32) and (5.30) imply that

$$|\ell(x)| \leq \sup_{x\in A}|\langle y,x\rangle| \tag{5.33}$$

The restriction of the $\sigma(Z,X)$-topology to $Y$ is the $\sigma(Y,X)$-topology (since both topologies are given by $\langle x,\,\cdot\,\rangle$) so since $A$ is compact in $Y$ in the $\sigma(Y,X)$-topology, it is closed in the $\sigma(Z,X)$-topology. Since $\ell \notin Y$ and $Z^*_{\sigma(Z,X)} = X$, there is, by Theorem 4.5, an $x \in X$ so $\sup_{y\in A}\langle y,x\rangle < \ell(x)$. Since $y$ is balanced, this says

$$\sup_{y\in A}|\langle y,x\rangle| < \ell(x)$$

This contradicts (5.33), so $\ell \in A \subset Y$, that is, $Z = X$. □

**Theorem 5.23** (Mackey–Arens Theorem) *Let $X,Y$ be a dual pair. The dual topologies on $X$ are precisely those locally convex topologies, $\mathcal{T}$, with $\sigma(X,Y) \subseteq \mathcal{T} \subseteq \tau(X,Y)$.*

*Proof* We need only show that any dual topology has $\mathcal{T} \subseteq \tau(X,Y)$, for $\sigma(X,Y)$ is, by definition, the weakest dual topology; and if $\sigma \subseteq \mathcal{T} \subseteq \tau$, $Y = X^*_\sigma \subseteq X^*_{\mathcal{T}} \subseteq X^*_\tau = Y$.

Let $A$ be a closed, balanced, convex $\mathcal{T}$-neighborhood of 0. Then, by the Bourbaki–Alaoglu theorem (Theorem 5.12), $A^\circ$ is $\sigma(Y,X)$-compact. Thus, $\rho_A(x) = \sup_{y\in A^\circ}|\langle y,x\rangle|$ (by Theorem 5.7) is continuous in the Mackey topology, that is, $A^{\text{int}} = \{x \mid \rho_A(x) < 1\}$ is $\tau(X,Y)$-open. Since the set of closed, balanced, convex $\mathcal{T}$-neighborhoods of 0 is a neighborhood base, $\mathcal{T} \subset \tau(X,Y)$. □

We turn to the $\beta$-topology. Recall that a barrel is a closed, absorbing, balanced, convex set and $X$ is called barreled if and only if every barrel is a neighborhood of 0.

**Theorem 5.24** *Let $X,Y$ be a dual pair. Then $X$ is barreled in the $\beta(X,Y)$-topology. Conversely, if $X$ is barreled and $Y$ is any set of continuous functionals on $X$ that separate points, then the $\beta(X,Y)$-topology is the original topology.*

*Remarks* 1. This shows that the $\beta(X,Y)$-topology is in some sense only dependent on $X$ and "explains" why $\beta(X^*,X^{**}) = \beta(X^*,X)$ for $X$ a Banach space.

2. In particular, by Proposition 3.23, every Fréchet space has the $\beta$-topology.

*Proof* This is an immediate consequence of Theorem 5.7, which describes topologies in terms of uniform convergence on polars, and Proposition 5.11, which says the polars of barrels are bounded sets and conversely. □

Given any locally convex space $X$, we can find $X^*$ and place the "intrinsic" topology $\beta(X^*,X)$ on $X^*$. This motivates the definition:

**Definition** A locally convex vector space $X$ is called *reflexive* if and only if
(i) $X$ has the $\beta(X, X^*)$-topology.
(ii) The dual of $X^*$ in the $\beta(X^*, X)$-topology is $X$.

By the discussion in Example 5.21, a Banach space is reflexive in the normed linear space sense if and only if it is reflexive in the sense just defined.

**Theorem 5.25** *Every Montel space is reflexive.*

*Proof* A Montel space, $X$, is, by definition, barreled so $X$ has the $\beta(X, X^*)$-topology by Theorem 5.24. Again by definition, every $\beta$-closed, bounded subset, $A$, of $X$ is compact. Any $\sigma(X, X^*)$-closed set is $\beta$-closed for $\sigma$ is weaker than $\beta$. By the lemma below, any $\sigma$-bounded set is $\beta$-bounded. Moreover, any $\beta$-compact set is $\sigma$-compact since the identity map of $X_\beta$ to $X_\sigma$ is continuous. Thus, the $\sigma(X, X^*)$-closed, bounded sets in $X$ and $\sigma(X, X^*)$-compact are the same, so $\tau(X^*, X) = \beta(X^*, X)$. Thus, the $\beta(X^*, X)$ dual of $X^*$ is $X$ by Theorem 5.22. □

**Lemma 5.26** *Let $X$ be a locally convex space and let $X^*$ be its dual. If $A \subset X$ is weakly bounded, that is, $\sup_{x\in A} |\ell(x)| < \infty$ for each $\ell \in X^*$, then $A$ is bounded, that is, $\sup_{x\in A} \rho_C(x) < \infty$ for each continuous seminorm, $\rho_C$, on $X$.*

*Proof* This result generalizes Proposition 3.2, which we will use in its proof. Use $\mathcal{T}$ to denote the topology on $X$ so, since $X^*_{\mathcal{T}} = X^*$, we have

$$\sigma(X, X^*) \subset \mathcal{T} \subset \tau(X, X^*) \tag{5.34}$$

by the Mackey–Arens theorem. Let $U$ be a closed, balanced, convex neighborhood of 0 in the $\mathcal{T}$-topology. We have to show that for some $\mu > 0$,

$$A \subset \mu U \tag{5.35}$$

Since $U$ is convex and closed, it is an intersection of half-spaces, so $\sigma$-closed, and thus, $(U^\circ)^\circ = U$. Since $\mathcal{T} \subset \tau$, $C \equiv U^\circ \subset X^*$ is $\sigma(X^*, X)$-compact.

Let $B$ be the subspace of $X^*$ generated by $C$ (algebraically, take no closures!), so

$$B = \bigcup_{n=1}^{\infty} nC \tag{5.36}$$

$C$ is absorbing for $B$, and for any $x \in B$, $\{\lambda \mid \lambda x \in C\}$ is bounded since $C$ is compact. Thus, $\rho_C$, the gauge of $C$, defines a norm on $B$. We claim $B$ is complete in the norm.

For, by Theorem 5.7,

$$\rho_C(\ell) = \sup_{x\in U} |\ell(x)|$$

So suppose $\ell_n$ is Cauchy in $\rho_C(\cdot)$. In particular, $\sup_n \rho_C(\ell_n) \equiv R < \infty$, so replacing $\ell_n$ by $\ell_n R^{-1}$, we can suppose $\sup_n \rho_C(\ell_n) = 1$, that is, $\{\ell_n\} \subset C$. Since $C$ is compact, by passing to a subsequence, we can suppose $\ell_n \to \ell_\infty$ in the $\sigma(X^*, X)$-topology. Now for $x \in U$,

$$|(\ell_\infty - \ell_m)(x)| = \lim_{n\to\infty} |(\ell_n - \ell_m)(x)| \le \lim_{n\to\infty} \rho_C(\ell_n - \ell_m)$$

so $\rho_C(\ell_\infty - \ell_m) \le \lim_{n\to\infty} \rho_C(\ell_n - \ell_m)$. Thus, since $\ell_n$ is Cauchy, $\ell_m \to \ell_\infty$ in $\rho_C$, that is, $B$ is complete.

Now we can appeal to Proposition 3.2. Each $x \in A$ defines a function $\varphi_x$ on $B$ by $\varphi_x(\ell) = \ell(x)$. By hypothesis, $A$ is weakly bounded, that is, $\sup_{x\in A} |\ell(x)| < \infty$ for each $\ell \in X^*$ and so for each $\ell \in B \subset X^*$, $\sup_{x\in A} |\varphi_x(\ell)| < \infty$. Thus, Proposition 3.2 says that

$$\sup_{x\in A} \|\varphi_x\|_{B^*} < \infty \tag{5.37}$$

But since $C$ is the unit ball in $B$,

$$\|\varphi_x\|_{B^*} = \sup_{\ell\in C} |\varphi_x(\ell)|$$

so (5.37) says

$$\sup_{x\in A} \sup_{\ell\in C} |\ell(x)| = \mu < \infty$$

or

$$\sup_{x\in A} \sup_{\ell\in C} |\ell(\mu^{-1}x)| = 1 \tag{5.38}$$

Since $C$ is balanced and convex, the definition of polar sets implies that (5.38) says if $x \in A$, then $\mu^{-1}x \subset C^\circ = U$, that is, $A \subset \mu U$ as we needed to prove. $\square$

# 6

# Monotone and convex matrix functions

In order for the definition of convexity of a function to make sense, the range space must have a notion of order. So it is natural to consider functions from self-adjoint operators to themselves and ask about convexity. In this chapter, we will focus on this question when $A \mapsto f(A)$ is generated by a function $f\colon \mathbb{R} \to \mathbb{R}$ via the functional calculus. Indeed, we will focus initially on $f(A)$ for $A$ an $n \times n$ matrix. $f(A)$ can then be defined by several equivalent methods:

(i) Find a polynomial $p$ with $p(\lambda_i) = f(\lambda_i)$ for all eigenvalues, $\lambda_i$, of $f$ and then let $f(A) = p(A)$ defined by $p(A) = \sum_{j=1}^N \alpha_j A^j$ if $p(\lambda) = \sum_{j=1}^N \alpha_j \lambda^j$.

(ii) Diagonalize $A$ by

$$A = U \begin{pmatrix} \lambda_1 & & 0 \\ & \ddots & \\ 0 & & \lambda_n \end{pmatrix} U^{-1}$$

and let

$$f(A) = U \begin{pmatrix} f(\lambda_1) & & 0 \\ & \ddots & \\ 0 & & f(\lambda_n) \end{pmatrix} U^{-1}$$

For scalar functions, $f$ is convex only if it is piecewise $C^1$ and $f'$ is monotone, so it is also natural to ask first about monotone functions on matrices, that is, functions $f\colon \mathbb{R} \to \mathbb{R}$ so $A \le B \Rightarrow f(A) \le f(B)$. The surprise is that this simple-looking question has a depth and richness one might not expect. But one has to ask a more restricted question, because later in this chapter, we will prove (see Corollary 6.27)

**Theorem 6.1** *Let $f\colon \mathbb{R} \to \mathbb{R}$ be monotone on all $2 \times 2$ self-adjoint matrices, that is, for all pairs of $2 \times 2$ self-adjoint matrices with $A \le B$, we have $f(A) \le f(B)$. Then $f$ is an affine function, that is, $f(x) = \alpha x + \beta$ with $\alpha \ge 0$.*

Thus, to get an interesting class, we do not require $f$ to be defined on all of $\mathbb{R}$ but only on an interval:

**Definition** Let $(a,b) \subset \mathbb{R}$ ($a$ or $b$ may be infinite). $M_n(a,b)$ is the set of all real-valued functions $f\colon (a,b) \to \mathbb{R}$ so that if $A, B$ self-adjoint $n \times n$ matrices with $\sigma(A) \cup \sigma(B) \subset (a,b)$ and $A \leq B$, then $f(A) \leq f(B)$.

$$M_\infty(a,b) = \bigcap_{n=1}^{\infty} M_n(a,b) \tag{6.1}$$

Since, given $A, B$, $(n-1) \times (n-1)$ matrices, we can consider $n \times n$ matrices $\tilde{A}, \tilde{B}$ with

$$\tilde{A} = \begin{pmatrix} A & 0 \\ 0 & \frac{1}{2}(a+b) \end{pmatrix}, \qquad \tilde{B} = \begin{pmatrix} B & 0 \\ 0 & \frac{1}{2}(a+b) \end{pmatrix}$$

it is easy to see that $M_{n-1}(a,b) \supset M_n(a,b) \supset \cdots$.

While we will focus on $(a,b)$, one can define for an arbitrary open set $\Omega \subset \mathbb{R}$, $M_n(\Omega)$ as the functions $f\colon \Omega \to \mathbb{R}$ so that $A, B$ self-adjoint with $\sigma(A) \cup \sigma(B) \subset \Omega$ and $A \leq B$ implies $f(A) \leq f(B)$. There is a second space $M_n^c(\Omega)$ where we add a condition on $A$ and $B$, namely, that there is a curve $\gamma(t)$, $0 \leq t \leq 1$, with $\gamma(0) = A$, $\gamma(1) = B$, $\gamma(t)$ is self-adjoint, and $\sigma(\gamma(t)) \subset \Omega$. One can see that $M_n^c(a,b) = M_n(a,b)$ (take $\gamma(t) = tB + (1-t)A$) and that such a curve $\gamma(t)$ exists if and only if for every connected component, $\Omega'$ of $\Omega$, $A$ and $B$ have the same number of eigenvalues in $\Omega'$.

**Example 6.2** We want to show the function $f(x) = x^2$ is not matrix monotone on $(0,\infty)$. We ask, in general, for two self-adjoint projections, $P$ and $Q$, when is it true that

$$(P+Q)^2 \geq P^2 \tag{6.2}$$

Clearly, (6.2) is equivalent to

$$C \equiv Q^2 + QP + PQ \geq 0 \tag{6.3}$$

so suppose (6.3) holds. For any smooth self-adjoint operator-valued function, $A(\varepsilon)$, we have

$$A(\varepsilon)CA(\varepsilon) \geq 0$$

so if $A(0)CA(0) = 0$, we have

$$\frac{dA(0)}{d\varepsilon}\, CA(0) + A(0)C\, \frac{dA}{d\varepsilon}(0) = 0 \tag{6.4}$$

Picking $A(\varepsilon) = (1-Q) + \varepsilon Q$, we see

$$\begin{aligned} 0 &= QC(1-Q) + (1-Q)CQ \\ &= QP(1-Q) + (1-Q)PQ \end{aligned} \tag{6.5}$$

Multiplying (6.5) on the right by $Q$, we see $QP(1-Q)=0$ or $PQ=QPQ$ so $QP=(PQ)^*=(QPQ)^*=QPQ=PQ$. Thus, (6.3) implies $[Q,P]=0$.

Conversely, if $[Q,P]=0$, then $QP=QP^2=PQP\geq 0$ so $Q^2+QP+PQ=Q^2+2PQP\geq 0$ and (6.3) holds. As a result, $P^2\leq(P+Q)^2$ if and only if $[P,Q]=0$.

If $n\geq 2$, there are lots of noncommuting projections, and so many $P,Q$ for which (6.2) fails, for example,

$$A=\begin{pmatrix}1&0\\0&0\end{pmatrix},\qquad B=\begin{pmatrix}1/2&1/2\\1/2&1/2\end{pmatrix}$$

Thus, $f(x)=x^2$ is not matrix monotone on $(0,\infty)$. We will see later (see the remark after Corollary 6.28) that $f(x)=x^\alpha$ is in $M_n(0,\infty)$, for $n\geq 2$, if and only if $0\leq\alpha\leq 1$. □

While we consider finite matrices, the following elementary argument shows that $M_\infty(a,b)$ also describes the infinite-dimensional case.

**Proposition 6.3** *Let $f\in M_\infty(a,b)$. Let $A,B$ be self-adjoint operators on a Hilbert space with* $\text{spec}(A)\cup\text{spec}(B)\subset(a,b)$. *Then*

$$A\leq B\Rightarrow f(A)\leq f(B)$$

*Remarks* 1. For simplicity, we will suppose $(a,b)$ is bounded, but using the notion of inequality for operators discussed, for example, in Kato [189], this theorem also holds for, say, $(a,b)=[0,\infty)$.

2. We will use the fact proven below (see Theorem 6.25) that any $f\in M_\infty$ is continuous.

*Proof* Let $P_1,P_2,\dots$ be a sequence of projections of dimension $1,2,\dots$ with $\text{s-lim}\,P_n=1$ (e.g., let $\{\varphi_j\}_{j=1}^\infty$ be an orthonormal basis and let $P_n$ be the projection onto the span of $\{\varphi_j\}_{j=1}^n$). If $A\leq B$ with $\sigma(A)\cup\sigma(B)\subset(a,b)$, define

$$\begin{aligned}A_n&=P_nAP_n+\tfrac{1}{2}(a+b)(1-P_n)\\B_n&=P_nBP_n+\tfrac{1}{2}(a+b)(1-P_n)\end{aligned}$$

Thus, $A_n\to A$ and $B_n\to B$ strongly, so by the continuity of the functional calculus ([303, Thm. VIII.20]), $f(A_n)\to f(A)$ and $f(B_n)\to f(B)$ strongly. Thus, to see $f(A)\leq f(B)$, it suffices that $f(A_n)\leq f(B_n)$. Since

$$f(A_n)=f(P_nAP_n)+f(\tfrac{1}{2}(a+b))(1-P_n)$$

and similarly for $B_n$ and $P_nAP_n\leq P_nBP_n$, $f\in M_\infty$ implies that $f(P_nAP_n)\leq f(P_nBP_n)$ so $f(A_n)\leq f(B_n)$. □

As a final preliminary, we note that to determine $M_n(a,b)$ for all $(a,b) \neq \mathbb{R}$, it suffices to determine it for one $(a,b)$.

**Proposition 6.4** (i) *Let* $(a,b)$ *be bounded. Let* $T\colon (0,1) \to (a,b)$ *by* $Tx = a + x(b-a)$. *Then* $f \in M_n(a,b)$ *if and only if* $f \circ T \in M_n(0,1)$.
(ii) *Let* $T\colon (0,\infty) \to (a,\infty)$ *by* $Tx = x + a$. *Then* $f \in M_n(a,\infty)$ *if and only if* $f \circ T \in M_n(0,\infty)$.
(iii) *Let* $T\colon (0,1) \to (1,\infty)$ *by* $Tx = x^{-1}$. *Then* $f \in M_n(1,\infty)$ *if and only if* $-f \circ T \in M_n(0,1)$.
(iv) *Let* $T\colon (0,\infty) \to (-\infty,0)$ *by* $Tx = -x$. *Then* $f \in M_n(-\infty,0)$ *if and only if* $-f \circ T \in M_n(-\infty,0)$.

*Proof* The maps $T$ and their inverses in (i)–(ii) are order preserving on operators, and in (iii)–(iv), $T$ and their inverses are order reversing on operators. □

Thus, one can focus on describing $M_n$ for a convenient $(a,b)$ and we will normally look at $(-1,1)$. The beautiful result in that case is

**Theorem 6.5** (Loewner's Theorem) *Let* $f$ *be a real-valued function on* $(-1,1)$. *It lies in* $M_\infty(-1,1)$ *if and only if there is a finite measure* $\mu$ *on* $[-1,1]$ *so that*

$$f(x) = f(0) + \int_{-1}^{1} \frac{x}{1+\lambda x}\, d\mu(\lambda) \tag{6.6}$$

We will provide a first proof of the "hard" half of this theorem in the next chapter, with another proof in Chapter 9 and further discussion in the Notes. For now, we will prove the "easy" half, namely, that (6.6) implies monotonicity.

*Proof that* (6.6) *implies* $f$ *is in* $M_\infty(-1,1)$ It clearly suffices to show that if $A \leq B$ with $\sigma(A) \cup \sigma(B) \subset (-1,1)$, then

$$\frac{A}{1+\lambda A} \leq \frac{B}{1+\lambda B} \tag{6.7}$$

For $\lambda = 0$, this is obvious. For $\lambda \neq 0$, write

$$\frac{x}{1+\lambda x} = \frac{1}{\lambda} - \frac{1}{\lambda^2}\, \frac{1}{\lambda^{-1}+x}$$

to see that (6.7) is equivalent to

$$-(A+\alpha)^{-1} \leq -(B+\alpha)^{-1} \tag{6.8}$$

for $\alpha \notin [-1,1]$, and this is just monotonicity of $x \to x^{-1}$ for strictly positive or strictly negative operators. □

We note immediately the strong consequences of the representation (6.6) for regularity of $f$ in $x$:

**Proposition 6.6** *Let $f$ defined on $(-1, 1)$ obey a representation (6.6). Then, $f$ is real analytic on $(-1, 1)$ and has an analytic continuation to all of $\mathbb{C}\backslash(-\infty, -1] \cup [1, \infty)$ that obeys*

$$\pm \operatorname{Im} f(z) > 0 \quad \text{if} \quad \pm \operatorname{Im} z > 0 \tag{6.9}$$

*Moreover, $d\mu$ is the weak limit of $\pi^{-1} \operatorname{Im} f((-\lambda - \varepsilon)^{-1})\, d\lambda$ in the sense that for any $g \in C_0^\infty[-1, 1]$,*

$$\pi^{-1} \int g(\lambda) \operatorname{Im} f((-\lambda - i\varepsilon)^{-1})\, d\lambda \underset{\varepsilon \downarrow 0}{\to} \int g(\lambda)\, d\mu(\lambda) \tag{6.10}$$

*Proof* For each $z \in \mathbb{C}\backslash(-\infty, -1] \cup [1, \infty)$, the integral in (6.6) converges, and if $d\mu$ is written as a weak limit of point measures, the convergence is uniform on compact subsets. Since $z \mapsto z(1 + \lambda_0 z)^{-1}$ is analytic for each $\lambda_0$, the integral defines an analytic function in the region in question. Moreover,

$$\operatorname{Im}\left[\frac{z}{1 + \lambda z}\right] = \operatorname{Im}\left[\frac{z + \lambda |z|^2}{|1 + \lambda z|^2}\right] = \frac{\operatorname{Im} z}{|1 + \lambda z|^2} \tag{6.11}$$

so (6.9) holds.

Finally, rewriting (6.6) as

$$f(x) = f(0) + \int_{-1}^{1} \frac{1}{\lambda + x^{-1}}\, d\mu(\lambda)$$

we see that

$$\begin{aligned} \frac{1}{\pi} \operatorname{Im} f((-\lambda_0 - i\varepsilon)^{-1}) &= \frac{1}{\pi} \int_{-1}^{1} \operatorname{Im}\left(\frac{1}{\lambda - \lambda_0 - i\varepsilon}\right) d\mu(\lambda) \\ &= \int_{-1}^{1} \frac{\varepsilon}{\pi} \frac{1}{(\lambda - \lambda_0)^2 - \varepsilon^2}\, d\mu(\lambda) \end{aligned}$$

and (6.10) follows from the fact that the integrand is an approximate identity. □

*Remark* The above explicit construction of $\mu$ shows $\mu$ is unique. This also follows from

$$\frac{x}{1 + \lambda x} = \sum_{n=0}^{\infty} (-1)^n x^{n+1} \lambda^n$$

which implies $f^{(n)}(0)/n! = \int_{-1}^{1} \lambda^{n-1}\, d\mu(\lambda)$ for $n \geq 1$. Since polynomials are dense in $C([-1, 1])$, the moments determine $\mu$.

There is a converse to Proposition 6.3, namely, if $F$ is analytic in $\mathbb{C}\backslash(-\infty, -1] \cup [1, \infty)$ with $\pm \operatorname{Im} F > 0$ if $\pm \operatorname{Im} z > 0$, then $F$ has a representation of the form (6.6).

Given Proposition 6.4 and the fact that the maps there preserve $\mathbb{C}_+$, we will have

**Proposition 6.7** *A real-valued function $f$ on $(a,b)$ lies in $M_\infty(a,b)$ if and only if $f$ has an analytic continuum to $\mathbb{C}_+ \cup \mathbb{C}_- \cup (a,b)$ with $\pm \operatorname{Im} f(z) > 0$ if $\pm \operatorname{Im} z > 0$.*

*Note* $a$ and/or $b$ can be infinite.

**Example 6.8** We can apply Proposition 6.7 to the function $f(x) = x^\alpha$ on $(0,\infty)$. $f(z) = z^\alpha$ and $\operatorname{Im} f(z) = |z|^\alpha \sin(\pi\alpha(\arg z))$. This is positive on all of $\mathbb{C}_+$ if and only if $0 \le \alpha \le 1$. Thus,

$$A \mapsto A^\alpha$$

for $A > 0$ is monotone on all finite matrices if and only if $0 \le \alpha \le 1$.

This can also be seen from an integral representation. By a contour integration (placing a cut for $x^\beta$ on $[0,\infty)$),

$$\int_0^\infty \frac{x^\beta}{x^2 + a}\, dx = \frac{\pi}{2}\, \frac{1}{\cos(\beta\pi/2)}\, a^{(\beta-1)/2}$$

which means that for any positive matrix $A$ and $\alpha = \frac{1}{2}(\beta - 1)$,

$$A^\alpha = \frac{2\sin(\pi\alpha)}{\pi} \int_0^\infty w^{\alpha-1}(A+w)^{-1} A\, dw$$

□

The analog of Loewner's theorem for the classes $M_n(\Omega)$ and $M_n^c(\Omega)$ defined before Example 6.2 is

**Theorem 6.9** *Let $\Omega$ be an open set in $\mathbb{R}$ and let $\tilde{\Omega}$ be the convex hull of $\Omega$, that is, $\tilde{\Omega} = (a,b)$ where $a = \inf_{x\in\Omega} x$ and $b = \sup_{x\in\Omega} x$. Then*

(i) *$f \in M_\infty^c(\Omega)$ if and only if $f$ is real analytic on $\Omega$ with an analytic continuation to $\Omega \cup \mathbb{C}_+ \cup \mathbb{C}_-$ with $\operatorname{Im} f > 0$ on $\mathbb{C}_+$.*

(ii) *$f \in M_\infty(\Omega)$ if and only if $f$ is the restriction to $\Omega$ of a function in $M_\infty(\tilde{\Omega})$, that is, if and only if $f$ has an analytic continuation to $\tilde{\Omega} \cup \mathbb{C}_+ \cup \mathbb{C}_-$ with $\operatorname{Im} f > 0$ on $\mathbb{C}_+$.*

See Rosenblum–Rovnyak [321, 322] for proofs.

Thus, if $a < b < c < d$ and $\Omega = (a,b) \cup (c,d)$, $M_\infty^c(\Omega) \ne M_\infty(\Omega)$. Functions in $M_\infty^c(\Omega)$ can have singularities in $[b,c]$ while functions in $M_\infty(\Omega)$ cannot.

Next, we will set up some preliminaries for the hard part of Loewner's theorem (Theorem 6.5), that is, that any monotone function in $M_\infty(a,b)$ has a representation of the form (6.6). A key will be to associate to any $f$ on $(a,b)$ and any distinct $x_1,\dots,x_n \in (a,b)$ an $n\times n$ matrix $L^{(n)}(x_1,\dots,x_n;f)$ by

$$L_{ij}^{(n)}(x_1,\dots,x_n;f) = \begin{cases} \frac{f(x_i)-f(x_j)}{x_i-x_j}, & \text{if } i \ne j \\ f'(x_i), & \text{if } i = j \end{cases} \tag{6.12}$$

Our immediate goal will be to prove $f$ is in $M_n(a,b)$ if and only if for all choices of distinct $x_1,\dots,x_n \in (a,b)$, $L^{(n)}(x_1,\dots,x_n;f)$ is a positive definite matrix.

As a preparatory step, we will study the object on the right of (6.12) and a generalization:

**Definition** Given any function $f$ on $(a,b)$ and $x_1,\dots,x_n$ distinct in $(a,b)$, we define the *divided difference* $[x_1,\dots,x_n;f]$ inductively by

$$[x_1;f] = f(x_1)$$

$$[x_1,\dots,x_n;f] = \frac{[x_1,\dots,x_{n-1};f] - [x_2,\dots,x_n;f]}{x_1 - x_n} \tag{6.13}$$

Often, if $f$ is understood, we will just use $[x_1,\dots,x_n]$.

**Proposition 6.10** *Fix $f$ on $(a,b)$.*

(i) *$[x_1,\dots,x_n]$ is a symmetric function of its arguments.*
(ii) *If $f$ is $C^k$, then $[x_1,\dots,x_{k+1}]$ can be extended continuously from $\{x_i \in (a,b) \mid x_i \neq x_j,\ \text{all } i \neq j\}$ to $(a,b)^{k+1}$.*

*Remark* More generally, if $f \in C^k$, $[x_1,\dots,x_\ell]$ for $\ell > k+1$ can be extended to the set of points where no more than $k+1$ $x$'s are identical.

*Proof* We will use a trick that will later be useful several times. We suppose $f$ is analytic in a neighborhood of $(a,b)$ and prove a formula for such $f$'s using the Cauchy integral formula. That plus a limiting argument proves the formula of the result in general.

This idea is especially useful because if $z \in \mathbb{C}\backslash(a,b)$ and $g_z(x) = 1/(z-x)$, we can compute $[x_1,\dots,x_n;g_z]$ easily, namely,

$$[x_1,\dots,x_n;g_z] = \prod_{j=1}^{n} \frac{1}{z - x_j} \tag{6.14}$$

This follows by induction from (6.13) and

$$\frac{1}{z-x_j} - \frac{1}{z-x_n} = \frac{x_j - x_n}{(z-x_j)(z-x_n)}$$

Basically, we will use the fact that $[x_1,\dots,x_n;f]$ is linear in $f$ and (6.14) is clearly symmetric on $x_1,\dots,x_n$. Explicitly, if $f$ is analytic and $C$ any contour in the domain of analyticity of $f$ that circles $(a,b)$, then

$$f(x) = \frac{1}{2\pi i} \oint_C \frac{f(\zeta)}{\zeta - x}\, d\zeta \tag{6.15}$$

so by (6.14),

$$[x_1,\dots,x_n;f] = \frac{1}{2\pi i} \oint_C f(\zeta) \prod_{j=1}^{n} \frac{1}{\zeta - x_j}\, d\zeta \tag{6.16}$$

$$= \sum_{j=1}^{n} \frac{f(x_j)}{\prod_{k\neq j}(x_j - x_k)} \tag{6.17}$$

by calculating the residues of $f(\zeta)/\prod_{j=1}^n(\zeta - x_j)$ at each of its poles. Since $[x_1,\dots,x_n;f]$ only depends on $f(x_1),\dots,f(x_n)$ and any $f$ agrees with $x = x_1,\dots,x_n$ with some polynomial, (6.17) holds for any $f$. The symmetry under permutations of the argument is obvious either from (6.16) or from (6.17).

(6.16) shows that if $f$ is analytic, $[x_1,\dots,x_n;f]$ can be continued to coincident points, and by using residue calculus if $(x_1,\dots,x_n)$ take the values $y_1, m_1$ times $y_2, m_2$ times and $y_\ell, m_\ell$ times (with $m_1+\dots+m_\ell = n$), then

$$[x_1,\dots,x_n;f] = \sum_{j=1}^{\ell} \frac{1}{(m_j-1)!}\left(\frac{d}{dx}\right)^{(m_j-1)}\left[f(x)\prod_{k\neq j}(x-y_k)^{-m_k}\right]\Bigg|_{x=y_j} \tag{6.18}$$

The result for general $f$ which are $C^q$ with $q \geq \max_j(m_j-1)$ follows from the lemma below. □

*Remark* In particular,

$$[x,\dots,x]_{k\text{-times}} = \frac{1}{(k-1)!}\, f^{(k-1)}(x) \tag{6.19}$$

**Lemma 6.11** *Let $f$ be a continuous function on a bounded interval $(a,b)$. Then there exists a sequence $f_n$ of entire functions on $\mathbb{C}$ so that for all $x \in (a,b)$, $f_n(x) \to f(x)$. The convergence is uniform on compact subsets of $(a,b)$. Moreover, if $f$ is $C^k$ in a neighborhood of some subset $[c,d] \subset (a,b)$, then $f_n$ can be chosen so that for $j = 0,1,\dots,k$, $f_n^{(j)}(x) \to f^{(j)}(x)$ uniformly in $x \in [c,d]$.*

*Proof* Define $g_\ell$ by

$$g_\ell(x) = \begin{cases} f(x), & a+\frac{1}{\ell} < x < b-\frac{1}{\ell} \\ 0, & \text{otherwise} \end{cases}$$

and

$$f_{m,\ell}(x) = \int_{-\infty}^{\infty} g_\ell(y)\exp(-m(x-y)^2)\left(\frac{\pi}{m}\right)^{-1/2} dy$$

Then $f_{m,\ell}$ is entire and the claimed convergence as first $m\to\infty$ and then $\ell\to\infty$ is easy to see. □

We will now make a formal definition of (6.12).

**Definition** If $f$ is a $C^1$ function on $(a,b) \subset \mathbb{R}$, we define the *Loewner matrix*, $L^{(n)}(x_1,\dots,x_n;f)$, for arbitrary points $x_1,\dots x_n$ in $(a,b)$ by

$$L_{ij}^{(n)}(x_1,\dots,x_n;f) = [x_i,x_j;f], \qquad i,j = 1,\dots,n \tag{6.20}$$

We will also need the notion of Schur product and one of its properties:

**Definition** If $A$ and $B$ are finite matrices, their *Schur product*, $A \odot B = C$, is defined by

$$C_{ij} = A_{ij} B_{ij}$$

*Remark* Schur product is *not* invariant under change of basis.

**Theorem 6.12** *If $A$ and $B$ are positive definite $n \times n$ matrices, their Schur product $A \odot B$ is positive definite.*

*Proof* If $\{\lambda_i\}_{i=1}^n$ are the eigenvalues of $A$ and $\{\varphi_i\}_{i=1}^n$ the eigenvectors, then

$$A = \sum \lambda_i P_{\varphi_i}$$

where $\lambda_i \geq 0$ and $P_{\varphi_i}$ is the projection onto $\varphi_i$. Since $A, B \mapsto A \odot B$ is bilinear, it suffices to show

$$P_\varphi \odot P_\psi$$

is positive definite. But

$$(P_\varphi \odot P_\psi)_{ij} = \overline{\varphi_i \psi_i}\, \varphi_j \psi_j$$

is the positive rank one operator $(\eta, \cdot\,)\eta$ where $\eta_i = \varphi_i \psi_i$. □

**Theorem 6.13** *Let $A$ be a diagonal $n \times n$ matrix*

$$A = \begin{pmatrix} x_1 & & 0 \\ & \ddots & \\ 0 & & x_n \end{pmatrix}$$

*Let $C$ be a Hermitian matrix and let $f$ be a $C^1$ function on $(a, b)$. Then $\lambda \mapsto f(A + \lambda C) \equiv \Phi(\lambda)$ is $C^1$ near $\lambda = 0$ and*

$$\Phi'(0) = L^{(n)}(x_1, \dots, x_n; f) \odot C \tag{6.21}$$

*More generally, if $f$ is $C^m$, then $\Phi$ is $C^m$ and*

$$\Phi^{(m)}_{i_1 i_{m+1}}(0) = m! \sum_{i_2, \dots, i_m} [x_{i_1}, \dots, x_{i_{m+1}}; f] \prod_{k=1}^{m} C_{i_k i_{k+1}} \tag{6.22}$$

*Proof* By Lemma 6.11, it suffices to prove the result for $f$ analytic in a neighborhood of $(a, b)$. Then write $f$ by (6.15), so

$$\lambda^{-1}[f(A) + \lambda C) - f(A)] = (2\pi i)^{-1} \oint_C f(\zeta)(\zeta - A - \lambda C)^{-1} C (\zeta - A)^{-1}\, d\zeta \tag{6.23}$$

Thus,

$$\Phi'(0) = (2\pi i) \oint f(\zeta)(\zeta - A)^{-1} C(\zeta - A)^{-1}\, d\zeta$$

and this has matrix elements

$$\begin{aligned}\Phi'(0)_{ij} &= (2\pi i)^{-1} \oint f(\zeta)(\zeta - x_i)^{-1} C_{ij} (\zeta - x_j)^{-1}\, d\zeta \\ &= [x_i, x_j; f] C_{ij} = [L^{(n)}(x_1, \dots, x_n; f) \odot C]_{ij}\end{aligned}$$

by (6.16). This proves (6.21).

Similarly, starting with

$$f(A + \lambda C) = (2\pi i)^{-1} \oint f(\zeta)(\zeta - A - \lambda C)^{-1}\, d\zeta$$

we get

$$\Phi^{(m)}(0) = (2\pi i) m! \oint f(\zeta)(\zeta - A)^{-1} [C(\zeta - A)^{-1}]^m\, d\zeta$$

so with $i_1 = i$ and $i_{m+1} = j$,

$$\Phi^{(m)}(0)_{ij} = (2\pi i)^{-1} m! \oint f(\zeta) \sum_{i_2, \dots, i_m} \prod_{k=1}^{n+1} (\zeta - x_{i_k})^{-1} \prod_{k=1}^{n} C_{i_k i_{k+1}}$$

which yields (6.22), given (6.16). □

We will need the following results to state a general theorem relating Loewner matrices to monotonicity.

**Proposition 6.14** *Let $A$ be an $n \times n$ self-adjoint matrix and let $\tilde{A}$ be the $(n-1) \times (n-1)$ matrix obtained by removing the last row and column. Let $\lambda_1 \le \lambda_2 \le \cdots \le \lambda_n$ be the eigenvalues of $A$ and $\mu_1 \le \mu_2 \le \cdots \le \mu_{n-1}$ of $\tilde{A}$. Then the eigenvalues interlace, that is,*

$$\lambda_1 \le \mu_1 \le \lambda_2 \le \mu_2 \le \cdots \le \mu_{n-1} \le \lambda_n \tag{6.24}$$

*First Proof* A standard application of the min-max principle. □

*Second Proof* Suppose first that $\lambda_1 < \lambda_2 < \cdots < \lambda_n$ and the corresponding eigenvectors $\varphi^{(1)}, \dots, \varphi^{(n)}$ obey $(\delta_n, \varphi^{(j)}) \neq 0$ for $j = 1, \dots, n$. Then by Cramer's rule,

$$\begin{aligned}m(z) \equiv (z - A)^{-1}_{nn} &= \frac{\det(z - \tilde{A})}{\det(z - A)} \\ &= \frac{\prod_{i=1}^{n-1} (z - \mu_i)}{\prod_{j=1}^{n} (z - \lambda_j)}\end{aligned}$$

so $m$ has zeros at $\mu_1, \dots, \mu_{n-1}$ and poles at $\lambda_1, \dots, \lambda_n$.

On the other hand, by an eigenfunction expansion,

$$m(z) = \sum_{k=1}^{n} |(\varphi^{(k)}, \delta_n)|^2 (z - \lambda_i)^{-1}$$

which shows that $m(x)$ is strictly monotone decreasing on each interval $(\lambda_i, \lambda_{i+1})$ and $m(x) \to +\infty$ as $x \downarrow \lambda_i$ and $m(x) \to -\infty$ as $x \uparrow \lambda_{i+1}$. It follows that $m$ has exactly one zero in each interval $(\lambda_i, \lambda_{i+1})$, that is, (6.24) holds.

Any $A$ is a limit of matrices which have the two stated properties. □

**Proposition 6.15** *An $n \times n$ self-adjoint matrix, $A$, is strictly positive definite if and only if for $m = 1, 2, \dots, n$,*

$$D_m = \det((A_{ij})_{1 \le i,j \le m})$$

*obeys $D_m > 0$.*

*Proof* A strictly positive matrix has all eigenvalues strictly positive and so its determinant is strictly positive. If $A$ is strictly positive, so is each $(A_{ij})_{1 \le i,j \le m}$ for $m \le n$. Thus, if $A$ is positive, each $D_m$ is positive.

We will prove the converse result inductively on $n$. For $n = 1$, the result is obvious. If we have it for $n-1$ by this hypothesis, the matrix $\tilde{A}$ of Proposition 6.14 is strictly positive definite. Thus, its eigenvalues obey

$$0 \le \mu_1 \le \cdots \le \mu_n$$

By Proposition 6.14, $\lambda_1 \le \mu_1 \le \lambda_2 \le \cdots \le \lambda_n$ so $\lambda_2 > 0, \dots, \lambda_n > 0$. Thus, $A$ is strictly positive definite if and only if $\lambda_1 > 0$. But

$$D_n = \lambda_1 \dots \lambda_n$$

so $D_n > 0$ implies $\lambda_1 > 0$. □

As the matrix $\left(\begin{smallmatrix} 0 & 0 \\ 0 & -1 \end{smallmatrix}\right)$ shows, this result is not true if strict positivity of the $D_n$'s is replaced by nonnegativity. However, one does have

**Proposition 6.16** *Let $A$ be an $n \times n$ self-adjoint matrix. For each subset $I \subset \{1, \dots, n\}$, let $A^I$ be the $\#I \times \#I$ matrix with matrix elements $\{A_{ij}\}_{i,j \in I}$, and let $D_I = \det(A^I)$. Then $A$ is a (not necessarily strictly) positive matrix if and only if $D_I \ge 0$ for all $I$.*

*Proof* That $A$ positive implies $D_I \ge 0$ is identical to the last proof, so we consider the converse. We will use alternating tensor products (see, e.g., [305, Sect. XIII.17]). Let $I = \{i_1, \dots, i_\ell\}$ with $i_1 < \cdots < i_\ell$ and $e^I = e_{i_1} \wedge \cdots \wedge e_{i_\ell}$ where $e_1, \dots, e_n$ is the normal basis for $\mathbb{R}^\nu$. Then, as noted in [305],

$$(e^I, \wedge^\ell(A) e^I) = \det(A^I) = D_I \ge 0$$

by hypothesis. Thus, since $\{e^I\}_{\#(I)=\ell}$ is an orthonormal basis for $\wedge^\ell(\mathbb{R}^\nu)$,

$$\operatorname{tr}(\wedge^\ell(A)) = \sum_{\#(I)=\ell} D_I \geq 0$$

It follows that if $\lambda_1, \dots, \lambda_n$ are the eigenvalues of $A$, then

$$c_\ell = \sum_{\#(I)=\ell} \left( \prod_{j\in I} \lambda_j \right) = \operatorname{tr}(\wedge^\ell(A)) \geq 0$$

Write the secular equation of $A$:

$$\begin{aligned} P(\lambda) &= \det(\lambda - A) = \prod_{i=1}^n (\lambda - \lambda_i) \\ &= \lambda^n - c_1\lambda^{n-1} + c_2\lambda^{n-2} + \cdots + (-1)^n c_n \end{aligned}$$

Since $c_j \geq 0$, if $\lambda < 0$,

$$(-1)^n P(\lambda) \geq |\lambda|^n > 0$$

and thus, $A$ has all nonnegative eigenvalues, so $A \geq 0$. □

Here is the main preliminary for Loewner's theorem:

**Theorem 6.17** *Let $f$ be a real-valued $C^1$ function on an interval $(a, b)$. Then the following are equivalent:*

(i) $f \in M_n(a, b)$

(ii) *For all distinct points $x_1, \dots, x_n \in (a, b)$, the Loewner's matrix $L^{(n)}(x_1, \dots, x_n; f)$ is positive definite.*

(iii) *For all $\ell \leq n$ and distinct points $x_1, \dots, x_\ell \in (a, b)$,*

$$\det(L^{(\ell)}(x_1, \dots, x_\ell; f)) \geq 0 \tag{6.25}$$

*Remarks* 1. We will see shortly (Theorem 6.25) that every $f \in M_n(a, b)$ with $n \geq 2$ is $C^1$.

2. This is just the analog of the fact that a $C^1$ function on $(a, b)$ is an ordinary monotone function if and only if $f'(x) \geq 0$ for all $x$.

*Proof* That (ii) ⇔ (iii) is Proposition 6.16.

Given $A$ and $B$ with $A \leq B$, let $C = B - A \geq 0$. Let $A(\lambda) = A + \lambda C = \lambda B + (1 - \lambda)A$. Let $U(\lambda)$ be a unitary matrix (not necessarily continuous!) which diagonalizes $A(\lambda)$ so

$$U(\lambda)A(\lambda)U(\lambda)^{-1} = \begin{pmatrix} x_1(\lambda) & & 0 \\ & \ddots & \\ 0 & & x_n(\lambda) \end{pmatrix}$$

Since $f(A(\lambda)) = U(\lambda_0)^{-1} f(U(\lambda_0)A(\lambda)U(\lambda_0)^{-1})U(\lambda_0)$, (6.21) implies

$$\frac{d}{d\lambda} f(A(\lambda))\bigg|_{\lambda=\lambda_0} \\ = U(\lambda_0)[L^{(n)}(x_1(\lambda_0), \dots, x_n(\lambda_0); f) \odot U(\lambda_0)CU(\lambda_0)^{-1}]U(\lambda_0)$$

Thus, if $L^{(n)}$ is positive, by Theorem 6.12, $\frac{d}{d\lambda} f(A(\lambda)) \geq 0$, and thus, $f(B) \geq f(A)$. That means (ii) $\Rightarrow$ (i).

Conversely, suppose $f \in M_n(a,b)$ and $x_1, \dots, x_n$ are distinct numbers in $(a,b)$. Let $C$ be the matrix $C_{ij} \equiv 1$ and

$$A = \begin{pmatrix} x_1 & & \\ & \ddots & \\ & & x_n \end{pmatrix} \tag{6.26}$$

By Theorem 6.13,

$$\frac{d}{d\lambda}(f(A + \lambda C))\bigg|_{\lambda=0} = L^{(n)}(x_1, \dots, x_n; f)$$

Thus, if $f$ is monotone so $\frac{d}{d\lambda} f(A + \lambda C) \geq 0$, $L^{(n)}$ is positive, that is, (i) implies (ii). $\square$

Our next subject is an aside on extended Loewner matrices, which can be skipped (jump ahead to Theorem 6.21). Loewner looked at more general matrices, namely, given $\mu_1, \mu_2, \dots, \mu_n, \lambda_1, \dots, \lambda_n \in (a,b)$ with $\mu_i \neq \lambda_j$ for all $i, j$, one defines the *extended Loewner matrix* by

$$[L_e^{(n)}(\mu_1, \dots, \mu_n, \lambda_1, \dots, \lambda_n; f)]_{ij} = [\lambda_i, \mu_j; f] \tag{6.27}$$

We are heading towards showing $\det(L_e^{(n)}) \geq 0$ if $f$ is monotone and $\mu_1 < \lambda_1 < \mu_2 < \lambda_2 < \cdots < \mu_n < \lambda_n$. By letting $\lambda_i = \mu_i + \varepsilon$ and taking $\varepsilon \downarrow 0$, the positivity of the determinants of the extended Loewner matrix implies the positivity of the determinants of the usual Loewner matrix and so, since the usual Loewner matrix is self-adjoint, its positivity follows by Proposition 6.16. This provides a second proof of the fact that Loewner's theorem implies the positivity of the Loewner matrix.

**Lemma 6.18** *Let $\mu_1 < \lambda_1 < \mu_2 < \cdots < \lambda_n$. Then there exist $n \times n$ matrices $A$ and $B$ and $2n+1$ vectors $\eta^{(1)}, \dots, \eta^{(n)}, \psi^{(1)}, \dots, \psi^{(n)}, \varphi$ so that*

(i) $(B - A)_{ij} = \varphi_i \varphi_j$ (*i.e., $B - A$ is positive, rank one*)
(ii) $A\eta^{(i)} = \mu_i \eta^{(i)}, \langle \eta^{(i)}, \eta^{(j)} \rangle = \delta_{ij}$
(iii) $B\psi^{(j)} = \lambda_j \psi^{(j)}, \langle \psi^{(i)}, \psi^{(j)} \rangle_{ij} = \delta_{ij}$
(iv) $\langle \varphi, \eta^{(i)} \rangle > 0, \langle \varphi, \psi^{(j)} \rangle > 0$
(v) $\det\langle \eta^{(i)}, \psi^{(j)} \rangle = 1$

*Proof* Let

$$P(z) = \frac{\prod_{j=1}^n (\lambda_j - z)}{\prod_{i=1}^n (\mu_i - z)} \tag{6.28}$$

$$= 1 + \sum_{i=1}^n \left( \frac{\alpha_i}{\mu_i - z} \right) \tag{6.29}$$

where

$$\alpha_i = \frac{\prod_{j=1}^n (\lambda_j - \mu_i)}{\prod_{k \neq i} (\mu_k - \mu_i)} > 0 \tag{6.30}$$

(6.29) follows from (6.28) since $P(z) \to 1$ as $z \to \infty$ and $P$ has simple poles only at $z = \mu_i$, so $P(z)$ minus the right side of (6.29) is an entire function going to zero at infinity and so zero. The value of $\alpha_i$ is obtained from (6.28) as $\lim_{z \to \mu_i;\, z \neq \mu_i} (\mu_i - z)P(z)$. That $\alpha_i > 0$ follows from the fact that both the numerator and denominator have $(i-1)$ negative factors.

We will pick

$$A = \begin{pmatrix} \mu_1 & & 0 \\ & \ddots & \\ 0 & & \mu_n \end{pmatrix}$$

and $(\eta^{(i)})_k = \delta_{ik}$ so (ii) holds. $\varphi$ can be chosen by

$$\varphi_k = \alpha_k^{1/2} \tag{6.31}$$

so the first part of (iv) is immediate.

Define the rank one matrix $C$ by

$$C_{ij} = \varphi_i \varphi_j \tag{6.32}$$

and

$$B(\alpha) = A + \alpha C \tag{6.33}$$

$$B(1) \equiv B$$

so (i) holds.

Now,

$$(B(\alpha) - z)^{-1} - (A - z)^{-1} = -\alpha (B(\alpha) - z)^{-1} C (A - z)^{-1}$$

so

$$\begin{aligned} &\langle \varphi, (B(\alpha) - z)^{-1} \varphi \rangle - \langle \varphi, (A - z)^{-1} \varphi \rangle \\ &\qquad = -\alpha \langle \varphi, (B(\alpha) - z)^{-1} \varphi \rangle \langle \varphi, (A - z)^{-1}) \varphi \rangle \end{aligned}$$

or

$$R_\alpha(z) \equiv \langle \varphi, (B(\alpha) - z)^{-1}\varphi\rangle = \frac{\langle \varphi, (A-z)^{-1}\varphi\rangle}{[1 + \alpha\langle \varphi, (A-z)^{-1}\varphi\rangle]} \tag{6.34}$$

Notice that, by (6.29),

$$\langle \varphi, (A-z)^{-1}\varphi\rangle = \sum_{i=1}^{n} \frac{\alpha_i}{\mu_i - z} = P(z) - 1$$

which is why we chose $\varphi$ as we did. Thus,

$$R_\alpha(z) = \frac{(P(z)-1)}{[1 + \alpha(P(z)-1)]} \tag{6.35}$$

$$= \frac{1}{\alpha} - \frac{1}{\alpha^2(P(z) - 1 + \alpha^{-1})} \tag{6.36}$$

It follows that $R_\alpha(z)$ has a pole at each root of $P(z) = 1 - \alpha^{-1}$. Taking $\alpha = 1$, we see $R_{\alpha=1}(z)$ has poles at $z = \mu_i$ so, by (6.28), $B$ has eigenvalues at $\{\mu_i\}_{i=1}^n$, and thus, those are all the eigenvalues. Moreover, since $R_{\alpha=1}(z)$ has a pole at each $\mu_i$, the corresponding eigenvectors are not orthogonal to $\varphi$.

By (6.29), $P(\alpha)$ is monotone increasing on $(\mu_i, \mu_{i+1})$ and on $(\mu_n, \infty)$, and it goes from $-\infty$ to $\infty$ on the former intervals and from $-\infty$ to 1 on the later intervals. It follows for each $\alpha \in (0,1)$, $P(z) = 1 - \alpha^{-1} < 0$ has a solution in $(\mu_i, \lambda_i)$ for $i = 1, \dots, n$. Thus, for $\alpha \in [0,1]$, $B(\alpha)$ has $n$ simple eigenvalues, and since (6.36) has poles, the eigenvectors are not orthogonal to $\varphi$.

Now, we use the fact that we have constructed real matrices so the eigenvectors are real. Moreover, by eigenvalue perturbation theory, the eigenvectors, $\psi^{(j)}(\alpha)$, can be chosen continuous in $\alpha$ with $\psi^{(j)}(\alpha = 0) = \eta_j$. Since $\langle \psi^{(j)}(\alpha), \varphi\rangle \neq 0$ for all $\alpha \in [0,1]$ and $\langle \psi^{(j)}(\alpha = 0), \varphi\rangle > 0$, we have $\langle \psi^{(j)}(\alpha), \varphi\rangle > 0$. Picking $\psi^{(j)} = \psi^{(j)}(\alpha = 1)$, we have (iii) and the second half of (iv).

Finally, by (ii), (iii), $M_{ij}(\alpha) = \langle \eta^{(i)}, \psi^{(j)}(\alpha)\rangle$ is a real orthogonal matrix, so $\det(M(\alpha)) = \pm 1$. Since $M(\alpha) = 1$ and $M(\alpha)$ is continuous in $\alpha$, $\det(M(\alpha)) \equiv 1$, and thus, taking $\alpha = 1$, (v) holds. □

*Remark* (6.34) is the fundamental formula in the theory of rank one perturbations; see [350].

**Proposition 6.19** *Let $A$ and $B$ be the matrices of Lemma 6.18 and let $f$ be a function defined on a neighborhood of $\{\mu_i\} \cup \{\lambda_j\}$. Then*

$$\det(f(B) - f(A)) = \left[\prod_j \langle \varphi, \eta^{(j)}\rangle \langle \varphi, \psi^{(j)}\rangle\right] \det([\lambda_i, \mu_j; f]) \tag{6.37}$$

*Remark* (6.30) gives an explicit formula for $\alpha_i = |\langle \varphi, \eta^{(i)} \rangle|^2$. By symmetry, there is a simple explicit formula for $B_j = |\langle \varphi, \psi^{(j)} \rangle|^2$. This allows one to rewrite (6.37) as

$$\begin{aligned} &\det(f(B) - f(A)) \\ &\quad = \left[ \prod_{i,j} |\mu_i - \lambda_j| \prod_{i<j} (\lambda_j - \lambda_i)^{-1/2} (\mu_j - \mu_i)^{-1/2} \right] \det([\lambda_i, \mu_j; f]) \end{aligned} \tag{6.38}$$

*Proof* Suppose $f$ is analytic in a neighborhood of $(a, b)$. Then

$$\begin{aligned} &\langle \psi^{(j)}, (f(B) - f(A))\eta^{(i)} \rangle \\ &\qquad = \frac{1}{2\pi i} \oint f(z) \left\langle \psi^{(j)}, \left( \frac{1}{z - B} - \frac{1}{z - A} \right) \eta^{(i)} \right\rangle dz \\ &\qquad = \frac{1}{2\pi i} \oint f(z) \langle \psi^{(j)}, (B - A)\eta^{(i)} \rangle \frac{1}{z - \lambda_j} \frac{1}{z - \mu_i} dz \\ &\qquad = [\lambda_j, \mu_i; f] \langle \psi^{(j)}, \varphi \rangle \langle \varphi, \eta^{(i)} \rangle \end{aligned} \tag{6.39}$$

Moreover,

$$\langle \eta^{(j)}, (f(B) - f(A)\eta^{(i)} \rangle = \sum_k \langle \eta^{(j)}, \psi^{(k)} \rangle \langle \psi^{(k)}, f(B) - f(A))\eta^{(i)} \rangle$$

so

$$\begin{aligned} \det(f(B) - f(A)) &= \det(\langle \eta^{(j)}, \psi^{(k)} \rangle \det([\lambda_j, \mu_i; f] \langle \psi^{(j)}, \varphi \rangle \langle \varphi, \eta^{(i)} \rangle) \\ &= \prod \langle \psi^{(j)}, \varphi \rangle \langle \varphi, \eta^{(k)} \rangle \det([\lambda_j, \mu_i; f]) \end{aligned}$$

since $\det(\langle \eta^{(j)}, \psi^{(k)} \rangle) = 1$ by Lemma 6.18(v). □

**Theorem 6.20** *Let $f$ be in $M_n(a, b)$ and $a < \mu_1 < \lambda_1 < \cdots < \mu_n < \lambda_n < b$. Then*

$$\det(L_e^{(n)}(\mu_1, \dots, \mu_n; \lambda_1, \dots, \lambda_n; f)) \geq 0$$

*Proof* Since $f$ is matrix monotone and $A \leq B$ (by (i) of Lemma 6.18), $f(A) \leq f(B)$ so $\det(f(B) - f(A)) \geq 0$. By (iv) of Lemma 6.18, $\langle \varphi, \eta^{(i)} \rangle > 0$ and $\langle \varphi, \psi^{(j)} \rangle > 0$. Thus, by (6.37), $\det([\lambda_i, \mu_j; f)]) \geq 0$. □

To prove Theorem 6.9, we will need extensions of the past two theorems:

**Theorem 6.21** *Let $\Omega$ be an open set in $\mathbb{R}$. Then $f \in M_n^c(\Omega)$ if and only if for all distinct $x_1, \dots, x_n \in \Omega$, the Loewner matrix $L(x_1, \dots, x_n; f)$ is positive.*

*Proof* Given that $M_n^c(\Omega)$ is defined by requiring the curve $\theta A + (1 - \theta)B$ have eigenvalues in $\Omega$, monotonicity is equivalent to positivity of the derivative, and so the proof of Theorem 6.17 extends. □

**Theorem 6.22** *Let $\Omega$ be an open set in $\mathbb{R}$. If $f \in M_n(\Omega)$, then*

$$\det(L_n(\lambda_1, \dots, \lambda_n; \mu_1, \dots, \mu_n; f)) \geq 0 \tag{6.40}$$

*for all $\lambda_1 < \mu_1 < \lambda_2 < \cdots < \lambda_n < \mu_n$ with $\{\lambda_j\}_{j=1}^n \cup \{\mu_j\}_{j=1}^n \subset \Omega$.*

*Proof* While we use interpolation to verify the properties of $A, B$ in Lemma 6.18, once one has $A, B$, we obtained (6.40) just from $f(B) \geq f(A)$ without needing eigenvalues of $\alpha A + (1 - \alpha)B$ to lie in $\Omega$. □

This concludes our discussion of extended Loewner matrices. We begin our analysis of the meaning of the positivity of the regular Loewner matrices by a complete analysis of the $2 \times 2$ case, that is, $M_2(a, b)$.

**Proposition 6.23** *Let $f \in M_2(a, b)$ be $C^3$. Then*

$$f''(x)^2 \leq \tfrac{2}{3} f'(x) f'''(x) \tag{6.41}$$

*for all $x \in (a, b)$.*

*Proof* We will eventually systematize going from

$$\det(L^{(n)}(x_1, \dots, x_n; f)) \geq 0$$

to derivative information, but for now let us do it by hand. Take $x_1 = x$ and $x_2 = x + \varepsilon$ and use

$$\begin{aligned} f'(x + \varepsilon) &= f'(x) + \varepsilon f''(x) + \tfrac{1}{2}\, \varepsilon^2 f'''(x) + o(\varepsilon^2) \\ \varepsilon^{-1}[f(x + \varepsilon) - f(x)] &= f'(x) + \tfrac{1}{2}\, \varepsilon f''(x) + \tfrac{1}{6}\, \varepsilon^2 f'''(x) + o(\varepsilon^2) \end{aligned}$$

in

$$\begin{aligned} \det(L^{(2)}(x_1, x_2; f)) &= f'(x) f'(x + \varepsilon) - (f(x + \varepsilon) - f(x))^2 \varepsilon^{-2} \\ &= \alpha + \beta\varepsilon + \gamma\varepsilon^2 + o(\varepsilon^2) \end{aligned}$$

where

$$\begin{aligned} \alpha &= f'(x)^2 - f'(x)^2 = 0 \\ \beta &= f'(x) f''(x) - 2(\tfrac{1}{2} f''(x)) f'(x) = 0 \\ \gamma &= \tfrac{1}{2} f'(x) f'''(x) - 2(\tfrac{1}{6} f'''(x)) f'(x) - \tfrac{1}{4}(f''(x))^2 \\ &= \tfrac{1}{6} f'(x) f'''(x) - \tfrac{1}{4} f''(x)^2 \end{aligned}$$

Since $\alpha = \beta = 0$ and $0 \leq \det(L^{(2)}(x_1, x_2; f))$, we have $4\gamma \geq 0$, which is (6.41). □

**Lemma 6.24** *Let $f \in M_n(a,b)$. Then there exist $C^\infty$ functions $f_j \in M_n(a + \frac{1}{j}, b - \frac{1}{j})$ so that $f_j(x) \to f(x)$ uniformly on each interval $(a+\varepsilon, b-\varepsilon)$. If $f$ is $C^\ell$, then*

$$\frac{d^m f_j}{dx^m} \to \frac{d^m f}{dx^m} \qquad \text{for } m = 0, 1, \dots, \ell$$

*Proof* Let $h_j$ be a $C^\infty$ approximate identity supported in $(-\frac{1}{j}, \frac{1}{j})$. Since $f(\cdot - \alpha)$ is in $M_n(a+\alpha, b+\alpha)$ and sums of monotone matrix functions are monotone matrix functions, $h_j * f \in M_n(a + \frac{1}{j}, b - \frac{1}{j})$. □

**Theorem 6.25** *If $f \in M_2(a,b)$ (in particular, if $f$ is in any $M_n(a,b)$), then $f$ is a $C^1$ function with $f'$ convex and*

$$\left| \frac{f(x) - f(y)}{x - y} \right|^2 \le f'(x) f'(y) \tag{6.42}$$

*for all $x, y \in (a,b)$. Moreover, if $f$ is nonconstant, $f'(x)$ is strictly positive for all $x \in (a,b)$.*

*Proof* By Lemma 6.24, $f$ is a limit of $C^\infty$ functions, $f_j$, in $M_2(a + \frac{1}{j}, b - \frac{1}{j})$. Since $f$ is monotone, $f_j' \ge 0$. (6.41) then implies $f_j''' \ge 0$ (for even if $f'(x_0) = 0$, $f'''(x_0)$ cannot be negative, for if it were, $f'(x) \ne 0$ for $x$ near $x_0$, so by (6.41), $f'''(x) \ge 0$ for $x$ near $x_1$ and so at $x_0$). Thus, Proposition 1.21 applies and $f$ is $C^1$, $f'$ is convex, and $f_j' \to f'$. Since $\det(L^{(2)}(x, y; f)) \ge 0$, (6.42) holds.

Finally, if $f'(x_0) = 0$ for any $x_0 \in (a,b)$, (6.42) implies $f$ is constant. □

**Theorem 6.26** (Dobsch–Donoghue Theorem) *Let $f$ be a nonconstant function on $(a,b)$. $f \in M_2(a,b)$ if and only if $f$ is $C^1$, $f' > 0$, and $(f')^{-1/2}$ is concave.*

*Proof* Let $f$ be $C^3$ and nonconstant. Then if $f \in M_2(a,b)$, $f' > 0$, and if $g = (f')^{-1/2}$, then

$$\begin{aligned} g'' &= \tfrac{3}{4}(f')^{-5/2}(f'')^2 - \tfrac{1}{2}(f')^{-3/2} f''' \\ &= \tfrac{3}{4}(f')^{-5/2}[(f'')^2 - \tfrac{2}{3} f' f'''] \le 0 \end{aligned}$$

by (6.41). By Lemma 6.24, we can find $C^\infty$ functions, $f_j \in M_2(a+\varepsilon, b-\varepsilon)$, so $f_j' \to f'$ and $f' > 0$. Thus, $(f_j')^{-1/2} \to (f')^{-1/2}$, so $(f')^{-1/2}$ is concave.

Conversely, if $f$ is $C^1$, $f' > 0$ and $g \equiv (f')^{-1/2}$ is concave, then for $x > y$,

$$\frac{f(x) - f(y)}{x - y} = \frac{1}{x - y} \int_x^y \frac{1}{g(z)^2}\, dz \tag{6.43}$$

$$= \int_0^1 \frac{1}{g(\theta x + (1-\theta) y)^2}\, d\theta \tag{6.44}$$

We get (6.44) from (6.43) by the change of variables $z = \theta x + (1-\theta)y$. Since $g$ is concave,

$$g(\theta x + (1-\theta)y) \geq \theta g(x) + (1-\theta)g(y)$$

so

$$\frac{f(x)-f(y)}{x-y} \leq \int_0^1 \frac{1}{(\theta g(x) + (1-\theta)g(y))^2}\, d\theta = \frac{1}{g(x)g(y)}$$

by an elementary integration. Thus,

$$\left(\frac{f(x)-f(y)}{x-y}\right)^2 \leq \frac{1}{g(x)^2}\,\frac{1}{g(y)^2} = f'(x)f'(y)$$

Thus, $\det(L^{(2)}(x,y;f)) \geq 0$. Since $\det(L^{(1)}(x,y;f)) = f'(x) \geq 0$, we conclude by Theorem 6.17 that $f \in M_2(a,b)$. □

**Corollary 6.27** ($\equiv$ Theorem 6.1) *If $f \in M_2(-\infty,\infty)$, then $f'$ is constant, that is, $f$ is affine.*

*Proof* $g = (f')^{-1/2}$ is a concave function on $\mathbb{R}$ which is nonnegative. Such a function must be constant, for if $(D^+g)(x_0) = a < 0$, then $(D^+g)(x) \leq a$ for $x \in (x_0,\infty)$ and so $g(x_0 + a^{-1}) \leq 0$, and if $(D^+g)(x_0) = a > 0$, then $(D^+g)(x_0) \geq a$ on $(-\infty, x_0)$ and $g(x_0 - a^{-1}) \leq 0$. Thus, $(D^+g)(x_0)$ is identically 0, so $g$ is constant and $f'$ is constant. □

**Corollary 6.28** *If $f \in M_2(0,\infty)$, then $f$ is a concave function.*

*Proof* As in the last proof, we have $(D^+g)(x) \geq 0$ so $g$ is increasing. Then $f' = 1/g^2$ is decreasing, that is, $D^+f' \leq 0$, which implies that $f$ is concave. □

*Remarks* 1. Below (see Theorem 6.38) we will generalize this corollary.

2. Given that for $0 \leq \alpha \leq 1$, $f(x) = x^\alpha$ is in $M_\infty(0,\infty)$ and for $\alpha > 1$, $f(x) = x^\alpha$ is not concave, and so not in $M_2(0,\infty)$, we see $f(x) = x^\alpha$ is in $M_n(0,\infty)$ if and only if $0 \leq \alpha \leq 1$ for any $n \geq 2$.

**Example 6.29** Let $g(x)$ on $(-1,1)$ be $1-|x|$ and $f(x) = \int_0^x g(y)^{-2}\, dy = x/(1-|x|)$. $f \in M_2(-1,1)$ but $f''$ is discontinuous at $x=0$, so $f \in M_2$ may not be $C^2$. □

We return to $M_n(a,b)$ and study suitable limits to generalize the fact that we have shown in case $n=2$ that

$$\begin{pmatrix} f'(x) & f''(x)/2! \\ f'(x)/2! & f'''(x)/3! \end{pmatrix}$$

is positive definite. We are heading towards showing that for any $n$, if $f \in M_n(a,b)$ and $x \in (a,b)$, the matrix $a_{ij} = f^{(i+j-1)}(x)/(i+j-1)!$ for $i,j = 1,\dots,n$ is positive definite. As a preliminary to this, we need

**Lemma 6.30** *Let $(c, d)$ be a bounded interval in $\mathbb{R}$, and for $x \notin [c, d]$, let*

$$f_0(x) = \int_c^d (y - x)^{-1}\, dy = \log\left(\frac{d - x}{c - x}\right)$$

*Let $x_1, \dots, x_n \in \mathbb{R}\backslash[c, d]$ and $A$ be the matrix*

$$a_{ij} = [x_1, \dots, x_i, x_1, \dots, x_j; f_0] \tag{6.45}$$

*Then $A$ is strictly positive.*

*Proof* By (6.14),

$$a_{ij} = \int_c^d \prod_{k=1}^{i} (y - x_k)^{-1} \prod_{\ell=1}^{j} (y - x_\ell)^{-1}\, dy$$

so that

$$\sum_{i,j=1}^{n} \bar{w}_i w_j a_{ij} = \int_c^d \left| \sum_{j=1}^{n} w_j \prod_{k=1}^{j} (y - x_k)^{-1} \right|^2 dy \tag{6.46}$$

$$> 0 \tag{6.47}$$

because $f(y) = \sum_{j=1}^{n} w_i \prod_{k=1}^{j} (y - x_k)^{-1}$ is a rational function which is, therefore, nonvanishing on $(c, d)$ except for a finite number of points. □

**Theorem 6.31** *Let $f$ be in $M_n(a, b)$. Then $f$ is $C^{(2n-3)}$ with $f^{(2n-3)}$ convex. Moreover, for any $x_1, \dots, x_n \in (a, b)$, the matrix*

$$a_{ij} = [x_1, \dots, x_i; x_1, \dots, x_j; f] \tag{6.48}$$

*is positive definite. If $f$ is $C^{(2n-1)}$, then for any $x_0 \in (a, b)$, the $n \times n$ matrix*

$$b_{ij} = \frac{f^{(i+j-1)}(x_0)}{(i + j - 1)!} \tag{6.49}$$

*$1 \le i, j \le n$ is positive definite.*

*Proof* Subtracting the first row from rows $2, 3, \dots, n$ in $\det(L^{(n)}(x_1, \dots, x_n; f))$ does not change the determinant, and shows

$$\det([x_i, x_j]) = \prod_{i=2}^{n} (x_i - x_1) \det(a_{ij}^{(1)})$$

where

$$a_{ij}^{(1)} = \begin{cases} [x_1, x_j], & \text{if } i = 1 \\ [x_1, x_i, x_j], & \text{if } i \ge 2 \end{cases}$$

Subtracting row 2 of $a_{ij}^{(1)}$ from rows $3,\ldots,n$ shows

$$\det([x_i, x_j]) = \prod_{i=2}^{n}(x_i - x_1)\prod_{j=3}^{n}(x_i - x_2)\det(c_{ij}^{(2)})$$

where

$$a_{ij}^{(2)} = \begin{cases} [x_1, x_j], & \text{if } i = 1 \\ [x_1, x_2, x_j], & \text{if } i = 2 \\ [x_1, x_2, x_i, x_j], & \text{if } i \geq 3 \end{cases}$$

Proceeding inductively and then doing the same thing with the columns shows that

$$\det([x_i, x_j]) = \prod_{i<j}(x_j - x_i)^2 \Delta_n(x_1, \ldots, x_n; f) \tag{6.50}$$

where

$$\Delta_n(x_1, \ldots, x_n; f) = \det([x_1, \ldots, x_i, x_1, \ldots, x_j; f]) \tag{6.51}$$

Since $\det([x_i, x_j]) \geq 0$, (6.50) implies that $\Delta_n(x_1, \ldots, x_n; f) \geq 0$ for every $f$ in $M_n(a, b)$. By Lemma 6.30, $\Delta_n(x_1, \ldots, x_n; f_0) > 0$. Define

$$g_n(\theta) = \Delta_n(x_1, \ldots, x_n; \theta f + (1-\theta) f_0) \tag{6.52}$$

$g_n$ is a polynomial in $\theta$, $g_n(\theta) \geq 0$ for $\theta \in [0,1]$ and $g_n(1) > 0$. Thus, $g_n(\theta) > 0$ on $[0,1]$ except for finitely many $\theta$'s for $j = 1, 2, \ldots, n$. Similarly, $g_j(\theta) > 0$ on $[0,1]$ except for finitely many $\theta$'s for $j = 1, 2, \ldots, n$. By Proposition 6.15, $A(\theta) > 0$ on $[0,1]$ for all but finitely many $\theta$'s. Picking $\theta_\ell$'s in $[0,1]$ with $\theta_\ell \geq 0$ and $\theta_\ell \to 0$, we see $A \geq 0$.

Now suppose $f$ is $C^{(2n-1)}$. Then, by Proposition 6.10 and (6.18), $[x_1, \ldots, x_i, x_1, \ldots, x_j]$ converges to $f^{(i+j-1)}(x_0)/(i+j-1)!$ as $x_1, \ldots, x_n \to x_0$, and so $A$ converges to $B$ and $B$ is positive definite.

Given a general $f$ in $M_n(a, b)$, by Lemma 6.24, we can approximate it with $C^\infty$ functions $f_j \in M_n(a + \varepsilon, b - \varepsilon)$. By the above, $f_j^{(2\ell+1)}(x) > 0$ for $\ell = 0, 1, 2, \ldots, n$. We claim this implies that $f$ is $C^{2n-3}$ and $f^{(2n-3)}$ is convex. Suppose $n \geq 3$. Since $f \in M_2(a, b)$, $f$ is $C^1$ and $f$ is convex (by Theorem 6.25). Let $g = D^- f'$. Then $g$ is monotone and the approximations $g_j = f_j''$ have $g_j'$ and $g_j'''$ nonnegative. By Theorem 1.22, $g$ is $C^1$ and $g' = f'''$ is convex. Proceeding inductively, we see $f$ is $C^{2n-3}$ and $f^{(2n-3)}$ is convex. $\square$

The final topic we will consider in this chapter is the analysis of matrix convex functions, given our analysis of matrix monotone functions. Not only is this subject of intrinsic interest, but our proof of Loewner's theorem in Chapter 9 will depend on this discussion. We will use Loewner's theorem in proving Theorem 6.33, but we will not use this result (but only Proposition 6.39 which depends only on Proposition 6.32 and Theorem 6.38) in Chapter 9.

**Definition** Let $(a,b)$ be an open interval in $\mathbb{R}$. We say a function $f$ on $(a,b)$ is convex on $n \times n$ matrices if and only if for all self-adjoint $n \times n$ matrices, $A, B$ with $\operatorname{spec}(A) \cup \operatorname{spec}(B) \subset (a,b)$ and $\theta \in [0,1]$, we have that

$$f(\theta A + (1-\theta)B) \leq \theta f(A) + (1-\theta) f(B) \tag{6.53}$$

The set of all such functions will be denoted $C_n(a,b)$.

Notice that $\operatorname{spec}(A) \cup \operatorname{spec}(B) \subset (a,b)$ implies $a \leq A \leq b$ and $a \leq B \leq b$ so $a \leq \theta A + (1-\theta)B \leq b$ and so, $\operatorname{spec}(\theta A + (1-\theta)B) \subset (a,b)$. As with monotone matrix functions, $C_n(a,b) \supset C_{n+1}(a,b)$ and we define

$$C_\infty(a,b) = \bigcap_n C_n(a,b)$$

We will mainly focus on $C_\infty$ but state some results for $C_n$ where the proofs naturally involve $n$.

**Proposition 6.32** *Let $f$ be a $C^2$ function on $(a,b)$. Define $f_x^{[1]}(y)$ by*

$$f_x^{[1]}(y) = [x, y; f] \tag{6.54}$$

*Then*
(i) *If $f_x^{[1]} \in M_n(a,b)$ for all $x \in (a,b)$, then $f \in C_n(a,b)$.*
(ii) *If $f \in C_n(a,b)$, then $f_x^{[1]} \in M_{n-1}(a,b)$ for all $x \in (a,b)$.*

*Proof* $g(\theta) \equiv f(\theta A + (1-\theta)B)$ is convex in operator sense if and only if $\langle \psi, g(\theta)\psi \rangle$ is convex for each $\psi$, if and only if $\frac{d^2}{d\theta^2}\langle \psi, g(\theta)\psi \rangle \geq 0$ for each $\psi$ if and only if $\frac{d^2}{d\theta^2} g(\theta) \geq 0$ for all $\theta$. Changing variables, we need $\frac{d^2}{d\theta^2}\Phi(\theta)\Big|_{\theta=0}$ is convex where $\Phi(\theta) = f(A + \theta C)$, $A$ is self-adjoint with $\operatorname{spec}(A) \subset (a,b)$, and $C$ is an arbitrary self-adjoint operator. Since $g$ and positivity are unitary invariant, we can suppose $A$ is diagonal of the form (6.26).

By Theorem 6.13,

$$\begin{aligned} \Phi_{ij}''(0) &= \sum_k [x_i, x_k, x_j; f] C_{ik} C_{kj} \\ &= \sum_k [x_i, x_j; f_{x_k}^{[1]}] C_{ik} C_{kj} \end{aligned} \tag{6.55}$$

Thus, $f \in C_n(a,b)$ if and only if (6.55) is positive definite for each choice of $\{x_1, \dots, x_n\}$ and self-adjoint $C$.

(i) If $f_{x_k}^{[1]} \in M_n(a,b)$ for each $k$, then by Theorem 6.21, each matrix $[x_i, x_j; f_{x_k}^{[1]}]$ is positive, and so

$$[x_i, x_j; f_{x_k}^{[1]}] = \sum_\ell \lambda_\ell^{(k)} \bar{\varphi}_i^\ell \varphi_j^\ell$$

with $\lambda_\ell^{(k)} \geq 0$. Thus, by (6.55),

$$\Phi''_{ij} = \sum_{k,\ell} \lambda_\ell^{(k)} \bar{C}_{ki} \bar{\varphi}_i^\ell C_{kj} \varphi_j^\ell$$

Since each matrix $\bar{\alpha}_i \alpha_j$ is positive definite, $\Phi''_{ij}(0)$ is positive definite, that is, $g \in C_n(a, b)$.

(ii) Fix $x_1, \dots, x_n \in (a, b)$. If (6.55) defines a positive matrix, that remains true for the $(n-1) \times (n-1)$ matrix obtained by restricting $i, j$ to $1, \dots, n-1$. Pick

$$\begin{aligned} C_{kl} &= 1, \qquad k = n,\ \ell \neq n \text{ or } k \neq n,\ \ell = n \\ &= 0, \qquad k \neq n,\ \ell \neq n \text{ or } k = \ell = n \end{aligned}$$

Since $i, j$ are restricted to $1, \dots, n-1$ in the sum only $k = n$ occurs and we see that $\{[x_i, x_j; f_{x_n}^{[1]}]\}_{1 \leq i,j \leq n-1}$ is positive. □

**Theorem 6.33** *Let $f$ be a function on an open interval $(a, b)$. Let $f_x^{[1]}$ be given by* (6.54). *The following are equivalent:*

(i) $f \in C_\infty(a, b)$
(ii) $f_x^{[1]} \in M_\infty(a, b)$ *for all* $x \in (a, b)$
(iii) $f_x^{[1]} \in M_\infty(a, b)$ *for one* $x \in (a, b)$

*In particular,* $f \in C_\infty(-1, 1)$ *if and only if $f$ is $C^1$ and there is a measure $\mu$ on* $[-1, 1]$ *so*

$$f(x) = f(0) + x f'(0) + \int_{-1}^{1} \frac{x^2}{1 + \lambda x}\, d\mu(\lambda) \tag{6.56}$$

*Proof* It follows from Proposition 6.32 that (i) is equivalent to (ii) and clearly (ii) implies (iii). Thus, we need only show (iii) implies (i).

We can suppose $a$ and $b$ are finite since, for example, $C_\infty(a, \infty) = \cap_{n=1}^\infty C_\infty(a, a+n)$. By translating, we can suppose the "one $x$" in (iii) is 0. By Loewner's theorem, $f_0^{[1]}$ then has the form

$$f_0^{[1]}(x) = \alpha + \int_{a^{-1}}^{b^{-1}} \frac{x}{1 + \lambda x}\, d\mu(\lambda)$$

for a measure $\mu$ on $[a^{-1}, b^{-1}]$. Since $f_0^{[1]}(x) = x^{-1}(f(x) - f(0))$, we see $f$ is $C_\infty$, and

$$f(x) = f(0) + f'(0)x + \int_{a^{-1}}^{b^{-1}} \frac{x^2}{1 + \lambda x}\, d\mu(\lambda)$$

Thus, to see $f \in C_\infty[a, b]$, we need only show each function

$$g_\lambda(x) = \frac{x^2}{(1 + \lambda x)}$$

is in $C_\infty(-\lambda^{-1},\infty)$ if $\lambda > 0$, in $C_\infty(-\infty,-\lambda^{-1})$ if $\lambda < 0$, and in $C_\infty(-\infty,\infty)$ if $\lambda = 0$. By reflection symmetry, we can suppose $\lambda \geq 0$. By (ii) $\Rightarrow$ (i), we need only show $g_{\lambda;y}(x) = [x,y;g_\lambda]$ is in $M_\infty(-\lambda^{-1},\infty)$ for all $y \in (-\lambda^{-1},\infty)$.

If $\lambda = 0$,

$$[x,y;g_\lambda] = \frac{x^2-y^2}{x-y} = x+y$$

is clearly matrix monotone in $x$ for any real $y$.

For $\lambda > 0$, a direct calculation shows that

$$g_{\lambda;y}(x) = \frac{x+y+\lambda xy}{(1+\lambda x)(1+\lambda y)} = \frac{x}{(1+\lambda x)(1+\lambda y)} + \frac{y}{1+\lambda y}$$

The second term is constant and the first is a positive multiple (since $1+\lambda y > 0$ on $(-\lambda^{-1},\infty)$) of $x/(1+\lambda x)$ which is matrix monotone, as shown in (6.7). □

**Example 6.34** By the above,

$$f(x) = \frac{x^2}{1+x}$$

in in $C_\infty(-1,1)$. But

$$f'(x) = \frac{2x}{1+x} - \frac{x^2}{(1+x)^2}$$

has a double pole at $x = -1$. It follows that $\mathrm{Im}\, f'(z)$ is not always positive on $\mathbb{C}_+$ so $f'$ is not matrix monotone. The natural conjecture that the derivative of a matrix convex function is matrix monotone is false! □

On the other hand,

**Proposition 6.35** *If $g$ is a $C^2$ function so $g'(x) \in M_\infty(a,b)$, then $g \in C_\infty(a,b)$.*

*Proof* Without loss of generality, we can suppose $0 \in (a,b)$ and $g(0) = 0$. If $g'(x) = f(x)$, then

$$\begin{aligned} x^{-1}g(x) &= x^{-1}\int_0^x f(y)\,dy \\ &= \int_0^1 f(xu)\,du \end{aligned} \tag{6.57}$$

by changing variable from $y$ to $u$ where $y = xu$. For $0 \leq u \leq 1$, $x \mapsto f(xu)$ is matrix monotone on $(a,b)$ since it is composition of monotone functions. Thus, by (6.57), $x^{-1}g(x)$ is in $M_\infty(a,b)$. Since $g(0) = 0$, this is $[x,0;g]$, so by Theorem 6.33, $g \in C_\infty(a,b)$. □

**Example 6.36** Let $f(x) = x^\alpha$ on $[0,\infty)$ for $\alpha > 0$. Then $f(0) = 0$ and $f_0^{[1]}(x) = x^{\alpha-1}$ is matrix monotone if and only if $0 \le \alpha - 1 \le 1$. Thus, $f$ is convex if and only if $1 \le \alpha \le 2$. We slightly cheated here by considering 0 an endpoint of $[0,\infty)$. The reader should check that the arguments go through in this slightly more general case. □

Finally, we have a vast generalization of Corollary 6.28. First, we will need a lemma:

**Lemma 6.37** *Let $\mathcal{H}$ be a Hilbert space which is a direct sum $\mathcal{H} = \mathcal{H}_1 \oplus \mathcal{H}_2$. Let $A\colon \mathcal{H} \to \mathcal{H}$ be a bounded self-adjoint operator with a decomposition*

$$A = \begin{pmatrix} A_{11} & A_{12} \\ A_{21} & A_{22} \end{pmatrix} \tag{6.58}$$

*where $A_{ij}\colon \mathcal{H}_j \to \mathcal{H}_i$ is $P_i A P_j$ with $P_i$ the orthogonal projection of $\mathcal{H}$ to $\mathcal{H}_i$. Then for any $\varepsilon$, there exists a $B_\varepsilon \ge A$ with*

$$B_\varepsilon = \begin{pmatrix} A_{11} + \varepsilon\mathbf{1} & 0 \\ 0 & C_\varepsilon \end{pmatrix} \tag{6.59}$$

*Proof* Let $C_\varepsilon = A_{22} + \varepsilon^{-1}\|A_{12}\|^2\mathbf{1}$. We need only show

$$\Delta_\varepsilon = \begin{pmatrix} \varepsilon\mathbf{1} & -A_{12} \\ -A_{12}^* & \varepsilon^{-1}\|A_{12}\|^2\mathbf{1} \end{pmatrix}$$

is positive definite. To see this, note that

$$\begin{aligned} \langle(\varphi_1,\varphi_2),&\Delta_\varepsilon(\varphi_1,\varphi_2)\rangle \\ &= \varepsilon\|\varphi_1\|^2 + \varepsilon^{-1}\|A_{12}\|^2\|\varphi_2\|^2 - 2\,\mathrm{Re}\langle\varphi_1, A_{12}\varphi_2\rangle \\ &\ge \varepsilon\|\varphi_1\|^2 + \varepsilon^{-1}\|A_{12}\|^2\|\varphi_2\|^2 - 2\|\varphi_1\|\,\|\varphi_2\|\,\|A_{12}\| \\ &= (\varepsilon^{1/2}\|\varphi_1\| - \varepsilon^{-1/2}\|A_{12}\|\,\|\varphi_2\|)^2 \ge 0 \end{aligned}$$

□

**Theorem 6.38** *If $f \in M_{2n}(a,\infty)$, then $-f \in C_n(a,\infty)$.*

*Remark* That the right endpoint is infinite is critical. While there is a conformal map that relates $M_n(a,\infty)$ to $M_n(0,1)$, it does not relate $C_n(a,\infty)$ to $C_n(0,1)$. For example, $f(x) = (1-x)^{-1}$ is in $M_n(0,1)$, but $f$ is not concave – it is convex, even operator convex.

*Proof* Given $n \times n$ matrices $A$ and $B$ and $\theta \in [0,1]$, write $\mathbb{C}^{2n} = \mathbb{C}^n \oplus \mathbb{C}^n$, pick $t$ so $\theta = \cos^2 t$, and define the $2n \times 2n$ matrices

$$U = \begin{pmatrix} \cos t\mathbf{1} & \sin t\mathbf{1} \\ -\sin t\mathbf{1} & \cos t\mathbf{1} \end{pmatrix}, \qquad C = \begin{pmatrix} A & 0 \\ 0 & B \end{pmatrix} \tag{6.60}$$

with $\mathbf{1}$ the $n \times n$ identity matrix. $U$ is unitary and $C$ self-adjoint.

By a direct calculation,

$$(UCU^{-1})_{11} = \theta A + (1-\theta)B \tag{6.61}$$

and similarly,

$$(Uf(C)U^{-1})_{11} = \theta f(A) + (1-\theta)f(B) \tag{6.62}$$

By Lemma 6.37, we can find $D_\varepsilon$ so

$$(UCU^{-1}) \le \begin{pmatrix} (UCU)_{11}^{-1} + \varepsilon\mathbf{1} & 0 \\ 0 & D_\varepsilon \end{pmatrix} \equiv Q_\varepsilon \tag{6.63}$$

Assume $\operatorname{spec}(A) \cup \operatorname{spec}(B) \subset [a,\infty)$. Then $\operatorname{spec}(C) \subset [a,\infty)$, and since $Q_\varepsilon \ge UCU^{-1}$, also $\operatorname{spec}(Q_\varepsilon) \subset [a,\infty)$ (here we use that the right endpoint is infinity). Since $f \in M_{2n}(a,b)$, $f(UCU^{-1}) \le f(Q_\varepsilon)$, and thus,

$$[Uf(C)U^{-1}]_{11} \le f(Q_\varepsilon)_{11} = f((UCU^{-1})_{11} + \varepsilon)$$

By (6.61) and (6.62), this says

$$\theta f(A) + (1-\theta)f(B) \le f(\theta A + (1-\theta)B + \varepsilon\mathbf{1})$$

Taking $\varepsilon \downarrow 0$, we see that $f$ is matrix concave, that is, $-f \in C_n(a,b)$. □

As noted in the above remark, the last theorem does not extend to direct information on $C_\infty(a,b)$ for $b < \infty$, but using invariance of $M_\infty(a,b)$ under the conformal map, we have the following:

**Proposition 6.39** *Let $f \in M_{2n+2}(-1,1)$ with $f(0) = 0$. Then $g_\pm$ defined by*

$$g_\pm(t) = \frac{t \pm 1}{t}\, f(t)$$

*lies in $M_n(-1,1)$.*

*Proof* Define $u(x) = (1+x)/(1-x) = 2(1-x)^{-1} - 1$, the conformal map of $(-1,1)$ to $(0,\infty)$ that takes $-1$ to $0$, $0$ to $1$, and $1$ to $\infty$. Let $h$ be defined by $h(u(x)) = f(x)$. By Proposition 6.4, $h \in M_{2n+2}(0,\infty)$. Thus, by Theorem 6.37, $-h \in C_{n+1}(0,\infty)$. Then, by Proposition 6.32, $\ell(u) = (-h(u) + h(1))/(u-1)$ is in $M_n(0,\infty)$. But by a direct calculation,

$$-\frac{1}{u-1} = \frac{x-1}{2x}$$

Using $u^{-1}$ to map $\ell$ back to a function on $(-1,1)$, we see that, since $h(1) = f(0) = 0$,

$$\frac{1}{2}\, g_-(t) = \frac{t-1}{2t}\, f(t)$$

lies in $M_n(-1,1)$.

Given any matrix monotone function $p$ on $(-1, 1)$, $(Rp)(t) = -p(-t)$ is also monotone of the same order. Since

$$Rg_{-}(t; Rf) = g_{+}(t)$$

we see that $g_{+}$ is also in $M_n(-1, 1)$. □

This innocent-looking result will be the key to one of the proofs of Loewner's theorem (see Chapter 9). We emphasize again that while Theorem 6.33 used Loewner's theorem, we did not use this result in the proof of Proposition 6.39.

# 7
# Loewner's theorem: a first proof

In this chapter, we will present the Bendat–Sherman proof of the hard part of Loewner's theorem, Theorem 6.5. This proof will rely on two theorems we state now and prove later in this chapter.

**Theorem 7.1** (Bernstein–Boas Theorem) *Let $f$ be a $C^\infty$ function on $(-1,1)$ so that $f^{(2n)}(x) \geq 0$ for $n = 0,1,2,\dots$ Then $f$ is the restriction to $(-1,1)$ of a function analytic in $\{z \mid |z| < 1\}$.*

**Theorem 7.2** (Hausdorff Moment Theorem) *Suppose $\{a_n\}_{n=0}^\infty$ is a sequence of real numbers so that*

(i)

$$|a_n| \leq CR^n, \qquad \textit{for some } C, R \tag{7.1}$$

(ii) *For each $n = 1,2,3,\dots$, the $n \times n$ matrix $\{a_{i+j-2}\}_{i,j=1,\dots,n}$ is positive definite.*

*Then there exists a finite Borel measure $\mu$ on $[-R,R]$ so that*

$$a_n = \int \lambda^n \, d\mu(\lambda) \tag{7.2}$$

*for all $n$.*

*Remarks* 1. Matrices like that arising in (ii) which are constant along diagonals that run from top-right to lower-left, that is,

$$\begin{pmatrix} a_0 & a_1 & a_2 & & \\ a_1 & a_2 & a_3 & & \\ a_2 & a_3 & a_4 & & \\ & & & \ddots & \\ & & & & \ddots \end{pmatrix}$$

are called *Hankel matrices*.

2. Since the polynomials are dense in $C([-R,R])$, the $\mu$ in (7.2) is unique.

3. (i) and (ii) are not only sufficient for there to be a $\mu$ on $[-R,R]$ obeying (7.2), they are also necessary. Obviously, if (7.2) holds, $|a_n| \leq a_0 R^n$, so (i) holds. Moreover, if (7.2) holds, then

$$\sum_{i,j=1}^{n} \bar{\alpha}_i \alpha_j a_{i+j-2} = \int \left| \sum_{j=1}^{n} \alpha_j x^{j-1} \right|^2 d\mu(x) \geq 0$$

so the matrices in (ii) are positive definite.

*Proof of Loewner's Theorem* (*Theorem 6.5*) Let $f \in M_\infty(-1,1)$. Let $g(x) = f'(x)$. By Theorem 6.31, $g^{(2n)}(x) = f^{(2n+1)}(x)$ is the diagonal matrix element of a positive matrix and so $g^{(2n)}(x) \geq 0$ for all $x \in (-1,1)$. By Theorem 7.1, $g$, and thus, $f$ are analytic in $\{z \mid |z| < 1\}$.

Let

$$h(x) \equiv \frac{[f(x) - f(0)]}{x} = \sum_{n=0}^{\infty} a_n x^n \tag{7.3}$$

where $a_n = f^{(n+1)}(0)/(n+1)!$. By the above for any $R > 1$,

$$|a_n| \leq C_R R^n$$

By Theorem 6.31, the matrix $\{a_{i+j-2}\}_{i,j=1,\dots,n}$ is positive, so by Theorem 7.2, (7.2) holds for a measure $d\mu$ on $[-R,R]$. Since $R$ is arbitrary with $R > 1$ and $d\mu$ is unique, $d\mu$ is supported on $[-1,1]$. By (7.3) and the fact that $f$ is analytic and so given by its Taylor series,

$$\begin{aligned} f(x) &= f(0) + \sum_{n=0}^{\infty} x^{n+1} \int_{-1}^{1} \lambda^n \, d\mu(\lambda) \\ &= f(0) + \int_{-1}^{1} \frac{x}{1 + x\lambda} \, d\mu(\lambda) \end{aligned}$$ □

We now turn to the proof of Theorem 7.1. As a preliminary, we need

**Proposition 7.3** *Let $h$ be $C^2$ in a neighborhood of $[-\varepsilon,\varepsilon]$. Then*

$$|h'(0)| \leq \varepsilon^{-1} \sup_{|y|\leq\varepsilon} |h(y)| + \varepsilon \sup_{|y|\leq\varepsilon} |h''(y)| \tag{7.4}$$

*Proof* By the mean value theorem, there exists $x_0 \in [-\varepsilon,\varepsilon]$, so

$$h'(x_0) = (2\varepsilon)^{-1}[h(\varepsilon) - h(-\varepsilon)]$$

and thus,

$$|h'(x_0)| \leq \varepsilon^{-1} \sup_{|y|\leq\varepsilon} |h(y)| \tag{7.5}$$

By the mean value theorem again, there is $y$ between 0 and $x_0$ so that

$$\frac{(h'(x_0) - h'(0))}{x_0} = h''(y)$$

Then

$$\begin{aligned} |h'(0)| &\leq |h'(x_0)| + |x_0|\,|h''(y)| \\ &\leq |h'(x_0)| + \varepsilon \sup_{|y|\leq\varepsilon} |f''(y)| \end{aligned} \tag{7.6}$$

(7.5) and (7.6) imply (7.4). □

*Proof of Theorem 7.1* Let $g(x) = \frac{1}{2}[f(x) + f(-x)]$, so

$$g^{(2n)}(x) = \tfrac{1}{2}[f^{(2n)}(x) + f^{(2n)}(-x)] \geq 0$$

and

$$g^{(n)}(0) = \begin{cases} f^{(n)}(0), & n \text{ even} \\ 0, & n \text{ odd} \end{cases}$$

Let $0 < x < 1$. Using $g^{(2n+1)}(0) = 0$, Taylor's theorem with remainder followed by the intermediate value theorem says that for some $y \in (0, x)$,

$$g(x) = \sum_{n=1}^{m} g^{(2n)}(0)\,\frac{x^{2n}}{2n!} + g^{(2m+2)}(y)\,\frac{x^{(2m+2)}}{(2m+2)!} \tag{7.7}$$

Since all terms on the right are positive,

$$\frac{g^{(2n)}(0)}{(2n)!} \leq x^{-2n} g(x)$$

so for all $\delta > 0$,

$$\frac{f^{(2n)}(0)}{(2n)!} \leq (1-\delta)^{-2n} \sup_{|y|\leq 1-\delta} |f(y)| \tag{7.8}$$

There was nothing special about 0 in this argument which shows

$$0 \leq \frac{f^{(2n)}(x)}{(2n)!} \leq (1 - |x| - \delta)^{-2n} \sup_{|y|\leq 1-\delta} |f(y)| \tag{7.9}$$

Applying (7.4) with $h = f^{(2n)}$,

$$\frac{|f^{(2n+1)}(0)|}{(2n)!} \leq [\delta^{-1}(1-2\delta)^{-2n} + \delta(1-2\delta)^{-2n-2}] \sup_{|y|\leq 1-\delta} |f(y)| \tag{7.10}$$

(7.9) and (7.10) show that the Taylor series

$$\tilde{f}(z) = \sum_{n=0}^{\infty} f^{(j)}(0)\,\frac{z^j}{j!}$$

converges for all $z \in D$ and so defines an analytic function there.

By the same argument that led to (7.7), if $x$ is real and if $0 < |x| < 1 - \delta$,

$$\begin{aligned}\left| f(x) - \sum_{n=1}^{2m+1} f^{(n)}(0)\,\frac{x^n}{n!} \right| &= |f^{(2m+2)}(y)|\,\frac{|x|^{2m+2}}{(2m+2)!} \\ &\leq (1+|x|-\delta)^{-(2m+2)} x^{2m+2} \sup_{|y|\leq 1-\delta} |f(y)| \quad (7.11)\end{aligned}$$

by (7.9). If $|x| < \frac{1}{2}$ and $\delta$ is taken small enough, $x(1-|x|-\delta)^{-1} < 1$ so (7.11) goes to zero as $m \to \infty$. Thus, $\tilde{f}(x) = f(x)$ if $|x| < \frac{1}{2}$. But the same argument shows that for any $y \in (-1,1)$, $f$ is equal to an analytic function on $(y - \frac{1}{2}(1 - |y|), y + \frac{1}{2}(1-|y|))$, so $f$ is real analytic and so equal to $f$ everywhere. □

We now turn to the proof of Theorem 7.2:

**Lemma 7.4** *Let $P(z)$ be a polynomial with $P(x) \geq 0$ for $x \in [-R, R]$. Then $P$ is a finite sum of polynomials of the form $Q^2$, $(R-z)Q^2$, $(R+z)Q^2$, and $(R^2-z^2)Q^2$ with $Q$ a real polynomial.*

*Proof* We will use induction on the degree of $P$. $\deg(P) = 0$ is immediate since $1 = 1^2$. So suppose the theorem is true of polynomials of degree $n-1$ and let $\deg(P) = n$. Pick a zero, $z_0$, of $P$. If $\operatorname{Im} z_0 \neq 0$, $\bar{z}_0$ is also a root, so

$$\begin{aligned}P(z) &= (z - z_0)(z - \bar{z}_0)\tilde{P}(z) \\ &= (z - \operatorname{Re} z_0)^2 \tilde{P}(z) + (\operatorname{Im} z_0)^2 \tilde{P}(z)\end{aligned}$$

and the induction step shows $P$ is of requisite form.

If $z_0 \in (-R, R)$, it must be a root of even order, hence at least a double zero. Then

$$P(z) = (z - z_0)^2 \tilde{P}(z)$$

and induction shows $P$ has the requisite form.

If $z_0 \geq R$, write

$$\begin{aligned}P(z) &= (z_0 - z)\tilde{P}(z) \\ &= (R - z)\tilde{P}(z) + (z_0 - R)\tilde{P}(z)\end{aligned}$$

where $\tilde{P} \geq 0$ on $[-R, R]$. Using the fact that $(R-z)(R-z)$ is a square, $(R-z)(R+z) = R^2 - z^2$, and $(R-z)(R^2-z^2) = (R+z)(R-z)^2$, induction shows $P$ is of the requisite form.

Similarly, if $z_0 \leq -R$,

$$\begin{aligned} P(z) &= (z - z_0)\tilde{P}(z) \\ &= (z + R)\tilde{P}(z) + (-R - z_0)\tilde{P}(z) \end{aligned}$$

and the above analysis carries over. □

*Proof of Theorem 7.2* Given two polynomials $P(z) = \sum_{j=0}^{n} \alpha_j z^j$ and $Q(z) = \sum_{k=0}^{m} \beta_k z^k$, define their inner product by

$$\langle P, Q\rangle = \sum_{\substack{j=0,\dots,n \\ k=0,\dots,m}} \bar{\alpha}_j \beta_k a_{j+k} \tag{7.12}$$

(this is motivated by the fact that if $d\mu$ exists, this will be the $L^2(d\mu)$ inner product). By hypothesis (ii), this is positive semidefinite and so defines a true inner product.

By the Schwarz inequality,

$$\begin{aligned} \langle xP, xP\rangle &= \langle P, x^2 P\rangle \\ &\leq \langle P, P\rangle^{1/2} \langle x^2 P, x^2 P\rangle^{1/2} \\ &\leq \langle P, P\rangle^{1-1/2^n} \langle x^{2^n} P, x^{2^n} P\rangle^{1/2^n} \end{aligned} \tag{7.13}$$

by iteration. If $P$ has degree $\ell$ and we use (7.12), the last inner product is a sum of $(\ell + 1)^2$ terms each at most $D^2 C R^{2^{n+1}} (1 + R)^{2\ell}$, where $D = \sup|\alpha_j|$ and $C, R$ are given by (7.1). Taking the $2^n$-th root and the limit $n \to \infty$ in (7.13), we see that

$$\langle xP, xP\rangle \leq R^2 \langle P, P\rangle$$

Put differently, if $P$ is a real polynomial,

$$\langle 1, (R^2 - x^2)P^2\rangle \geq 0$$

Moreover, if $P$ is a real polynomial,

$$\begin{aligned} \langle 1, P^2\rangle &\geq 0 \\ \langle 1, (R \pm x)P^2\rangle &\geq 0 \end{aligned}$$

where the latter comes from

$$|\langle xP, P\rangle| \leq \langle xP, xP\rangle^{1/2} \langle P, P\rangle \leq (R^2)^{1/2} \langle P, P\rangle$$

By the lemma, if $Q$ is any polynomial, positive on $[-R, R]$, then

$$\langle 1, Q\rangle \geq 0 \tag{7.14}$$

For any polynomial $\|Q\|_\infty 1 \pm Q \geq 0$ on $[-R, R]$ with $\|Q\|_\infty = \sup_{|x| \leq R} |Q(x)|$, and thus, (7.14) implies

$$|\langle 1, Q\rangle| \leq \|Q\|_\infty \langle 1, 1\rangle = a_0 \|Q\|_\infty$$

Thus, since polynomials are dense, $Q \mapsto \langle 1, Q \rangle$ extends to a linear functional $C([-R, R])$. Since any positive function is a limit of positive polynomials, this extension defines a measure $\mu$ on $[-R, R]$ with

$$\langle 1, Q \rangle = \int_{-R}^{R} Q(x)\, d\mu(x)$$

In particular,

$$a_n = \langle 1, x^n \rangle = \int_{-R}^{R} x^n\, d\mu(x) \qquad \square$$

For a discussion of moment problems when (7.1) is not assumed, see, for example, Simon [353, Sect. 3.8].

# 8

# Extreme points and the Krein–Milman theorem

The next four chapters will focus on an important geometric aspect of compact sets, namely, the role of extreme points where:

**Definition** An *extreme point* of a convex set, $A$, is a point $x \in A$, with the property that if $x = \theta y + (1-\theta)z$ with $y, z \in A$ and $\theta \in [0,1]$, then $y = x$ and/or $z = x$. $\mathcal{E}(A)$ will denote the set of extreme points of $A$.

In other words, an extreme point is a point that is not an interior point of any line segment lying entirely in $A$. This chapter will prove a point is a limit of convex combinations of extreme points and the following chapters will refine this representation of a general point.

**Example 8.1** The $\nu$-simplex, $\Delta_\nu$, is given by (5.3) as the convex hull in $\mathbb{R}^{\nu+1}$ of $\{\delta_1, \dots, \delta_{\nu+1}\}$, the coordinate vectors. It is easy to see its extreme points are precisely the $\nu+1$ points $\{\delta_j\}_{j=1}^{\nu+1}$. The hypercube $C_0 = \{x \in \mathbb{R}^\nu \mid |x_i| \le 1\}$ has the $2^\nu$ points $(\pm 1, \pm 1, \dots, \pm 1)$ as extreme points. The ball $B^\nu = \{x \notin \mathbb{R}^\nu \mid |x| \le 1\}$ has the entire sphere as extreme points, showing $\mathcal{E}(A)$ can be infinite.

An interesting example (see Figure 8.1) is the set $A \subset \mathbb{R}^3$, which is the convex hull of

$$A = \text{ch}(\{(x,y,0) \mid x^2+y^2=1\} \cup \{(1,0,\pm 1)\}) \tag{8.1}$$

Its extreme points are

$$\mathcal{E}(A) = \{(x,y,0) \mid x^2+y^2=1,\ x \ne 1\} \cup \{(1,0,\pm 1)\}$$

$(1,0,0) = \frac{1}{2}(1,0,1) + \frac{1}{2}(1,0,-1)$ is not an extreme point. This example shows that even in the finite-dimensional case, the extreme points may not be closed. In the infinite-dimensional case, we will even see that the set of extreme points can be dense! □

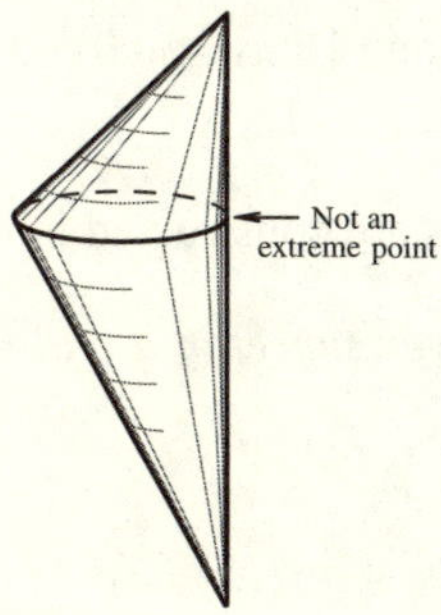


Figure 8.1 An example of not closed extreme points

If a point, $x$, in $A$ is not extreme, it is an interior point of some segment

$$[y, z] = \{\theta y + (1 - \theta)z \mid 0 \leq \theta \leq 1\} \tag{8.2}$$

with $y \neq z$. If $y$ or $z$ is not an extreme point, we can write them as convex combinations and continue. (If $A$ is compact and in $\mathbb{R}^\nu$, and if one extends the line segment to be maximal, one can prove this process will stop in finitely many steps. Indeed, that in essence is the method of proof we will use in Theorem 8.11.) If one thinks about writing $y, z$ as convex combinations, one "expects" that any point in $A$ is a convex linear combination of extreme points of $A$ – and we will prove this when $A$ is compact and finite-dimensional. Indeed, if $A \subset \mathbb{R}^\nu$, we will prove that at most $\nu + 1$ extreme points are needed. This fails in infinite dimension, but we will find a replacement, the Krein–Milman theorem, which says that any point is a limit of convex combinations of extreme points. These are the two main results of this chapter.

Extreme points are a special case of a more general notion:

**Definition** A *face* of a convex set is a nonempty subset, $F$, of $A$ with the property that if $x, y \in A$, $\theta \in (0, 1)$, and $\theta x + (1 - \theta)y \in F$, then $x, y \in F$. A face, $F$, that is strictly smaller than $A$ is called a *proper face*.

Thus, a face is a subset so that any line segment $[xz] \subset A$, with interior points in $F$ must lie in $F$. Extreme points are precisely one-point faces of $A$. (*Note*: See the remark before Proposition 8.6 for a later restriction of this definition.)

**Example 8.2** (Example 8.1 continued) $\Delta_\nu$ has lots of faces; explicitly, it has $2^{\nu+1} - 2$ proper faces, namely, $\nu + 1$ extreme points, $\binom{\nu+1}{2}$ facial lines, ..., $\binom{\nu+1}{\nu}$ faces of dimension $(\nu - 1)$. The hypercube $C_\nu$ has $3^\nu - 1$ faces, namely, $2^\nu$ extreme points, $\nu 2^{\nu-1}$ facial lines, $\binom{\nu}{2}2^{\nu-2}$ facial planes, ..., $2\binom{\nu}{\nu-1}$ faces of dimension $(\nu - 1)$. The only faces on the ball are its extreme points. The faces of the set $A$ of (8.1) are its extreme points, the line $\{(1, 0, y) \mid |y| \leq 1\}$, and the lines

$\{\theta(x_0, y_0, 0)+(1-\theta)(1, 0, 1)\}$ and $\{\theta(x_0, y_0, 0)+(1-\theta)(1, 0, -1)\}$, where $x_0, y_0$ are fixed with $x_0^2 + y_0^2 = 1$ and $x_0 \neq 1$. □

A canonical way proper faces are constructed is via linear functionals.

**Theorem 8.3** *Let $A$ be a convex subset of a real vector space. Let $\ell\colon A \to \mathbb{R}$ be a linear functional with*
(i)

$$\sup_{x \in A} \ell(x) = \alpha < \infty \tag{8.3}$$

(ii) *$\ell \restriction A$ is not constant.*
*Then*

$$\{y \mid \ell(y) = \alpha\} = F \tag{8.4}$$

*if nonempty, is a proper face of $A$.*

*Remark* If $A$ is compact and $\ell$ is continuous, of course, $F$ is nonempty.

*Proof* Since $\ell$ is linear, $F$ is convex. Moreover, if $y, z \in A$ and $\theta \in (0,1)$ and $\theta y + (1-\theta)z \in F$, then $\theta\ell(y) + (1-\theta)\ell(z) = \alpha$ and $\ell(y) \leq \alpha$, $\ell(z) \leq \alpha$ implies $\ell(y) = \ell(z) = \alpha$, that is, $y, z \in F$. By (ii), $F$ is a proper subset of $A$. □

The hyperplane $\{y \mid \ell(y) = \alpha\}$ with $\alpha$ given by (8.3) is called a *tangent hyperplane* or *support hyperplane*. The set (8.4) is called an *exposed set*. If $F$ is a single point, we call the point an *exposed point*.

**Example 8.4** We have just seen that every exposed set is a face so, in particular, every exposed point is an extreme point. I'll bet if you think through a few simple examples like a disk or triangle in the plane or a convex polyhedron in $\mathbb{R}^3$, you'll conjecture the converse is true. But it is not! Here is a counterexample in $\mathbb{R}^2$ (see Figure 8.2):

$$A = \{(x,y) \mid -1 \leq x \leq 1,\ -2 \leq y \leq 0\} \cup \{(x,y) \mid x^2 + y^2 \leq 1\}$$

The boundary of $A$ above $y = -2$ is a $C^1$ curve, so there is a unique supporting hyperplane through each such boundary point. The supporting hyperplane through the extreme point $(1,0)$ is $x = 1$ so $(1,0)$ is not an exposed point, but it is an extreme point. □

**Proposition 8.5** *Any proper face $F$ of $A$ lies in the topological boundary of $A$. Conversely, if $A \subset X$, a locally convex space (and, in particular, in $\mathbb{R}^\nu$), and $A^{\text{int}}$ is nonempty, then any point $x \in A \cap \partial A$ lies in a proper face.*

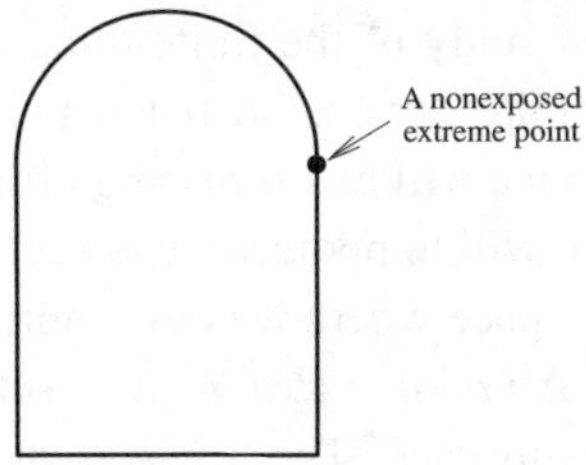


Figure 8.2 A nonexposed extreme point

*Proof* Let $x \in F$ and pick $y \in A\backslash F$. The set of $\theta \in \mathbb{R}$ so $z(\theta) \equiv \theta x + (1-\theta)y \in A$ includes $[0,1]$, but it cannot include any $\theta > 1$ for if it did, $\theta = 1$ (i.e., $x$) would be an interior point of a line in $A$ with at least one endpoint in $A\backslash F$. Thus, $x = \lim_{n\downarrow 0} z(1+n^{-1})$ is a limit point of points not in $A$, that is, $x \in \bar{A} \cap \overline{X\backslash A} = \partial A$.

For the converse, let $x \in A \cap \partial A$ and let $B = A^{\text{int}}$. Since $B$ is open, Theorem 4.1 implies there exists a continuous $L \neq 0$ with $\alpha = \sup_{y\in B} L(y) \leq L(x)$. Since $x \in A$, $L(x) = \alpha$. Since $B$ is open, $L[B]$ is an open set (Lemma 4.2), so the supporting hyperplane $H = \{y \mid L(y) = \alpha\}$ is disjoint from $B$ and so $H \cap A$ is a proper face. □

To have lots of extreme points, we will need lots of boundary points, so it is natural to restrict ourselves to closed convex sets. The convex set $\mathbb{R}^\nu_+ = \{x \in \mathbb{R} \mid x_i \geq 0 \text{ all } i\}$ has a single extreme point, so we will also restrict to bounded sets. Indeed, except for some examples, we will restrict ourselves to compact convex sets in the infinite-dimensional case. Convex cones are interesting but can normally be treated as suspensions of compact convex sets; see the discussion in Chapter 11. So we will suppose $A$ is a compact convex subset of a locally convex space. As noted in Corollary 4.9, $A$ is weakly compact, so we will suppose henceforth that we are dealing with the weak topology.

*Remark* Henceforth, we will also restrict the term "face" to indicate a *closed* set.

**Proposition 8.6** *Let $F \subset A$ with $A$ a compact convex set and $F$ a face of $A$. Let $B \subset F$. Then $B$ is a face of $F$ if and only if it is a face of $A$. In particular, $x \in F$ is in $\mathcal{E}(F)$ if and only if it is also in $\mathcal{E}(A)$, that is,*

$$\mathcal{E}(F) = F \cap \mathcal{E}(A)$$

*Proof* If $B$ is a face of $A$, $x \in B$, and $x$ is an interior point of $[y,z] \subset F$, it is an interior point of $[y,z] \subset A$. Thus, $y, z \in A$, so $y, z \in B$, and thus, $B$ is a face of $F$.

Conversely, if $B$ is a face of $F$, $x \in B$, and $x \in [y,z] \subset A$, since $x \in F$, the fact that $F$ is a face implies $y, z \in F$ so $[y,z] \subset F$. Thus, since $B$ is a face of $F$, $y, z \in B$ and so $B$ is a face of $A$. □

We turn next to a detailed study of the finite-dimensional case. We begin with some notions that involve finite dimension but which are useful in the infinite-dimensional case also. Since we will be discussing affine subspaces, affine spaces, affine independence, etc., we will temporarily use vector subspaces, etc. to denote the usual notions in a vector space where we don't normally include "vector."

Let $X$ be a vector space. An *affine subspace* is a set of the form $a + W$ where $a \in X$ and $W$ is a vector subspace. The *affine span* of a subset $A \subset X$ is the smallest affine subspace containing $A$. If $A = \{e_1, \dots, e_n\}$, then its affine span is just

$$S(e_1, \dots, e_n) = \left\{ \theta_1 e_1 + \cdots + \theta_n e_n \;\middle|\; \theta \in \mathbb{R}^n, \sum_{i=1}^n \theta_i = 1 \right\} \tag{8.5}$$

as is easy to see since $\sum_{i=1}^n \theta_i = 1$ implies that

$$\theta_1 e_1 + \cdots + \theta_n e_n = e_1 + \sum_{j=2}^n \theta_j (e_j - e_1) \tag{8.6}$$

so the right-hand side of (8.5) is $e_1$ plus the vector span of $\{e_j - e_1\}_{j=2}^n$. The convex hull of $\{e_1, \dots, e_n\}$ is, of course,

$$\mathrm{ch}(e_1, \dots, e_n) = \left\{ \theta_1 e_1 + \cdots + \theta_n e_n \mid \theta \in \mathbb{R}^n, \sum_{i=1}^n \theta_i = 1, \, \theta_i \ge 0 \right\} \tag{8.7}$$

We call $\{e_1, \dots, e_n\}$ *affinely independent* if and only if $\sum_{i=1}^n \theta_i e_i = 0$ and $\sum_{i=1}^n \theta_i = 0$ implies $\theta \equiv 0$. By (8.6) this is true if and only if $\{e_j - e_1\}_{j=2}^n$ are vector independent.

**Proposition 8.7** $\mathrm{ch}(e_1, \dots, e_n)$ *always has a nonempty interior as a subset of* $S(e_1, \dots, e_n)$.

*Proof* By successively throwing out dependent vectors from $P = \{e_j - e_1\}_{j=2}^n$, find a maximal independent subset of $P$. By relabeling, suppose it is $P' \equiv \{e_j - e_1\}_{j=1}^k$ so $\{e_1, \dots, e_k\}$ are affinely independent, and each $e_\ell - e_1$ with $\ell > k$ is a linear combination of $P'$. Then $S(e_1, \dots, e_n) = S(e_1, \dots, e_k)$.

Since $\mathrm{ch}(e_1, \dots, e_k) \subset \mathrm{ch}(e_1, \dots, e_n)$, it suffices to prove the result when $e_1, \dots, e_n$ are affinely independent. In that case, $\varphi\colon \Delta_{n-1} \to \mathrm{ch}(e_1, \dots, e_k)$ is a bijection and continuous, so a homeomorphism. Since $\Delta_{n-1}$ has a nonempty interior ($\{(\theta_1, \dots, \theta_n) \mid \sum_{i=1}^n \theta_i = 1, \, 0 < \theta_i\}$), so does $\mathrm{ch}(e_1, \dots, e_k)$. □

*Remark* The $\theta$'s are called *barycentric coordinates* for $S(e_1, \dots, e_\ell)$ and $\mathrm{ch}(e_1, \dots, e_\ell)$.

**Theorem 8.8** *Let $A \subset \mathbb{R}^\nu$ be a convex set. Then there is a unique affine subspace $W$ of $\mathbb{R}^\nu$ so that $A \subset W$, and as a subset of $W$, $A$ has a nonempty interior.*

*Proof* Pick $e_1 \in A$ and consider $B = A - e_1 \ni 0$. Let $W$ be the subspace generated by $B$, that is, let $f_1, \dots, f_{\ell-1}$ be a maximal linear independent subset of $B$, and let $X$ be the vector span of $\{f_j\}_{j=1}^{\ell-1}$. Let $e_j = f_{j-1} + e_1$ for $j = 2, \dots, \ell$ so $e_1 + X \equiv W$ is the affine span of $\{e_j\}_{j=1}^{\ell}$. By construction $B \subset X$ so $A \subset W = S(e_1, \dots, e_\ell)$. By Proposition 8.7, $\text{ch}(e_1, \dots, e_\ell) \subset A$ is open in $S$, so $A$ has no nonempty interior as a subset of $S$.

$W$ is unique because any affine subspace containing $A$ must contain $e_1, \dots, e_\ell$ and so $S(e_1, \dots, e_\ell)$. If its dimension were larger than $W$, $W$ would have empty interior in it and so would $A$. Thus, the condition that $A$ have nonempty interior uniquely determines $W$. □

**Definition** The *dimension* of a convex set $A \subset \mathbb{R}^\nu$ is the dimension of the unique affine subspace given by Theorem 8.8. The interior of $A$ as a subset of $W$ is written $A^{\text{iint}}$ and called the *intrinsic interior* of $A$. $\partial^i A$, the *intrinsic boundary* of $A = \bar{A} \backslash A^{\text{iint}}$.

**Proposition 8.9** *Let $A$ be a compact convex subset of $\mathbb{R}^\nu$. Then*

(i) *$\partial^i A$ is the union of all the proper faces of $A$.*

(ii) *If $x \in \partial^i A$ and $y$ is any point in $A^{\text{iint}}$, $\{\theta \mid (1-\theta)x + \theta y \in A\} = [0, \alpha]$ for some $\alpha > 1$.*

(iii) *If $x \in A^{\text{iint}}$ and $y \in A$, $\{\theta \mid (1-\theta)x + \theta y \in A\} \cap (-\infty, 0) \neq \emptyset$.*

*Remark* This gives us an intrinsic definition of $A^{\text{iint}}$. $x \in A^{\text{iint}}$ if and only if for any $y \in A$, the line $[y, x]$ continued past $x$ lies in $A$ for at least a while. Similarly, $\partial^i A$ is determined by the condition that any line that intersects $A$ in more than one point enters and leaves $A$ at points in $\partial^i A$ and any $x \in \partial^i A$ lies on such a line as an endpoint.

*Proof* (i) This follows from Proposition 8.5 if we view $A$ as a subset of $W$.

(ii) We know $x \in \partial^i A$ lies in some face $F$. Since $A^{\text{iint}}$, viewed as a subset of $W$, is disjoint from the boundary, $y \notin F$. As in the proof of Proposition 8.5, $\{\theta \mid (1-\theta)x + \theta y \in A\} \cap (-\infty, 0) = \emptyset$. Since this set is connected and compact and contains $[0, 1]$, it must be the requisite form. That $\alpha > 1$ and $\alpha \neq 1$ follow from (iii).

(iii) $[x, y]$ lies in $A$, so in $W$, so since $A^{\text{iint}}$ is open in $W$, $\{\theta \mid (1-\theta)x + \theta y \in A^{\text{iint}}\}$ is open. Since it contains 0, it must contain an interval $(-\varepsilon, \varepsilon)$ about 0. □

**Proposition 8.10** *Let $A \subset \mathbb{R}^\nu$ be a compact convex set. Let $\ell = \dim(A)$ and let $F$ be a proper face of $F$. Then $\dim(F) < \ell$.*

*Proof* Let $A \subset W$ where $W$ is the unique $\ell$-dimensional space containing $A$. If $\dim(F) = \ell$, then $W$ must also be the unique $\ell$-dimensional space containing $F$, and so $F$ has not empty interior. But as a set in $W$, $F \subset \partial A$, which contradicts $F^{\text{int}} \neq \emptyset$. Thus, $\dim(F) < \ell$. □

We are now ready for the main finite-dimensional result:

**Theorem 8.11** (Minkowski–Carathéodory Theorem) *Let $A$ be a compact convex subset of $\mathbb{R}^\nu$ of dimension $n$. Then any point in $A$ is a convex combination of at most $n+1$ extreme points. In fact, for any $x$, one can fix $e_0 \in \mathcal{E}(A)$ and find $e_1, \dots, e_n \in \mathcal{E}(A)$ so $x$ is a convex combination of $\{e_j\}_{j=0}^n$. If $x \in A^{\text{iint}}$, then $x = \sum_{j=0}^n \theta_j e_j$ with $\theta_0 > 0$. In particular,*

$$A = \text{ch}(\mathcal{E}(A)) \tag{8.8}$$

*Remarks* 1. It pays to think of the square in $\mathbb{R}^2$ which has four extreme points, but where any point is in the convex hull of three points (indeed, for most interior points in exactly two ways).

2. The example of the $n$ simplex $\Delta_n$ shows that for general $A$'s, one cannot do better than $n+1$ points. Of course, for some sets, one can do better. No matter what value of $\nu$, the ball $B^\nu$ has the property that any point is a convex combination of at most two extreme points.

*Proof* We use induction on $n$. $n = 0$, that is, single-point sets, is trivial. Suppose we have the result for all sets, $B$, with $\dim(B) \le n-1$. Let $A$ have dimension $n$ and $x \in A$ and $e_0 \in \mathcal{E}(A)$. Take the line segment $[e_0, x]$ and extend it – $\{\theta \mid (1-\theta)e_0 + \theta x \in A\} = [0, \alpha]$ for some $\alpha$ by Proposition 8.9. Let $y = (1-\alpha)e_0 + \alpha x$. Since $\alpha \ge 1$,

$$x = \theta_0 e_0 + (1-\theta_0) y \tag{8.9}$$

where $\theta_0 = 1 - \alpha^{-1} \ge 0$.

By construction, $y \in \partial^i A$ and so, by Proposition 8.9, $y \in F$, some proper face of $A$. By Proposition 8.10, $\dim(F) \le n-1$, so by the induction hypothesis, $y = \sum_{j=1}^n \varphi_j e_j$ where $\varphi_j \ge 0$, $\sum_{j=1}^n \varphi_j = 1$, and $\{e_1, \dots, e_n\} \subset \mathcal{E}(F)$. By Proposition 8.6, $\mathcal{E}(F) \subset \mathcal{E}(A)$. Thus,

$$x = \sum_{j=0}^n \theta_j e_j$$

where $\theta_j = (1-\theta_0)\varphi_j$ for $j = 1, \dots, n$.

If $\theta_0 = 0$, by (8.9), $x = y$ and $x \in \partial^i A$. Thus, if $x \in A^{\text{iint}}$, $\theta_0 \ne 0$. □

We will have more to say about extreme points of finite-dimensional convex sets in Chapter 15 when we discuss a particular convex set, the set of all doubly stochastic matrices. In particular, we will show that a compact, convex set, $K$, in $\mathbb{R}^\nu$ has finitely many extreme points if and only if it is a finite intersection of closed half-spaces (Corollary 15.3).

In the infinite-dimensional case, it is not clear that $\mathcal{E}(A)$ is nonempty – we will go through the main construction in two phases. We will first show that $\mathcal{E}(A) \neq \emptyset$ for $A$ a compact convex subset of a locally convex space and then, fairly easily, we will be able to show that

$$A = \mathrm{cch}(\mathcal{E}(A))$$

which is the Krein–Milman theorem. The following illustrates that the infinite-dimensional case is subtle.

**Example 8.12** Let $A$ be the closed unit ball in $L^1(0,1)$. Let $f \in A$ with $f \neq 0$. Then $H_f(s) = \int_0^s |f(t)|\,dt$ is a continuous function with $H_f(0) = 0$ and $H_f(1) = \alpha \leq 1$. Thus, there exists $s_0$ with $H_f(s_0) = \alpha/2$. Let

$$g = 2f\chi_{(0,s_0)}$$
$$h = 2f\chi_{(s_0,1)}$$

Then $\|g\|_1 = \|h\|_1 = \|f\|_1 = \alpha \leq 1$ and $f = \frac{1}{2}h + \frac{1}{2}g$. Since $h \neq g$, $f$ is not an extreme point. Clearly, $0 = \frac{1}{2}(f - f)$ is not extreme either. Thus, $A$ has no extreme points!

We will show below that any compact convex subset, $A$, of a locally convex space has $\mathcal{E}(A) \neq \emptyset$. This means that the unit ball in $L^1(0,1)$ cannot be compact in any topology making it into a locally convex space. In particular, because of the Bourbaki–Alaoglu theorem, $L^1(0,1)$ cannot be the dual of any Banach space. This is subtle because $\ell^1(\mathbb{Z})$ is a dual (of $c_0(\mathbb{Z})$, the bounded sequences vanishing at infinity). Of course, the unit ball in $\ell^1(\mathbb{Z})$ has lots of extreme points in each $\pm\delta_n$. □

**Proposition 8.13** *Let A be a compact convex subset of a locally convex space, X. Then* $\mathcal{E}(A) \neq \emptyset$.

*Proof* Extreme points are one-point faces. We will find them as minimal faces. So let $\mathfrak{F}$ be the family of proper faces of $A$ with $F_1 > F_2$ if $F_1 \subset F_2$. This is a partially ordered set and it has the chain property, that is, if $\{F_\alpha\}_{\alpha\in I}$ is linearly ordered, then it has an "upper" bound ("upper" here means small since a "larger than" means contained in), namely, $\cap_{\alpha\in I} F_\alpha$. This is closed, a face (by a simple argument), and nonempty because of the intersection property for compact sets.

Thus, by Zorn's lemma, there exist minimal faces. Suppose $F$ is such a minimal face and $F$ has at least two distinct points $x$ and $y$. By Corollary 4.6, there is a linear functional on $X$ and so on $F$ with $\ell(x) \neq \ell(y)$. Since $F$ is compact,

$$\tilde{F} = \{z \in F \mid \ell(z) = \sup_{w\in F} \ell(w)\}$$

is nonempty. It is a face of $F$ and so, by Proposition 8.6, $\tilde{F}$ is a face of $A$. Since $\ell(x) \neq \ell(y)$, it cannot be that both $x$ and $y$ lie in $\tilde{F}$, so $\tilde{F} \subsetneq F$, violating minimality. It follows that $F$ has a single point and that point must be an extreme point. □

*Remark* In $L^1(0,1)$, $F_\alpha = \{f \in L^1 \mid \|f\|_1 = 1, f \geq 0, \text{ and } f(x) = 0 \text{ on } (0,\alpha)\}$ is a face and it is linearly ordered (since $\alpha > \beta \Rightarrow F_\alpha \subset F_\beta$), but $\cap_\alpha F_\alpha$ is empty. This proves the lack of compactness directly.

**Theorem 8.14** (The Krein–Milman Theorem) *Let $A$ be a compact convex subset of a locally convex vector space, $X$. Then*

$$A = \text{cch}(\mathcal{E}(A)) \tag{8.10}$$

*Proof* Since $\mathcal{E}(A) \subset A$ and $A$ is closed and convex, $B \equiv \text{cch}(\mathcal{E}(A)) \subset A$. Suppose $B \neq A$ so there exists $x_0 \in A \backslash B$. Since $B$ is closed and convex, by Theorem 4.5, there exists $\ell \in X^*$ so

$$\ell(x_0) > \sup_{y \in B} \ell(y) \tag{8.11}$$

Let $F = \{x \in A \mid \ell(x) = \sup_{z \in A} \ell(z)\}$. Then $F$ is nonempty since $A$ is compact, a face, and by (8.11),

$$F \cap B = \emptyset \tag{8.12}$$

By Proposition 8.13, $F$ has an extreme point, $y_0$, and then, by Proposition 8.6, $y_0 \in \mathcal{E}(A)$. Thus, $y_0 \in B$, contradicting (8.12). □

*Remark* In the next chapter (see Theorem 9.4), we will prove a sort of converse of this theorem.

**Example 8.15** Let $X = C_{\mathbb{R}}([0,1])$ and let $A$ be the unit ball in $\|\cdot\|_\infty$. If $|f(x)| < 1$ for some $x_0$ in $[0,1]$, then by continuity for some $\varepsilon$, $|f(y)| < 1$ for $|y - x_0| < \varepsilon$ and we can find $g \neq 0$ supported in $(x_0 - \varepsilon, x_0 + \varepsilon)$, so both $f + g$ and $f - g$ lie in $A$. Since $f = \frac{1}{2}(f+g) + \frac{1}{2}(f-g)$, $f$ is not an extreme point. Thus, extreme points have $|f(x)| = 1$. By continuity and reality, $A$ has precisely two extreme points $f \equiv \pm 1$. $\text{cch}(\mathcal{E}(A))$ is the constant functions in $A$ so $A \neq \text{cch}(\mathcal{E}(A))$. Thus, $C_{\mathbb{R}}([0,1])$ is not a dual space. □

**Example 8.16** This is an important example. Let $X$ be a compact Hausdorff space and let $A = \mathcal{M}_{+1}(X)$ be the set of regular Borel probability measures on $X$. The extreme points of $A$ are precisely the single-point pure points, $\delta_x$, since if $C \subset X$ has $0 < \mu(C) < 1$ and

$$\mu_C(B) = \mu(C)^{-1}\mu(B \cap C)$$
$$\mu_{X \backslash C} = \mu(X \backslash C)^{-1}\mu(B \backslash C)$$

then with $\theta = \mu(C)$, $\mu = \theta\mu_C + (1-\theta)\mu_{X\backslash C}$ so $\mu$ is not an extreme point.

Suppose $\mu$ has the property that $\mu(A)$ is 0 or 1 for each $A \subset X$. If $x \neq y$ are both in $\mathrm{supp}(\mu)$, we can find disjoint open sets $B, C$ with $x \in B$ and $y \in C$. By the $0, 1$ law, either $\mu(B) = 0$ or $\mu(C) = 0$ or both. But that would mean $x$ and $y$ cannot both be in $\mathrm{supp}(\mu)$. Thus, $\mathrm{supp}(\mu)$ is a single point and $\mu = \delta_x$ for some $x$, that is, the only extreme points are among the $\{\delta_x\}$. But each $\delta_x$ is an extreme point since $\delta_x = \frac{1}{2}\mu + \frac{1}{2}\nu$ implies $\mathrm{supp}(\mu) \subset \{x\}$ so $\mu = \delta_x$. Thus, $\mathcal{E}(A) = \{\delta_x \mid x \in X\}$.

$\mathrm{ch}(\mathcal{E}(A))$ is the pure point measures. $A$ is compact in the $\sigma(\mathcal{M}(X), C(X))$-topology and so the Krein–Milman theorem says that the pure point measures are weakly dense – something that is easy to prove directly. □

**Example 8.17** In some ways, this is an extension of the last example. Let $X$ be a compact Hausdorff space and let $T\colon X \to X$ be a continuous bijection. A regular Borel probability measure $\mu$ on $X$ is called *invariant* if and only if $\mu(T^{-1}[A]) = A$ for all $A \subset X$. This is equivalent to

$$\int f(Tx)\, d\mu(x) = \int f(x)\, d\mu(x) \tag{8.13}$$

for all $f \in C(X)$. An invariant measure, $\mu$, is called *ergodic* if and only if $\mu(A \triangle T[A]) = 0$ (i.e., $A = T[A]$ $\mu$ a.e.) implies $\mu(A)$ is 0 or 1.

Let $T^*$ map $\mathcal{M}_{+,1}(X) \to \mathcal{M}_{+,1}(X)$ by

$$\int f(x)\, d(T^*\mu)(x) = \int f(Tx)\, d\mu(x)$$

Pick any $\mu \in \mathcal{M}_{+,1}(X)$ and let

$$\mu_n = \frac{1}{n} \sum_{j=0}^{n-1} (T^*)^j(\mu)$$

Then for any $f \in C(X)$,

$$|\mu_n(f) - \mu_n(Tf)| = \left| \frac{1}{n} [((T^*)^n \mu)(f) - \mu(f)] \right| \leq \frac{2}{n} \|f\|_\infty \tag{8.14}$$

Thus, if $\mu_\infty$ is any weak-$*$ limit point of $\mu_n$, $\mu_\infty(Tf) = \mu_\infty(f)$ for all $f$, that is, $T^*\mu_\infty = \mu_\infty$. Since $\mathcal{M}_{+,1}(X)$ is compact in the weak-$*$ topology, we conclude

$$\mathcal{M}^I_{+,1}(T) = \{\mu \in \mathcal{M}_{+,1} \mid T^*\mu = \mu\}$$

is not empty.

We claim $\mu \in \mathcal{M}^I_{+,1}(T)$ is ergodic if and only if $\mu \in \mathcal{E}(\mathcal{M}^I_{+,1}(T))$. Suppose $\mu$ is not ergodic. Then there exists an almost invariant set $A$ with $0 < \mu(A) < 1$. $\mu$ can be decomposed $\mu = \theta\mu_A + (1-\theta)\mu_{X\setminus A}$ with $\theta = \mu(A)$ and $\mu_C(B) = \mu(C)^{-1}\mu(B \cap C)$.

Conversely, suppose $\mu$ is ergodic. Then in $L^2(X, d\mu)$, define $(Uf)(x) = f(Tx)$. Then $U$ is unitary. Since as functions on $\partial D$,

$$\frac{1}{n}\sum_{j=0}^{n-1} e^{in\theta} \to \begin{cases} 1, & \theta = 0 \\ 0, & \theta \in (0, 2\pi) \end{cases}$$

the continuity of the functional calculus (see [303, Thm. VIII.20]) implies

$$\frac{1}{n}\sum_{j=1}^{n-1} U^n f \xrightarrow{L^2} P_{\{1\}} f \tag{8.15}$$

where $P_{\{1\}}$ is the projection onto the invariant functions, that is, those $g$ with $Ug = g$. (This is essentially a version of the von Neumann ergodic theorem.) We claim that, since $\mu$ is ergodic, any such $g$ is constant. For clearly, $\operatorname{Re} g$ and $\operatorname{Im} g$ obey $Ug = g$ so we can suppose $g$ is real. But then, for all rational $(\alpha, \beta)$, $\{x \mid \alpha < g(x) < \beta\}$ is almost $T$-invariant and so it has measure 0 or 1. This implies $g$ is a.e. constant. Since $\langle 1, U^n f\rangle = \langle 1, f\rangle = \mu(f)$, we see the constant must be $\mu(f) = \int f(x)\, d\mu(x)$.

We have thus shown that if $\mu$ is ergodic, then

$$\int \left| \frac{1}{n}\sum_{j=0}^{n-1} f(T^{n-1}x) - \mu(f) \right|^2 d\mu(x) = 0 \tag{8.16}$$

Suppose now $\mu = \theta\nu + (1-\theta)\eta$ with $0 < \theta < 1$. Since (8.16) has a positive integrand, we see that (8.16) holds if $d\mu$ is replaced by $d\nu$ or $d\eta$ (but $\mu(f)$ is left unchanged). Thus,

$$\int \frac{1}{n}\sum_{j=0}^{n-1} f(T^{n-1}x)\, d\nu(x) \to \mu(f) \tag{8.17}$$

But since $\nu$ is invariant, the left side of (8.17) is $\nu(f)$ for any $n$. Thus, $\nu(f) = \mu(f)$, and similarly, $\eta(f) = \mu(f)$. It follows that $\nu = \eta = \mu$, that is, $\mu$ is an extreme point.

We have therefore shown that ergodic measures are precisely the extreme points of $\mathcal{M}_{+,1}^I(T)$. The Krein–Milman theorem therefore implies the existence of ergodic measures. If $\mathcal{M}_{+,1}^I(T)$ has more than one point, there must be multiple extreme points.

Now suppose that $\{T_\alpha\}_{\alpha\in I}$ is an arbitrary family of commuting maps of $X$ to $X$. Invariant measures for all the $T_\alpha$'s at once are defined in the obvious way, and $\mu$ is called ergodic if $\mu(A\Delta T_\alpha[A]) = 0$ for all $\alpha$ implies $\mu(A)$ is 0 or 1. Since the $T$'s commute, $T_\alpha^*$ maps each $\mathcal{M}_{+,1}^I(T_\beta)$ to itself, and so by repeating the proof that $\mathcal{M}_{+,1}(X)$ has invariant measures, we see $\mathcal{M}_{+,1}^I(T_\beta)$ has a $T_\alpha^*$-invariant point. By induction, there are invariant measures for any finite set $\{T_{\alpha_i}^*\}_{i=1}^{\ell}$, and then by compactness and the fact that invariant measures are closed, invariant measures for all $\{T_\alpha\}_{\alpha\in I}$. We summarize in the following theorem. This example is discussed further in Example 9.7. □

**Theorem 8.18** *Let $X$ be a compact Hausdorff space and let $\{T_\alpha\}_{\alpha\in I}$ be a family of commuting bijections of $X$ to itself. Then $\mathcal{M}_{+,1}^I(\{T_\alpha\})$, the set of common invariant measures, is nonempty. The ergodic measures are precisely $\mathcal{E}(\mathcal{M}_{+,1}^I(\{T_\alpha\}))$, the extreme points, and are therefore also nonempty.*

As an example, if $X$ is a compact abelian group, and for each $x \in X$, $T_x\colon X \to X$ by $T_x(y) = xy$, then there is an invariant measure. We have therefore constructed a Haar measure in this case, which is known to be unique. Similar ideas can be used to construct what are invariant means on noncompact abelian groups. See the Notes.

(8.15) provides a useful criterion for ergodicity.

**Theorem 8.19** *Let $\mu$ be an invariant measure for a continuous bijection $T$ on a compact Hausdorff space. For any function $f \in L^2(X, d\mu)$ and $n = 0, 1, \dots,$ define*

$$(Av_n f)(x) = \frac{1}{2n+1} \sum_{j=-n}^{n} f(T^j x) \tag{8.18}$$

*Then $\mu$ is ergodic if and only if*

$$\lim_{n\to\infty} \mu(|Av_n f|^2) = |\mu(f)|^2 \tag{8.19}$$

*For* (8.19) *to hold, it suffices that it holds for a dense set, $S$, in $L^2(X, d\mu)$.*

*Proof* (8.19) is equivalent to weak operator convergence as operators on $L^2(X, d\mu)$,

$$(Av_n)^*(Av_n) \xrightarrow{w} (1, \cdot\,)1$$

the projection onto 1, so since $\|Av_n\| \le 1$, it suffices to prove it for a dense set.

If $T$ is ergodic, then (8.15) implies (8.19). Conversely, if (8.19) holds, $A$ is an invariant set, and $\chi_A$ is its characteristic function, then $Av_n(\chi_A) = \chi_A$ so (8.19) implies $\mu(A) = \mu(A)^2$, that is, $\mu(A)$ is 0 or 1. Thus, $\mu$ is ergodic. □

**Example 8.20** Let $X = \partial D$, the unit circle. Let $\alpha$ be an irrational number and let

$$T(e^{i\theta}) = e^{i(\theta + 2\pi\alpha)}$$

Let $d\mu = d\theta/2\pi$ and $f_m = e^{im\theta} \in L^2(\partial D, d\mu)$. Then, for $m \ne 0$,

$$\begin{aligned} Av_n(f_m) &= (2n+1)^{-1}\biggl(\sum_{j=-n}^{j} e^{2\pi i j\alpha m}\biggr) f_m \\ &= (2n+1)^{-1}\,\frac{\sin(2\pi(n+\frac{1}{2})m\alpha)}{\sin(\pi m\alpha)}\, f_m \end{aligned}$$

hence $\|Av_n(f_m)\| \to 0$ if $m \neq 0$. Since $\{f_m\}_{m=0,\pm 1,\dots}$ are a basis of $L^2(\partial D, d\mu)$, (8.19) holds, so $\mu$ is ergodic. Notice $A = \{e^{2\pi i \alpha m}\}_{m=-\infty}^{\infty}$ is an invariant set but it has measure 0. It can be shown that $\mu$ is the only invariant measure in this case. □

**Example 8.21** Given a locally compact group, $G$, a unitary representation is a continuous map $U$ taking $G$ to the unitary operators on a Hilbert space, $\mathcal{H}$. Given such a representation, one can form the functions $F_{\varphi,U}(g) = \langle \varphi, U(g)\varphi\rangle$ for each $U \in \mathcal{H}$. One can show that as $\varphi$ runs over all unit vectors and $U$ over all representations, $\{F_{\varphi,U}\}$ forms a compact convex subset in $C(G)$ in the $\|\cdot\|_\infty$-topology. Its extreme points will correspond to what are called irreducible representations, and one can use the Krein–Milman theorem to prove the existence of such representations. □

Just the existence of extreme points in compact convex sets is powerful. The penultimate topic in this chapter provides proofs of two analytic results that would seem to have no direct connection to the Krein–Milman theorem. First, we provide a proof of the Stone–Weierstrass theorem; see, for example, [303, Appendix to Sect. IV.3] for the "usual" proof.

**Theorem 8.22** (Stone–Weierstrass Theorem) *Let $X$ be a compact Hausdorff space. Let $A$ be a subalgebra of $\mathbb{C}_{\mathbb{R}}(X)$, the real-valued function on $X$, so that for any $x, y \in X$ and $\alpha, \beta \in \mathbb{R}$, there exists $f \in A$ so $f(x) = \alpha$ and $f(y) = \beta$. Then $A$ is dense in $\mathbb{C}_{\mathbb{R}}(X)$ in $\|\cdot\|_\infty$.*

*Proof* $\mathcal{M}(X) = \mathbb{C}_{\mathbb{R}}(X)^*$ is the space of real signed measures on $X$ with the total variation norm, that is, for any $\mu$, there is a set, unique up to $\mu$-measure zero sets, $B \subset X$ so $\mu \restriction B \geq 0$, $\mu \restriction X\backslash B \leq 0$, and $\|\mu\| = \mu(B) + |\mu(X\backslash B)|$.

Define

$$L = \{\mu \in \mathcal{M}(X) \mid \|\mu\| \leq 1;\ \mu(f) = 0 \text{ for all } f \in A\}$$

Then, since the unit ball in $\mathcal{M}(X)$ is compact in the weak-$*$ topology, $L$ is a compact convex set. If $L$ is larger than $\{0\}$, $L$ has an extreme point which necessarily has $\|\mu\| = 1$ since, if $0 < \|\mu\| \leq 1$, $\mu$ is a nontrivial concave combination of $\mu/\|\mu\|$ and 0.

If $g \in A$ and $\|g\|_\infty \leq 1$, then $g\, d\mu \in L$ for any $\mu \in L$ since $\int f(g\, d\mu) = \int (fg)\, d\mu = 0$ and $\|g\, d\mu\| \leq \|g\|_\infty \|\mu\|$. If $0 \leq g \leq 1$, then

$$\begin{aligned}&\|g\, d\mu\| + \|(1-g)\, d\mu\| \\ &\quad = \int_B g\, d\mu + \int_B (1-g)\, d\mu - \int_{X\backslash B} g\, d\mu - \int_{X\backslash B} (1-g)\, d\mu = \|\mu\|\end{aligned}$$

so

$$\mu = \frac{g\, d\mu}{\|g\, d\mu\|} + \frac{(1-g)\, d\mu}{\|(1-g)\, d\mu\|}$$

is a convex combination of elements in $L$. Thus,

$$\begin{aligned}&\mu \text{ extreme and } 0 \le g \le 1\\ &\qquad \text{with } g \in A \Rightarrow g = 0 \text{ a.e. } d\mu \text{ or } (1-g) = 0 \text{ a.e. } d\mu\end{aligned} \tag{8.20}$$

If $\text{supp}(d\mu)$ has two points $x, y$, we can pick $f \in A$ with $f(x) = 1$, $f(y) = 2$. Thus, $g = f^2/\|f\|_\infty^2$ has $0 \le g \le 1$ and $0 < g(x) < \frac{1}{4}$, and so $g \in (0, \frac{1}{4})$ in a neighborhood $U$ of $x$. Since $\mu(U) \ne 0$, (8.20) fails. We conclude $\text{supp}(d\mu)$ is a single point, $x$. But then $\mu \in L$ implies $f(x) = 0$ for all $f \in A$, violating the assumption about $f(x) = \alpha$ can have any real value $\alpha$.

This contradiction implies $L = \{0\}$ which, by the Hahn–Banach theorem, implies that $A$ is dense. □

*Remark* The Stone–Weierstrass theorem does not hold if $\mathbb{C}_\mathbb{R}(X)$ is replaced by $\mathbb{C}(X)$, the complex-valued function. The canonical example of a nondense subalgebra of $\mathbb{C}(X)$ with the $\alpha, \beta$ property is the analytic functions on $\mathbb{D}$. It is a useful exercise to understand why the above proof breaks down in this case.

The second application concerns vector-valued measures, that is, measures with values in $\mathbb{R}^\nu$, equivalently, $n$-tuples of signed real measures. Given such a measure, $\vec{\mu}$, one can form the scalar measure $\tilde{\mu} = \sum_{i=1}^N |\mu_i|$, which we suppose is finite. Then $d\mu_i = f_i \, d\tilde{\mu}$ with $f_i \in L^1$.

**Definition** A scalar measure, $d\mu$, is called *weakly nonatomic* if and only if for any $A$ with $\mu(A) > 0$, there exists $B \subset A$ so $\mu(B) > 0$ and $\mu(A \backslash B) > 0$.

In much of the literature, what we have called weakly nonatomic is called nonatomic, but we defined nonatomic in Chapter 2 as a measure obeying Corollary 8.24 below. That corollary shows the definitions are equivalent, so one can drop "weakly" once one has the theorem.

**Theorem 8.23** (Lyapunov's Theorem) *Let $\tilde{\mu}$ be a weakly nonatomic finite measure on $(M, \Sigma)$, a space with countably generated sigma algebra, and $\vec{f} \in L^1(M, \tilde{\mu}; \mathbb{R}^\nu)$ fixed. Then*

$$\left\{ \int_A \vec{f} \, d\mu \;\middle|\; A \subset M, \textit{ measurable} \right\} \subset \mathbb{R}^\nu$$

*is a compact convex subset of $\mathbb{R}^\nu$.*

Before proving this, we note a corollary and make some remarks:

**Corollary 8.24** *Let $\mu$ be a $\sigma$-finite scalar positive measure which is weakly nonatomic. Then $\mu$ is nonatomic, that is, for any $A$ and any $\alpha \in (0, \mu(A))$, there is $B \subset A$ with $\mu(B) = \alpha$.*

*Proof* By a simple approximation argument, we can suppose $\mu(A) < \infty$. Applying the theorem to $\mu \restriction A$, we see $\{\mu(B) \mid B \subset A, \text{ measurable}\} \subset \mathbb{R}$ is convex. Since $\mu(\emptyset) = 0$ and $\mu(A) = \mu(A)$, we see this convex set must be $[0, \mu(A)]$. □

The main remark that helps us understand the proof is that the extreme points of $\{f \in L^\infty(M, d\mu) \mid 0 \le f \le 1\}$ are precisely the characteristic functions.

*Proof of Theorem 8.23* Let $Q = \{g \in L^\infty \mid 0 \le g \le 1\}$. Then $Q$ is a convex set, compact in the $\sigma(L^\infty, L^1)$-topology, and $F\colon Q \to \mathbb{R}^\nu$ by

$$F(g) = \int g \vec{f} \, d\mu$$

is a continuous linear function, so $\{F(g) \mid g \in Q\}$ is a compact convex set, $S$. We will show that for any $\vec{\alpha} \in S$, there is $g = \chi_A \in Q$ with $F(g) = \alpha$, so $\{\int_A \vec{f} \, d\mu \mid A \subset M\}$ is $S$, and so convex.

Let $Q_{\vec{\alpha}} = \{g \in Q \mid F(g) = \vec{\alpha}\}$. $Q_{\vec{\alpha}}$ is a closed subset of $S$ and so a compact convex subset. By the Krein–Milman theorem, $Q_{\vec{\alpha}}$ has an extreme point $g$. We will prove $g = \chi_A$ using the fact that $\mu$ is weakly nonatomic.

Suppose for some $\varepsilon > 0$, $A = \{x \mid \varepsilon < g < 1 - \varepsilon\}$ has $\mu(A) > 0$. By induction, we can find $B_1, \dots, B_{n+1}$ disjoint, so $\mu(B_j) > 0$ and $\cup_{j=1}^{n+1} B_j = A$. Let $\vec{\alpha}_j = \int_{B_j} \vec{f} \, d\mu$. Since $\mathbb{R}^n$ has dimension $n$, we can find $(\beta_1, \dots, \beta_{n+1}) \in \mathbb{R}^{n+1}$, so that $\sum_{j=1}^{n+1} \beta_j \vec{\alpha}_j = 0$, some $\beta_j \neq 0$ and $|\beta_j| < \varepsilon$ for all $j$. Let

$$g_\pm = g \pm \sum \beta_j \chi_{B_j}$$

Since $|\beta_j| < \varepsilon$ and $\varepsilon < g < 1 - \varepsilon$ on $B_j$, we have that $0 \le g_\pm \le 1$. Since some $\beta_j \neq 0$, $g_+ \neq g_-$. Since $\sum \beta_j \vec{\alpha}_j = 0$, $g_\pm \in Q_{\vec{\alpha}}$. Clearly, $g = \frac{1}{2} g_+ + \frac{1}{2} g_-$, violating the fact that $g$ is an extreme point of $Q_{\vec{\alpha}}$. It follows that $g$ is 0 or 1 for a.e. $x$, that is, $g = \chi_A$ for some $A$. Thus, $\int_A \vec{f} \, d\mu = \vec{\alpha}$. □

We end this chapter with a few results relating extreme points and linear or affine maps between spaces and sets. These will be needed in the next chapter.

**Proposition 8.25** *Let $X$ and $Y$ be locally convex spaces and let $A, B$ be compact convex subsets of $X$ and $Y$, respectively. Let $T\colon X \to Y$ be a continuous linear map. Then if $T[\mathcal{E}(A)] \subset B$, we have that $T[A] \subset B$.*

*Proof* Since $T$ is linear and $B$ is convex, each $T(\sum_{i=1}^n \theta_i x_i)$ with $\sum_{i=1}^n \theta_i = 1$ and $x_i \in \mathcal{E}(A)$ lies in $B$. Then, since $B$ is closed and $T$ is continuous, the same is true of limits. Since $A = \text{cch}(\mathcal{E}(A))$, we see $T[A] \subset B$. □

**Definition** Let $X$ and $Y$ be locally convex spaces and let $A, B$ be convex subsets of $X$ and $Y$, respectively. A map $T\colon A \to B$ is called *affine* if and only if for all $x, y \in A$ and $\theta \in [0, 1]$, $T(\theta x + (1 - \theta) y) = \theta T(x) + (1 - \theta) T(y)$.

**Proposition 8.26** *Let $A$ and $B$ be compact convex subsets of locally convex spaces and let $T\colon A \to B$ be a continuous affine map. Then for any face, $F$, of $B$, $G \equiv T^{-1}[F]$, if nonempty, is a face of $A$.*

*Proof* $G$ is closed since $F$ is closed and $T$ is continuous. If $x \in G$, $y, z \in A$, and $x = \theta y + (1-\theta)z$ with $\theta \in (0,1)$, then $T(x) \in F$, $T(y), T(z) \in B$, and $T(x) = \theta T(y) + (1-\theta)T(z)$. Since $F$ is a face, $T(y), T(z) \in F$, that is, $y, z \in G$. Thus, $G$ is a face. □

# 9

# The Strong Krein–Milman theorem

The representation theorem for points in a compact convex set in terms of extreme points is clean in the finite-dimensional case – a point is a convex combination of finitely many extreme points. But in the form we have it so far, the infinite-dimensional case is murky – points are only limits of convex sums of extreme points. An attractive thought is that somehow this limit of sums is just an integral. We will take a first stab at this idea in this chapter, a stab that is often fine and which we will raise to high art in the next chapter.

While the main result (Theorem 9.2) in this chapter is somewhat mathematically unsatisfying since it only asserts that any point in $A$, a compact convex subset, is an integral of points in $\overline{\mathcal{E}(A)}$ (and we will show lots of examples where $\overline{\mathcal{E}(A)}$ is all of $A$!), it is powerful and includes many classical integral representation theorems in cases where $\mathcal{E}(A)$ is closed. Indeed, we will prove Bernstein's, Bochner's, and Loewner's theorems in this chapter.

We first need to define what we even mean by an integral of points. Consider a probability measure, $\mu$, of bounded support on $\mathbb{R}^\nu$. What do we mean by $\int \vec{x}\, d\mu(\vec{x})$? Obviously, it is the point, $p$, whose coordinates are given by

$$p_i = \int x_i \, d\mu(\vec{x}), \qquad i = 1, \dots, \nu$$

If $d\mu = \sum_{\alpha=1}^{m} \theta_\alpha \delta(\vec{x} - \vec{x}^{(\alpha)})$, then $p = \sum \theta_\alpha \vec{x}^{(\alpha)}$, so this generalizes convex combinations. In infinite dimensions, linear functions play the role of coordinates which justifies:

**Theorem 9.1** *Let $A$ be a compact convex subset of a real locally convex vector space, $X$. Let $\mu \in \mathcal{M}_{+,1}(A)$ be a Baire probability measure on $A$. Then there is a unique point $r(\mu) \in A$, called the* barycenter *or* resultant *of $\mu$ so that for any $\ell \in X^*$,*

$$\ell(r(\mu)) = \int_A \ell(x)\, d\mu(x) \tag{9.1}$$

*The map $r$ is a continuous affine map of $\mathcal{M}_{+,1}(A)$ (with the weak-$*$ topology) onto $A$ and is the unique such map with $r(\delta_x) = x$. More generally, if $B \subset A$ is any closed subset and $\nu(A\backslash B) = 0$, then $r(\nu) \in \text{cch}(B)$ and*

$$r[\{\nu \mid \nu(A\backslash B) = 0\}] = \text{cch}(B) \tag{9.2}$$

*Remarks* 1. Since $X^*$ separates points in $X$, (9.1) can hold not only for a unique point in $A$ but even for a unique point in $X$.

2. If $X$ is a complex vector space, view it as a real space with real linear maps.

*Proof* Let $|||\ell||| = \sup_{x\in A} |\ell(x)| < \infty$, since $\ell$ is continuous on the compact set $A$. Let $B$ be the infinite product

$$B = \mathop{\times}_{\ell \in X^*} \{\lambda \mid |\lambda| \le |||\ell|||\} \tag{9.3}$$

which is compact by Tychonoff's theorem. Let $r_0 : \mathcal{M}_{+,1}(A) \to B$ by

$$r_0(\mu)_\ell = \int_A \ell(x)\, d\mu(x) \tag{9.4}$$

and let $I\colon A \to B$ by

$$I(x)_\ell = \ell(x) \tag{9.5}$$

Both maps are clearly continuous. The range of $I$, that is, $I[A]$, is a closed convex subset of $B$. $I$ is a bijection of $A$ and $I[A]$ since $X^*$ separates points and so, since $A$ is compact, a homeomorphism, so $I^{-1} : I[A] \to A$ is continuous.

We can view $B$ as a subset of the huge vector space $Y = \times_{\ell\in X^*} \mathbb{R}$, given the weak topology (i.e., $y_\alpha \to y_\infty$ if and only if $(y_\alpha)_\ell \to (y_\infty)_\ell$ for each $\ell$). If we let $Z = C(A)^*$, the vector space of finite signed measures on $A$, $r_0$ clearly maps $Z$ to $Y$. As noted in Example 8.16, $\mathcal{E}(\mathcal{M}_{+,1}(A)) = \{\delta_x\}_{x\in X}$. Clearly, for such a $\delta_x$, $r_0(\delta_x) = I(x)$, so $r_0[\mathcal{E}(\mathcal{M}_{+,1}(X))] \subset I(A)$. It follows by Proposition 8.25 that $r_0[\mathcal{M}_{+,1}(X)] \subset I(A)$ so the map $r = I^{-1} \circ r_0$ is a continuous map of $\mathcal{M}_{+,1}$ to $A$.

Let $\tilde{r}$ be another continuous affine map of $\mathcal{M}_{+,1}(A)$ to $A$ with $\tilde{r}(\delta_x) = x$. Since $\tilde{r}$ and $r$ are continuous, $\{\mu \mid r(\mu) = \tilde{r}(\mu)\}$ is closed in $\mathcal{M}_{+,1}(A)$. Since $r$ and $\tilde{r}$ are affine, this set is convex. Since it contains $\mathcal{E}(\mathcal{M}_{+,1}(A))$, it must be all of $\mathcal{M}_{+,1}(A)$, proving uniqueness.

Let $B \subset A$ be closed. Then $\{\nu \mid \nu(A\backslash B) = 0\}$ is a face of $\mathcal{M}_{+,1}(A)$ so it is precisely $\text{cch}(\{\delta_x \mid x \in B\})$, and thus, $\delta_x$ are its extreme points. By Proposition 8.25 again, since $r(\delta_x) \in \text{cch}(B)$ for any $x \in B$, $r$ takes $\{\nu \mid \nu(A\backslash B) = 0\}$ to $\text{cch}(B)$. The range is closed and convex and contains $B$. Hence, it is all of $\text{cch}(B)$. □

The following corollary is so significant we call it a theorem:

**Theorem 9.2** (The Strong Krein–Milman Theorem) *Let $A$ be a compact convex subset of a locally convex vector space. Any point in $A$ is the barycenter of a measure on $\overline{\mathcal{E}(A)}$.*

*Proof* In the last theorem, take $B = \overline{\mathcal{E}(A)}$ and note that by the Krein–Milman theorem, $\text{cch}(\,\overline{\mathcal{E}(A)}\,) = A$. □

Before giving some examples of this important result, we will explore the general theory of barycenters a little more and provide some examples that show the potential weakness of Theorem 9.2 since these examples will have $\overline{\mathcal{E}(A)} = A$ (!).

**Theorem 9.3** (Bauer's Theorem) *Let $x \in \mathcal{E}(A)$ and let $r(\mu) = x$ for a $\mu \in \mathcal{M}_{+,1}(A)$. Then $\mu = \delta_x$. More generally, if $F$ is a face of $A$ and $r(\mu) \in F$, then $\mu(A\backslash F) = 0$. Conversely, if $F$ is a closed set so that $r(\mu) \in F \Rightarrow \mu(A\backslash F) = 0$, then $F$ is a face (and if $\{x\}$ is a point so $r(\mu) = x$ for a unique $\mu$, then $x \in \mathcal{E}(A)$).*

*Proof* Since $x \in \mathcal{E}(A)$ if and only if $\{x\}$ is a face, the results for faces imply the results for extreme points.

Let $F$ be a face. Then $r$ takes $\{\mu \mid \mu(A\backslash F) = 0\}$ to $\text{cch}(F) = F$ by Theorem 9.1. Thus, $r^{-1}[F]$ is nonempty and so, by Proposition 8.26, $r^{-1}[F]$ is a face, $\tilde{F}$, of $\mathcal{M}_{+,1}(A)$. Thus, $\mathcal{E}(\tilde{F}) \subset \mathcal{E}(\mathcal{M}_{+,1}(A)) = \{\delta_x \mid x \in A\}$. Since $r(\delta_x) = x$ is in $F$ if and only if $x \in F$, we conclude $\mathcal{E}(\tilde{F}) = \{\delta_x \mid x \in F\}$. Since $\tilde{F}$ is a compact convex set, the Krein–Milman theorem applied to it implies $\tilde{F} = \text{cch}(\{\delta_x \mid x \in F\}) = \{\mu \mid \mu(A\backslash F) = 0\}$. We have thus proven that if $F$ is a face, then $r(\mu) \in F$ if and only if $\mu(A\backslash F) = 0$.

The converse is essentially trivial. If $F$ is a closed set so that $r(\mu) \in F \Rightarrow \mu(A\backslash F) = 0$ and if $\theta \in (0,1)$ and $x, y \in A$ with $\theta x + (1-\theta)y \in F$, then $r(\theta\delta_x + (1-\theta)\delta_y) \in F$ which implies $(\theta\delta_x + (1-\theta)\delta_y)(A\backslash F) = 0$, which means $x, y \in F$, so $F$ is a face. □

The following proof shows the power of using barycenters:

**Theorem 9.4** (Milman's Theorem) *Let $A$ be a compact convex subset of a locally convex space, $X$. Let $B \subset A$ so that* $\text{cch}(B) = A$. *Then $\mathcal{E}(A) \subset \bar{B}$.*

*Proof* Since $\text{cch}(B) = A$, we have $\text{cch}(\bar{B}) = A$. By Theorem 9.1, $r[\{x \mid \nu(A\backslash \bar{B}) = 0\}] = A$, so for any $x \in \mathcal{E}(A)$, there is a measure $\mu$ with $\mu(A\backslash \bar{B}) = 0$ so that $r(\mu) = x$. By Bauer's theorem, since $x \in \mathcal{E}(A)$ and $r(\mu) = x$, $\mu$ must be $\delta_x$, that is, $\delta_x(A\backslash \bar{B}) = 0$, that is, $x \in \bar{B}$. □

As we will see shortly, there are many interesting cases where $\mathcal{E}(A)$ is closed. However, it is instructive to see a collection of examples where the opposite extreme holds: where $\mathcal{E}(A)$ is dense in $A$. This cannot happen in finite dimensions (except for the one-point set) since $\mathcal{E}(A) \subset \partial^i A$ which is a closed set disjoint from $A^{\text{iint}}$ which is nonempty if $\dim(A) \geq 1$. The reader interested in seeing Theorem 9.2 in action can skip past Example 9.8.

**Example 9.5** (Unit balls in Hilbert space and in $L^p$) Let $\mathcal{H}$ be a (separable) Hilbert space and $B = \{x \in \mathcal{H} \mid \|x\| \le 1\}$. Then $B$ is weakly compact and

$$\mathcal{E}(B) = \{x \mid \|x\| = 1\} \tag{9.6}$$

by the parallelogram rule. For

$$\frac{1}{2}\left(\|x\|^2 + \|y\|^2\right) = \left\|\frac{x+y}{2}\right\|^2 + \left\|\frac{x-y}{2}\right\|^2$$

shows if $\|x\| \le 1$, $\|y\| \le 1$, and $x - y \ne 0$, then $\|\frac{x+y}{2}\| < 1$, that is, if $\|z\| = 1$, $z$ is not $\frac{1}{2}x + \frac{1}{2}y$ with $x, y \in B$ and $x \ne y$. Thus, $\{x \mid \|x\| = 1\} \subset \mathcal{E}(B)$. If $\|x\| = \theta < 1$,

$$x = \theta\left[\frac{x}{\|x\|}\right] + (1-\theta)0$$

is not an extreme point of $B$, so (9.6) holds.

We claim $\mathcal{E}(B)$ is dense in $B$ in the weak topology. For given $x \in B$, let $e_1, e_2, \dots$ be an orthonormal basis for $\{x\}^\perp$ and let $x_n = x + \sqrt{1 - \|x\|^2}\, e_n$. Then $x_n \to x$ weakly and $\|x_n\| = 1$ means $x_n \in \mathcal{E}(B)$.

Notice that $A_n = \{x \mid \|x\| \le 1 - n^{-1}\}$ is closed so $C_n = \{x \mid 1 - n^{-1} < \|x\| \le 1\}$ is open in $B$, and thus, $\mathcal{E}(B) = \bigcap_n C_n$ is a dense $G_\delta$ in $B$.

For $1 < p < \infty$, $L^p$ also has the property that the extreme points of the unit ball are all the points in the unit sphere (see the discussion of uniform convexity in the Notes) and so $\mathcal{E}(\{f \mid \|f\|_p \le 1\})$ is dense in $\{f \mid \|f\|_p \le 1\}$ in the $\sigma(L^p, L^q)$-topology. □

**Example 9.6** (Lipschitz functions) One might think that the issue in the last example is that the topology is the weak topology which has so many convergent sequences that it isn't surprising that $\mathcal{E}(B)$ is closed. Of course, any compact convex subset has a topology identical to the weak topology, but some sets are also compact in some norm or other "strong" topology. That is not true of the last example, but here is one where it is.

Let $X = C([0,1])$ the continuous functions on $[0,1]$. Let $L = \{f \mid \forall x, y \in [0,1],\ |f(x) - f(y)| \le |x - y| \text{ and } f(0) = 0\}$. $L$ is clearly closed and since the functions are uniformly equicontinuous, $L$ is compact (see [303, Sect. I.6]). We claim $\mathcal{E}(L)$ is dense in $L$. A sawtooth function is one for which there are $x_0 = 0 < x_1 < x_2 < \cdots < x_n = 1$ so that $f$ is affine on each interval $[x_{j-1}, x_j]$ with slope $+1$ or $-1$ (different slopes allowed on each interval). We claim each sawtooth function is in $\mathcal{E}(L)$ (there are other functions in $\mathcal{E}(L)$) and the sawtooth functions are dense.

Let $f$ be a sawtooth function and let $g, h \in L$ so $f = \frac{1}{2}g + \frac{1}{2}h$. On $[0, x_1]$, since $f(0) = 0$, $f(x) = \pm x$ so $\frac{1}{2}|g(x) + h(x)| = |x|$. Since $|g(x)| \le |x|$, $|h(x)| \le |x|$, we must have equality and so $g = h = f$ on $[0, x_1]$. An induction proves $g = h = f$, that is, $f \in \mathcal{E}(L)$.

Next, let $f \in L$. For each $n$, we will find a sawtooth function $f_n$ with $\|f_n - f\|_\infty \le n^{-1}$ so the sawtooth functions are dense. Let

$$x_{2j} = \frac{j}{n} \tag{9.7}$$

for $j = 0, 1, 2, \dots, n$. We will pick $x_1, x_3, \dots, x_{2n-1}$ shortly. $f_n$ will be picked so that

$$f_n\left(\frac{j}{n}\right) = f\left(\frac{j}{n}\right) \tag{9.8}$$

Since $|f(x) - f(y)| \le |x - y|$, it follows that $|f_n(x) - f(x)| \le 2(2n)^{-1}$ on each $[\frac{j-1}{n}, \frac{j}{n}]$. Since $f \in L$, $f_n(\frac{j}{n}) - f_n(\frac{j-1}{n}) = \alpha_j/n$ with $|\alpha_j| \le 1$. Let $\theta_j = \frac{1}{2}(\alpha_j + 1) \in [0, 1]$ and let

$$x_{2j-1} = \frac{j-1}{n} + \frac{\theta_j}{n} \tag{9.9}$$

$$f_n(x_{2j-1}) = f\left(\frac{j-1}{n}\right) + \frac{\theta_j}{n} \tag{9.10}$$

Finally, make $f_n$ piecewise linear on each $[x_{j-1}, x_j]$. By construction, the slope is $+1$ on each $[x_{2j}, x_{2j+1}]$ and $-1$ on $[x_{2j-1}, x_{2j}]$ so $f$ is a sawtooth function and (9.8) holds, so $\|f_n - f\|_\infty \le n^{-1}$. □

**Example 9.7** (The Poulsen simplex as states of a spin system) In some ways this is a continuation of Example 8.17. Let $X = \{0, 1\}^{\mathbb{Z}}$ (where $\{0, 1\}$ is the two-point set), that is, $X$ is sequences $\{x_n\}_{n=-\infty}^{\infty}$ with each $x_n$ either 0 or 1. The analysis below works if $\{0, 1\}$ is replaced by any compact set $Y$, but it is traditional to take $Y = \{0, 1\}$ (or $Y = \{-1, 1\}$) because of the origin in the mathematical physics of classical spin systems. Let $T\colon X \to X$ be the shift, that is,

$$(Tx)_n = x_{n-1} \tag{9.11}$$

Give $X$ the weak product topology, that is, $x^{(m)} \to x^{(\infty)}$ if and only if $x_n^{(m)} \to x_n^{(\infty)}$.

$X$ can be realized as the set of all subsets of $\mathbb{Z}$ (under $x \mapsto A = \{n \mid x_n = 1\}$), topologized by $A_m \to A_\infty$ if and only if for all finite sets $F$, $A_m \cap F = A_\infty \cap F$ for $m$ large, and $T$ is then just set translation.

Inside $C(X)$, we define for each finite interval $[a, b] \subset \mathbb{Z}$, $C_{[a,b]}$ to be those functions of $X$ only dependent on $\{x_n\}_{n\in[a,b]}$. We can think of $C_{[a,b]}$ either as a set of functions in $C(X)$ or as the $2^{\#[a,b]}$-dimensional set of functions on $\{0, 1\}^{[a,b]}$. The adjoint of the embedding defines a map $i_{[a,b]}^*\colon \mathcal{M}_{+,1} \Rightarrow \mathcal{M}_{+,1}(\{0, 1\}^{[a,b]})$.

Given a measure $\nu$ on $\{0,1\}^{[a,b]}$, we can define a measure $j_{[a,b]}(\nu)$ on $\Omega$ by letting $m = 1+b-a$, defining $\nu^{(n)}$ on $\{0,1\}^{[a+nm,\, b+nm]}$ via translation and then

$$j_{[a,b]}(\nu) = \bigotimes_{n=-\infty}^{\infty} \nu^{(n)} \tag{9.12}$$

based on the product decomposition $X = \times_{m=-\infty}^{\infty}\{0,1\}^{[a+nm,\, b+nm]}$.

One has that $i^*_{[a,b]} \cdot j_{[a,b]}$ is the identity although, of course,

$$j_{[a,b]} \cdot i^*_{[a,b]} \equiv P_{[ab]} : \mathcal{M}_{+,1}(X) \to \mathcal{M}_{+,1}(X) \tag{9.13}$$

is not. Now consider the interval $[-n,n]$. If $\mu$ is a $T$-invariant measure, $\nu \equiv P_{[-n,n]}\mu$ is *not* in general an invariant measure, but it is only periodic, that is, $\nu(T^{2n+1}[A]) = \nu([A])$. To get an invariant measure, we define

$$Q_n(\mu) = \frac{1}{2n+1} \sum_{j=-n}^{n} (P_{[-n,n]}\mu)(T^j[\,\cdot\,]) \tag{9.14}$$

We claim that each $Q_n(\mu)$ is ergodic and that as $n \to \infty$,

$$Q_n(\mu) \to \mu \tag{9.15}$$

in the weak (i.e., $\sigma(\mathcal{M}(X), C(X))$) topology. This will prove that the extreme points of $\mathcal{M}^I_{+,1}(T)$ are dense in $\mathcal{M}^I_{+,1}(T)$. When we discuss simplexes in Chapter 11, we will return to this and see how counterintuitive it is.

To prove (9.15), note that by the Stone–Weierstrass theorem, $\cup_{[a,b]\subset\mathbb{Z},\,\text{finite}} C([a,b])$ is dense in $C(X)$, so to prove (9.15), it suffices to prove

$$\lim_{n\to\infty} Q_n(\mu)(f) \to \mu(f) \tag{9.16}$$

for each $f$ in some $C([a,b])$. Let $m = 1+d-c$. Then if $\mu$ is translation invariant,

$$[j_{[c,d]} \cdot i^*_{[c,d]}(\mu)](f) = \mu(f) \tag{9.17}$$

so long as for some $k$,

$$[a,b] \subset [c+km, d+km] \tag{9.18}$$

for $j \cdot i^*(\mu)$ is a product measure, $[a,b]$ lies in a single factor in the infinite product. Let $\ell = 1+b-a$ be the length of $[a,b]$. Suppose $m \geq \ell$. Then $m-\ell+1$ translates of $T^j[a,b]$ out of $m$ successive translates have (9.18) holding.

Thus, if $f \in C([a,b])$, $\ell = 1+b-a$, and $\ell \leq 2n+1$, then

$$|Q_n(\mu)(f) - \mu(f)| \leq \frac{1}{2n+1}\, 2\ell \|f\|_\infty$$

goes to zero as $n \to \infty$, proving (9.16) for $f$, and so proving (9.15).

To prove ergodicity, we use Theorem 8.19. Let $\eta \equiv Q_n(\mu)$. Let $\eta_0 = P_{[-n,n]}(\mu)$ so

$$\eta = \frac{1}{2n+1} \sum_{j=-n}^{n} \eta_0(T^j[\,\cdot\,]) \tag{9.19}$$

Since $\cup C[a,b]$ is dense in $C(X)$, it is dense in $L^2(X, d\eta)$, so it suffices to prove (8.19) for real-valued $f \in C[a,b]$. Because $\eta_0$ is a product measure, for all $|j - \ell|$ sufficiently large,

$$\eta_0(f \circ T^j \, f \circ T^\ell) = \eta_0(f \circ T^j)\eta_0(f \circ T^\ell) \tag{9.20}$$

Noting that, by (9.19),

$$\frac{1}{2m+1} \sum_{j=-m}^{m} \eta_0(f \circ T^j) = \eta(f)$$

we see that (9.20) implies that

$$\lim_{m \to \infty} \eta_0(Av_m(f)^2) = \eta(f)^2 \tag{9.21}$$

and, similarly for each fixed $j$,

$$\lim_{m \to \infty} \eta_0([Av_m(f) \circ T^j]^2) = \eta(f)^2 \tag{9.22}$$

By (9.19), this implies

$$\lim_{m \to \infty} \eta(Av_m(f)^2) = \eta(f)^2$$

so $\eta$ is ergodic by Theorem 8.19. □

**Example 9.8** In Simon [353, Sect. 3.8], the moment problem is discussed, that is, given a set of numbers $\{a_n\}_{n=0}^\infty$, look for positive measures on $\mathbb{R}$ obeying

$$a_n = \int x^n \, d\mu \tag{9.23}$$

The condition that the Hankel matrices $\{a_{i+j-2}\}_{i,j=1}^n$ be positive definite for each $n$ is a necessary and sufficient condition for there to exist such a $\mu$ (see [353, Thm. 3.8.4]). However, $\mu$ may not be unique. For example, it is not unique if

$$a_n = \begin{cases} \Gamma((2n+1)/\alpha)/\Gamma(1/\alpha) & \\ 0, & \text{if } n \text{ is odd} \end{cases} \tag{9.24}$$

for $\alpha < 1$ (see [353, Example 3.8.1]), nor if

$$a_n = \exp(\tfrac{1}{4}\,n^2 + \tfrac{1}{2}\,n) \tag{9.25}$$

(see Stieltjes [363]).

In a suitable space, the set of $\mu$ solving (9.23) for a fixed set of moments is a compact convex set which, as we have just explained, may contain multiple points. In that case, it is infinite-dimensional and its extreme points are dense. □

The power of the Strong Krein–Milman theorem is seen in the proof of a classical theorem of Bernstein, to which we now turn:

**Definition** Given a function, $f$, on $[0,\infty)$, we define

$$(\Delta_n^h f)(x) = \sum_{k=0}^{n} (-1)^{n-k} \binom{n}{k} f(x+kh) \tag{9.26}$$

This is interpreted as $(\Delta_0^h f)(x) = f(x)$ for $n = 0$.

**Definition** A function $f$ on $(0,\infty)$ is called *completely monotone* if and only if for all $x, h > 0$ and $n = 0, 1, 2, \dots$,

$$(-1)^n (\Delta_n^h f)(x) \geq 0 \tag{9.27}$$

*Remark* Often (9.27) is replaced with the assumption $f$ is $C^\infty$ and

$$(-1)^n f^{(n)}(x) \geq 0 \tag{9.28}$$

We will see shortly that if $f$ is $C^\infty$, (9.27) and (9.28) are equivalent, and eventually that (9.27) implies that $f$ is $C^\infty$.

Given $f$ on $(0,\infty)$ and $x_1, \dots, x_n$, define the Bernstein matrix by

$$B(x_1, \dots, x_n; f) = f(x_i + x_k) \tag{9.29}$$

It is, of course, a Hankel matrix if $x_{i+1} - x_i = \alpha$, a constant.

**Example 9.9** Simple examples of completely monotone functions (check (9.28)) include

$$f(x) = e^{-\alpha x}, \qquad \alpha \geq 0$$

and

$$f(x) = (x+1)^{-\beta}, \qquad \beta \geq 0$$

**Theorem 9.10** (Bernstein's Theorem) *Let $f$ be a bounded function on $(0,\infty)$. Then the following are equivalent:*

(i) *$f$ is completely monotone.*
(ii) *$f$ is $C^\infty$ and* (9.28) *holds.*
(iii) *For any $x_1, \dots, x_n \in (0,\infty)$, $B(x_1, \dots, x_n; f)$ is positive definite.*
(iv) *For any $x_1, \dots, x_n \in (0,\infty)$, $\det(B(x_1, \dots, x_n; f)) \geq 0$.*

(v) *There exists a finite measure $d\mu$ on $[0,\infty)$ so that*

$$f(x) = \int_0^\infty e^{-\alpha x}\, d\mu(\alpha) \tag{9.30}$$

*In particular, if one, and hence all, of these conditions holds, $f$ has an analytic continuation to a function analytic on $\{z \mid \operatorname{Re} z > 0\}$. Moreover, $d\mu$ is unique and $\mu([0,\infty)) = \|f\|_\infty$.*

*Remarks* 1. Since $\Delta_1^h f \le 0$ if $f$ is completely monotone, $f$ is positive and decreasing, so bounded on any interval $(a,\infty)$ with $a > 0$. Below (see Theorem 9.16), we discuss what holds if the condition that $f$ be bounded is dropped.

2. (iii) is a continuum analog of the condition on the moments (that $\{a_{i+j-2}\}_{i,j=1}^n$ be positive definite) that we had for the moment problem to be solvable. (9.30) is a kind of continuous moment condition. This is one of many indications that Laplace transforms are the continuum equivalent to power series.

3. The analogy between power series and Laplace transforms can be made clearer by noting the following analog of Bernstein's theorem (see the Notes for a proof and further discussion). Note that by mapping $x \to -x$, Bernstein's theorem can be rephrased to say a function on $f$ on $(-\infty,0)$ has $f^{(n)}(x) \ge 0$ for all $x$ if and only if $f(x) = \int_0^\infty e^{\alpha x}\, d\mu(\alpha)$ for a measure $\mu$. The analog of this result is that a bounded $C^\infty$ function $f$ on $(0,1)$ has $f^{(n)}(x) \ge 0$ for all $x$ if and only if $f(x) = \sum_{n=0}^\infty a_n x^n$ for $a_n \ge 0$ with $\sum_{n=0}^\infty a_n < \infty$.

*Start of Proof* We will prove

$$\begin{array}{ccccc} & & \text{(ii)} & & \\ & \nearrow & & \searrow & \\ \text{(v)} & & & & \text{(i)} \Rightarrow \text{(v)} \Rightarrow \text{(analyticity)} \\ & \searrow & & \nearrow & \\ & & \text{(iii)} \Leftrightarrow \text{(iv)} & & \end{array}$$

(ii) $\Rightarrow$ (i), (iii) $\Rightarrow$ (i), (i) $\Rightarrow$ (v), and the uniqueness will require some machinery and their proof will appear below. Here are the other steps:

(iii) $\Leftrightarrow$ (iv) This follows from Proposition 6.16.

(v) $\Rightarrow$ (iii) If $f$ has the representation (9.30), $x_1,\dots,x_n \in (0,\infty)$, and $\zeta_1,\dots,\zeta_n \in \mathbb{C}$, then

$$\sum_{i,j=1}^n \bar\zeta_i \zeta_j f(x_i + x_j) = \int_0^\infty \Big| \sum_{j=1}^n \zeta_j e^{-\alpha x_j} \Big|^2 d\mu(\alpha) \ge 0$$

so the matrices $B(x_1,\dots,x_n; f)$ are positive.

(v) $\Rightarrow$ (analyticity) This is obvious for any sum of $e^{-\alpha x}$'s and follows by a simple limiting argument for general $\mu$'s. Alternately, we can define $f(z)$ for $\operatorname{Re} z > 0$

by replacing $x$ in (9.30) by $z$, and it is easy to see the resulting $f$ has a complex derivative.

<u>(v) $\Rightarrow$ (ii)</u> As in the last argument, $f$ is $C^\infty$. Moreover,

$$f^{(n)}(x) = (-1)^n \int x^n e^{-\alpha x}\, d\mu(x)$$

clearly has $(-1)^n f^{(n)}(x) \geq 0$ (indeed, unless $f$ is constant, $(-1)^n f(x) > 0$). $\square$

To start the other parts of the proof, we need to study the operators $\Delta_n^h$:

**Proposition 9.11** (a)

$$[\Delta_n^h(\Delta_1^h f)](x) = (\Delta_{n+1}^h f)(x) \tag{9.31}$$

(b)

$$\Delta_n^h = (\Delta_1^h)^n \tag{9.32}$$

(c) *Let $A^{(n)}$ be the $n \times n$ Bernstein matrix*

$$A^{(n)} = B(x, x+h, x+2h, \dots, x+(n-1)h; f) \tag{9.33}$$

*and let $\alpha^{(n)}$ be the $n$ component vector*

$$\alpha_j^{(n)} = (-1)^j \binom{n-1}{j-1} \tag{9.34}$$

*Then*

$$\langle \alpha^{(n)}, A^{(n)} \alpha^{(n)} \rangle = (\Delta_{2(n-1)}^h f)(x) \tag{9.35}$$

(d) *$\Delta_1^h$ and $d/dx$ commute applied to $C^1$ functions.*

(e) *If $f$ is $C^n$,*

$$(\Delta_n^h f)(x) = \int_0^h ds_1 \int_0^h ds_2 \dots \int_0^h ds_n\, f^{(n)}(x + s_1 + \dots + s_n) \tag{9.36}$$

(f) *If $f$ is $C^n$,*

$$\lim_{h \downarrow 0} \frac{(\Delta_h^n f)(x)}{h^n} = f^{(n)}(x) \tag{9.37}$$

(g) *If $\ell_1, \dots, \ell_n$ are positive integers, then*

$$(\Delta_1^{\ell_1 h} \dots \Delta_1^{\ell_n h} f)(x) = \sum_{j_1=0}^{\ell_1 - 1} \sum_{j_2=0}^{\ell_2 - 1} \cdots \sum_{j_n=0}^{\ell_n - 1} (\Delta_n^h f)(x + (j_1 + \dots + j_n)h) \tag{9.38}$$

*Proof* (a) Since

$$(\Delta_1^h f)(x) = f(x+h) - f(x) \tag{9.39}$$

we have

$$
\begin{aligned}
(\Delta_n^h \Delta_1^h f)(x) &= \sum_{k=0}^{n} (-1)^{n-k} \binom{n}{k} [f(x+(k+1)h) - f(x+kh)] \\
&= \sum_{k=0}^{n+1} (-1)^{n+1-k} \left[ \binom{n}{k} + \binom{n}{k-1} \right] f(x+kh)
\end{aligned}
$$

(where $\binom{n}{n+1} \equiv 0$). The standard formula

$$
\binom{n}{k} + \binom{n}{k-1} = \binom{n+1}{k} \tag{9.40}
$$

proves (9.31).

(b) follows from (a).

(c) We have

$$
\begin{aligned}
\langle \alpha^{(n)}, A^{(n)} \alpha^{(n)} \rangle &= \sum_{i,j=1}^{n} f(x+(i+j-2)h)(-1)^{1-j} \binom{n-1}{i-1} \binom{n-1}{j-1} \\
&= \sum_{k=0}^{2(n-1)} f(x+kh)(-1)^k \left( \sum_{j=0}^{k} \binom{n-1}{j} \binom{n-1}{k-j} \right)
\end{aligned}
$$

so that (9.35) follows from $(-1)^{2(n-1)-k} = (-1)^k$, and for $k \le 2m$,

$$
\sum_{j=0}^{k} \binom{m}{j} \binom{m}{k-j} = \binom{2m}{k} \tag{9.41}
$$

which follows if we note that to pick $k$ elements from $\{1, \dots, 2m\}$, we must pick $j \le k$ elements from $\{1, \dots, m\}$ and then $k-j$ elements from $\{m+1, \dots, 2m\}$.

(d) This is immediate from the definition.

(e) We will prove this by induction. Clearly, if $f$ is $C^1$,

$$
(\Delta_1^h f)(x) = f(x+h) - f(x) = \int_0^h f'(x+s_1)\, ds_1 \tag{9.42}
$$

Supposing (9.36) for $n-1$, we have

$$
\begin{aligned}
(\Delta_n^h f)(x) &= (\Delta_{n-1}^h (\Delta_1^h f))(x) \\
&= \int_0^h ds_1 \int_0^h ds_2 \dots \int_0^h ds_{n-1} (\Delta_1^h f)^{(n-1)}(x+s_1+\cdots+s_{n-1})
\end{aligned} \tag{9.43}
$$

By (d) and (9.42),

$$
(\Delta_1^h f)^{(n-1)}(y) = (\Delta_1^h (f^{(n-1)}))(x) = \int_0^h ds_n f^{(n)}(y+s_n)\, ds_n \tag{9.44}
$$

(9.43) and (9.44) imply (9.36).

(f) This is immediate from (9.36).

(g) We have for any positive integer $\ell$,

$$(\Delta_1^{\ell h} f)(x) = \sum_{j=0}^{\ell-1} (\Delta_1^h f)(x + jh)$$

Thus, (b) and the fact that if $f_b(x) = f(x+b)$, then $\Delta_n^h(f_b) = (\Delta_n^h f)_b$ implies (9.38). □

**Corollary 9.12** (Includes (ii) ⇒ (i) in Theorem 9.10) *Let $f$ be $C^\infty$ on $(a,b)$. Then $f$ is completely monotone if and only if $(-1)^n f^{(n)}(x) \geq 0$ on $(a,b)$.*

*Proof* Immediate from (9.36) and (9.37). □

**Proposition 9.13** (This is (iii) ⇒ (i) in Theorem 2.9) *Let $f$ be a bounded function on $(0,\infty)$ so that for all $x, h > 0$ and $n$, the $n \times n$ Bernstein matrix, $B(x_1, x + h, \dots, x + (n-1)h)$ is positive. Then $f$ is completely monotone.*

*Proof* By (9.35), under the assumption on $B$, we have

$$(\Delta_{2m}^h f)(x) \geq 0 \tag{9.45}$$

for $m = 0, 1, 2, \dots$

Now suppose $g$ is bounded on $(0,\infty)$ and that $g(x) \geq 0$, $(\Delta_2^h g)(x) \geq 0$ for all $x, h$. We claim $(\Delta_1^h g)(x) \leq 0$ for suppose $(\Delta_1^h g)(x) \equiv \alpha > 0$. By hypothesis,

$$(\Delta_2^h g)(x) = \Delta_1^h(\Delta_1^h g)(x) \geq 0$$

so by induction,

$$(\Delta_1^h g)(x + \ell h) \geq \alpha$$

for all $\ell = 0, 1, 2, \dots$ so

$$g(x + nh) \geq g(x) + n\alpha$$

which implies that $g(x+nh) \to \infty$ as $n \to \infty$, contradicting the assumption that $g$ is bounded.

Since $|(\Delta_n^h f)(x)| \leq 2^n \|f\|_\infty$, we can apply this argument to $g(x) = (\Delta_{2n}^h f)(x)$ for $n = 0, 1, 2, \dots$ to see that $(\Delta_{2n-1}^h f)(x) \leq 0$. Thus, $f$ is completely monotone. □

We are then left with the heart of the proof of Bernstein's theorem, which is that (i) ⇒ (v). We will show that $\{f \mid f \text{ is completely monotone and } \|f\|_\infty \leq 1\}$ is compact in the topology of local uniform convergence, that its extreme points are $\{e^{-\alpha x} \mid \alpha \in [0,\infty]\}$, and then apply the Strong Krein–Milman theorem. One might be tempted to take the set where $\|f\|_\infty = 1$, but that set is not closed in this topology since $f_\alpha(x) \equiv e^{-\alpha x}$ has $f_\alpha \to 0$ as $\alpha \to \infty$. Indeed, this solves a puzzle you might have about this application of the Strong Krein–Milman theorem. The measures it produces are on a compact set but the measure in (9.30) is on $[0, \infty)$ which is not compact. By throwing in a point at $\infty$ corresponding to the 0 function, we will get a compact set of extreme points!

The abstract theorem we will use to help us identify the extreme points is

**Proposition 9.14** *Let $C$ be a cone in a vector space, $V$. Let $\ell\colon C \to [0, \infty)$ be a function that obeys*
(i) $\ell(x) = 0$ *if and only if* $x = 0$,
(ii) $\ell(\lambda x) = \lambda \ell(x)$ *for all* $\lambda \geq 0$, $x \in C$,
(iii) $\ell(x + y) = \ell(x) + \ell(y)$ *for all* $x, y \in C$.
*Let*

$$C_1 = \{x \in C \mid \ell(x) \leq 1\} \tag{9.46}$$

*and*

$$H_1 = \{x \in C \mid \ell(x) = 1\} \tag{9.47}$$

*Then*
(a) $C_1$ *and* $C \backslash C_1$ *are both convex sets.*
(b) $\mathcal{E}(C_1) = \mathcal{E}(H_1) \cup \{0\}$
(c) *If* $x, y, z \in C_1$ *with* $x \in \mathcal{E}(C_1)$ *and* $x = y + z$, *then* $y = \theta x$, $z = (1 - \theta)x$ *for* $\theta \in [0, 1]$.

*Remarks* 1. A *ray* in a cone, $C$, is a set of the form $R = \{\lambda x \mid \lambda \in (0, \infty)\}$ for some $x \in C$. The ray is called an *extreme ray* if $R$ is a face of $C$, that is, $y, z \in C$ and $y + z \in R$ implies $y = \theta x$, $z = (1 - \theta)x$. This result says that rays through extreme points of $H_1$ generate extreme rays of $C$ and they are all the extreme rays. If $C_1$ is compact in a locally convex topology, we learn from the Krein–Milman theorem that any point in $C$ is a limit of sums of points in extreme rays.

2. We have said nothing in the proposition about topology, but in cases of interest, $C_1$ is compact. A compact convex subset of a cone so that $C \backslash C_1$ is also convex is called a *cap* of $C$. We claim in that case, we can take $\ell = \rho_C$, the gauge of $C$, and conversely, in the situation discussed, $\ell$ is the gauge of $C$. Under hypotheses (i)–(iii), clearly $\lambda x \in C$ if and only if $\ell(x) \leq \lambda^{-1}$ so $\ell(x) = \inf\{\lambda^{-1} \mid \lambda x \in C\} = \rho_C$. Conversely, if $C$ is a cap since $C$ is compact, any $x \neq 0$, $\lambda x \neq C$ for large $\lambda$, so $\rho_C(x) = 0$ for $x$ in $C$ implies $0 \in C$. $x_\varepsilon \equiv (1 + \varepsilon)x/\rho_C(x)$ and $y_\varepsilon \equiv (1 + \varepsilon)y/\rho_C(y)$ are both not in $C$, so

$$(1 + \varepsilon)\, \frac{x + y}{\rho_C(x) + \rho_C(y)} = \frac{\rho_C(x)}{\rho_C(x) + \rho_C(y)}\, x_\varepsilon + \frac{\rho_C(y)}{\rho_C(x) + \rho_C(y)}\, y_\varepsilon$$

is not in $C$, that is,

$$\rho_C(x + y) \geq \frac{[\rho_C(x) + \rho_C(y)]}{1 + \varepsilon}$$

since $\varepsilon$ is arbitrary and $\rho_C$ is convex, we see that $\rho_C$ obeys (iii).

3. We have said nothing in the proposition about topology but in applications, $\ell$ may not be continuous, although if $C_1$ is compact, $\ell$ must be lsc. In the Bernstein and Bochner theorem examples, $\ell$ will not be continuous and neither $H_1$ nor $\mathcal{E}(H_1)$ will be closed.

*Proof* (a) Since

$$\ell(\theta x + (1-\theta)y) = \theta\ell(x) + (1-\theta)\ell(y) \tag{9.48}$$

it follows immediately that both $C_1$ and $C\backslash C_1$ are convex.

(b) Since $\ell$ obeys (9.48), $H_1$ has the same property as a face does (except it may not be closed), so $\mathcal{E}(H_1) \subset \mathcal{E}(C_1)$, and since $\ell(x) = 0$ only if $x = 0$, $\{0\} \in \mathcal{E}(C_1)$. If $\ell(x) = \theta < 1$, then $x = \theta(\theta^{-1}x) + (1-\theta)0$ is not an extreme point.

(c) If $y$ or $z$ is zero, the conclusion is obvious. If not, $\ell(x) \neq 0$ so $\ell(x) = 1$ by (b). Moreover, $\ell(x) = \ell(y) + \ell(z)$, so if $\theta \equiv \ell(y)$, then $\ell(\theta^{-1}y) = 1 = \ell((1-\theta)^{-1}z)$, and thus,

$$x = \theta(\theta^{-1}y) + (1-\theta)((1-\theta)^{-1}z)$$

implies $\theta^{-1}y = (1-\theta)^{-1}z = x$ since $x$ is extreme. □

We return now to Bernstein's theorem. We need a final set of preliminaries. Let $\mathcal{B}$ be the cone of bounded completely monotone functions on $(0,\infty)$.

**Proposition 9.15** *Let $f \in \mathcal{B}$. Then*

(i) $\|f\|_\infty = \lim_{x\downarrow 0} f(x)$ *so* $f \mapsto \|f\|_\infty$ *is an additive function on* $\mathcal{B}$.

(ii) $f$ *is a continuous convex function.*

(iii) *If* $x, y \geq a$, *then*

$$|f(x) - f(y)| \leq a^{-1}\|f\|_\infty |x-y| \tag{9.49}$$

(iv) *For any* $x, h_1, \dots, h_n$, *we have*

$$(-1)^n(\Delta_1^{h_1} \dots \Delta_1^{h_n} f)(x) \geq 0 \tag{9.50}$$

*Proof* (i) $(\Delta_1^h f)(x) \leq 0$ means $f(x)$ increases as $x$ decreases. Since $f$ is bounded, $\lim_{x\downarrow 0} f(x)$ exists and clearly equals $\|f\|_\infty$ since $f \geq 0$. Since each $f \mapsto f(x)$ is additive, so is $\|f\|_\infty$.

(ii) $(\Delta_2^h f)(x) \geq 0$ says precisely that $f$ is midpoint-convex in the sense discussed in Proposition 1.3. In generality, midpoint convexity does not imply $f$ is continuous, but we have something else going for us, namely, $f$ is monotone, so for any $x$, $f(x \pm 0) \equiv \lim_{\varepsilon\downarrow 0} f(x \pm \varepsilon)$ exists. By monotonicity,

$$f(x+0) \leq f(x-0) \tag{9.51}$$

On the other hand, since

$$x - \varepsilon = \tfrac{1}{2}(x - 3\varepsilon) + \tfrac{1}{2}(x+\varepsilon)$$

midpoint convexity implies

$$f(x-\varepsilon) \le \tfrac{1}{2} f(x-3\varepsilon) + \tfrac{1}{2} f(x+\varepsilon)$$

(i.e., $(\Delta_2^{2\varepsilon} f)(x-\varepsilon) \ge 0$) which means

$$f(x-0) \le \tfrac{1}{2} f(x+0) + \tfrac{1}{2} f(x-0) \tag{9.52}$$

(9.51) and (9.52) imply that $f$ is continuous and so, by Proposition 1.3, it is convex.

(iii) Since $f$ is convex and monotone for $0 < z < a < x < y$, we have

$$0 \le (y-x)^{-1}[f(x)-f(y)] \le (a-z)^{-1}(f(z)-f(a))$$

by Theorem 1.24. Taking $z \downarrow 0$ and using $|f(a)-f(z)| \le \|f\|_\infty$ since $0 < f(a) < f(z)$, we obtain (9.49).

(iv) If $h_1, \dots, h_n$ are integral multiples of a single $h_0$, this follows from (9.38). Continuity then implies the result for arbitrary $h_1, \dots, h_n$. □

*Remark* By taking $h_1 \downarrow 0$ in (9.50), after dividing by $h_1$ and $h_2 = \dots = h_n = h$, we see $-(Df^+)(x)$ is completely monotone (and bounded on subsets $[a, \infty)$ for $a > 0$). Thus, by (ii), $D^+ f$ is convex, so $f$ is $C^1$ and $-f'$ is completely monotone. By induction, $f$ is $C^\infty$. This provides a direct proof that (i) ⇒ (ii) in Theorem 9.10 without going through the representation (9.30).

*Completion of the Proof of Theorem 9.10* Let $X$ be the set of bounded continuous functions on $(0, \infty)$ with the topology generated by the seminorms

$$\rho_n(f) = \sup_{n^{-1} \le x \le n} |f(x)|$$

which is metrizable. (While $X$ is not separable, the set $\mathcal{B}_1$ below is separable since it is a compact metric space.)

Let $\mathcal{B}_1 = \{f \in \mathcal{B} \mid \|f\|_\infty \le 1\}$. If $f_n \in B_1$ and $f_n \to f$ is the topology of $X$, then $(\Delta_n^h f)(x) \ge 0$ since we are taking pointwise limits in (9.26), and for any $x$, $|f(x)| = \lim |f_n(x)| \le \liminf \|f_n\|_\infty \le 1$ so $f \in \mathcal{B}_1$, that is, $\mathcal{B}_1$ is closed. Because of (9.49), $\{f \in \mathcal{B}_1\}$ is uniformly equicontinuous, so $\mathcal{B}_1$ is compact in the $X$-topology (see [303, Sect. I.6]). By Proposition 9.15(i), $\ell(f) = \|f\|_\infty$ is a degree 1, homogeneous, additive function on $\mathcal{B}_1$, so Proposition 9.13 applies.

Suppose $f \in \mathcal{E}(\mathcal{B}_1)$, $f \ne 0$. Fix $b > 0$. Then $f_b(x) = f(x+b)$ obeys $(\Delta_n^h f)(x) = (\Delta_n^h f)(x+b)$ so $f_b \in \mathcal{B}$, and since $f$ is monotone,

$$0 \le f_b \le f \tag{9.53}$$

Moreover,

$$\Delta_n^h (f - f_b)(x) = -(\Delta_n^h \Delta_1^b f)(x)$$

so, by (9.50), $f - f_b \in \mathcal{B}$ also. By (9.53), $\|f_b\|_\infty, \|f - f_b\|_\infty \le 1$, so $f_b, f - f_b \in \mathcal{B}_1$, and thus, by Proposition 9.13(b),

$$f_b(x) = \theta(b) f(x) \tag{9.54}$$

where $0 \le \theta(b) \le 1$. Since $f$ is continuous and $\|f\|_\infty = \lim_{x\downarrow 0} f(x) > 0$, we know $f(x) > 0$ for $x \in (0, \delta)$ for some $\delta > 0$. By (9.54), $\theta(\frac{\delta}{2}) = f(\frac{5\delta}{6})/f(\frac{\delta}{3}) > 0$. But then, by (9.54) again, $f(x) > 0$ in $(0, \frac{3\delta}{2})$ and, by induction and (9.54), on all of $(0, \infty)$.

Thus, for any $x, b > 0$,

$$\theta(b) = \frac{f(x+b)}{f(x)}$$

It follows that $\theta$ is continuous in $b$ and

$$\theta(b_1 + b_2) = \theta(b_1)\theta(b_2) \tag{9.55}$$

This implies $\theta(b) = \theta(1)^b$, first for $b$ rational (from (9.55)) and then for all $b$ by continuity. Writing $\theta(1) = e^{-\alpha}$, we have

$$f(x) = \lim_{y\downarrow 0} f(x+y) = e^{-\alpha x} \lim_{y\downarrow 0} f(y) = e^{-\alpha x}$$

since $\|f\|_\infty = \lim_{y\downarrow 0} f(y) = 1$.

Thus, any extreme point must be among $\{0\} \cup \{1\} \cup \{e^{-\alpha x}\}_{0<\alpha<\infty}$. Since $\mathcal{B}_1$ is more than one-dimensional, it must have at least one extreme point of the form $e^{-\alpha_0 x}$. For any $\lambda > 0$, the map $f \mapsto M_\lambda f = f(\lambda\,\cdot\,)$ is an affine, invertible map of $\mathcal{B}_1$ to itself, so it takes extreme points to extreme points. It follows that $e^{-\alpha_0 \lambda x}$ is an extreme point for all $\lambda > 0$. It follows that every $e^{-\alpha x}$, $\alpha \neq 0$ is an extreme point. Since 1 is the unique point in $\mathcal{B}_1$ with $\lim_{x\to\infty} f(x) = 1$, 1 is also an extreme point.

We have thus proven that $\mathcal{E}(\mathcal{B}_1) = \{f_\alpha \mid \alpha \in [0,\infty)\} \cup \{0\}$ where

$$f_\alpha(x) = e^{-\alpha x} \tag{9.56}$$

$\alpha \mapsto f_\alpha(x)$ is clearly continuous and in the topology on $X$, $\lim_{\alpha\to\infty} f_\alpha = 0$. So writing $f_\infty \equiv 0$, $\mathcal{E}(\mathcal{B}_1)$ is a continuous image of $[0,\infty]$ and so closed.

By the Strong Krein–Milman theorem, any $f \in \mathcal{B}_1$ can be written

$$f = \int_{[0,\infty]} f_\alpha \, d\mu(\alpha)$$

in the sense that for any continuous $L$ on $X$,

$$L(f) = \int_{[0,\infty]} L(f_\alpha)\, d\mu(\alpha)$$

Since $f \mapsto f(x)$ is continuous, (9.30) holds for $f$, except $\mu$ is a probability measure and the integral is over $[0, \infty]$. But $\mu(\{\infty\})$ contributes nothing to the integral, so we can drop $\infty$ and get a measure $\mu$ on $[0, \infty)$ with $\mu([0, \infty)) \leq 1$. By the monotone convergence theorem and (9.30),

$$\|f\|_\infty = \lim_{x \downarrow 0} f(x) = \mu([0, \infty))$$

The Stone–Weierstrass theorem shows that the span of $\{e^{-\alpha x}\}_{\alpha \in (0,\infty)}$ is dense in $C_0([0, \infty))$, the continuous functions vanishing at infinity, so $\{\int e^{-\alpha x}\, d\mu(\alpha)\}_{\text{all } x \in (0,\infty)}$ determine the measure $\mu$ on $(0, \infty)$, that is, $\mu$ is unique, as claimed. □

Finally, in our consideration of Bernstein's theorem, we note and prove the result in case $f$ is not assumed bounded.

**Theorem 9.16** (Bernstein's Theorem: Unbounded Case) *Let $f$ be a real-valued function on $(0, \infty)$. Then the following are equivalent:*

(i) *$f$ is completely monotone.*
(ii) *$f$ is $C^\infty$ with $(-1)^n f^{(n)}(x) \geq 0$.*
(iii) *For any $(x_1, \dots, x_n) \in (0, \infty)$, $B(x_1, \dots, x_n; f)$ is a positive matrix and $f$ is bounded on $[1, \infty)$.*
(iv) *For any $(x_1, \dots, x_n) \in (0, \infty)$, $\det(B(x_1, \dots, x_n; f))$ is a positive matrix and $f$ is bounded on $[1, \infty)$.*
(v) *There exists a measure $d\mu$ on $[0, \infty)$ with $\int e^{\ \varepsilon x}\, d\mu(x)$ for all $\varepsilon > 0$ so that (9.30) holds.*

*Proof* In case (i), (ii), (v), this follows by applying Theorem 9.10 to $f(x + \varepsilon)$ for each $\varepsilon > 0$. For (iii), (iv), we note we only used $f$ bounded at $+\infty$ to go from (iii) to (i). □

*Remark* Our proof shows that for (iii) $\Rightarrow$ (i), it suffices that $\lim_{x \to \infty} \frac{f(x)}{|x|} = 0$. By using higher differences, it actually suffices that $\limsup_{x \to \infty} \frac{f(x)}{|x|^m} < \infty$ for some $m$.

That completes our discussion of Bernstein's theorem. We note an unsatisfactory aspect of the proof we gave of Bernstein's theorem. Behind the scenes, there is a spectacular result being proven, namely, that if $(-1)^n f^{(n)}(x) \geq 0$, then $f$ is the restriction to $(0, \infty)$ of a function analytic in $\{z \mid \operatorname{Re} z > 0\}$. There is no "explanation" of why this is true. Since it is true for extreme points, it holds in general. Of course, the Bernstein–Boas theorem (Theorem 7.1) and its proof provide an understanding, but we did not use it in this argument. We now turn to Bochner's theorem, a result which has other, more "standard" proofs; see, for example, [304, Sect. IX.2].

**Definition** A (weakly) *positive definite function* on $\mathbb{R}^\nu$ is a function $f \in L^\infty(\mathbb{R}^\nu)$ so that for all $g \in L^1(\mathbb{R}^\nu)$,

$$\int f(x-y)g(x)\,\overline{g(y)}\,d^\nu x\,d^\nu y \geq 0 \tag{9.57}$$

A positive definite function in classical sense (*pdfcs*) is a continuous function $f$ on $\mathbb{R}^\nu$ so for all $\zeta_1, \dots, \zeta_n \in \mathbb{C}$ and $x_1, \dots, x_n \in \mathbb{R}^\nu$,

$$\sum_{i,j=1}^{n} \bar{\zeta}_i \zeta_j\, f(x_j - x_i) \geq 0 \tag{9.58}$$

**Theorem 9.17** (Bochner's Theorem) *Let $f\colon \mathbb{R}^\nu \to \mathbb{C}$. The following are equivalent:*

(i) *$f$ is a positive definite function.*
(ii) *$f$ is equal a.e. to a pdfcs.*
(iii) *There is a finite measure $d\mu$ on $\mathbb{R}^\nu$ so for a.e. $x \in \mathbb{R}^\nu$,*

$$f(x) = \int e^{ik\cdot x}\, d\mu(k) \tag{9.59}$$

*The measure $d\mu$ in* (9.59) *is unique.*

We will eventually prove (iii) ⇒ (ii) ⇒ (i) ⇒ (iii). As a preliminary, we need some critical connections between positive definite functions and pdfcs that go beyond (ii) ⇒ (i).

**Proposition 9.18** (a) *If $f$ is a pdfcs, then $f$ is bounded and*

$$\overline{f(x)} = f(-x), \qquad |f(x)| \leq f(0) \tag{9.60}$$

(b) *If $f$ is a pdfcs, then $f$ is positive definite.*
(c) *Let $\varphi_n(x)$ be an approximate identity* (*i.e.*, $\varphi_n(x) = n^\nu \varphi(nx)$, $\int \varphi(y)\,d^\nu y = 1$, $\varphi \geq 0$, $\varphi \in C_0^\infty(\mathbb{R}^\nu)$) *and define for any $f \in L^\infty$,*

$$[\Phi_n(f)](x) = \int \varphi_n(y)\varphi_n(z) f(x+z-y)\,d^\nu y\,d^\nu z \tag{9.61}$$

*Then if $f$ is a positive definite function, for each $n$, $\Phi_n(f)$ is a pdfcs and*

$$\|f\|_\infty = \lim_{n\to\infty} \|\Phi_n(f)\|_\infty \tag{9.62}$$

(d) *$f \mapsto \|f\|_\infty$ on positive definite functions is additive.*

*Proof* (a) For $f$ to be a pdfcs, the matrix ($x_1 = 0$, $x_2 = x$),

$$\begin{pmatrix} f(0) & f(x) \\ f(-x) & f(0) \end{pmatrix}$$

must be a positive matrix which implies (9.60). The inequality follows from the fact that the trace, $2f(0)$, and determinant, $f(0)^2 - |f(x)|^2$, are both nonnegative.

(b) If $g$ in (9.57) is a continuous function of compact support, (9.57) follows from (9.58) by approximating the integral by Riemann sums since we know $f$ is continuous. By (a), $f$ is bounded, so by an approximating argument, (9.57) for continuous functions, $g$, of compact support implies it for all $g \in L^1$.

(c) That $\Phi_n(f)$ is continuous is a standard argument since $w \mapsto f(\cdot - w)$ is continuous from $\mathbb{R}$ to $L^1$. To see that $\Phi_n(f)$ obeys (9.58), note that

$$
\begin{aligned}
&\sum_{i,j=1}^{m} \bar{\zeta}_i \zeta_j \Phi_n(f)(x_j - x_i) \\
&\qquad = \sum_{i,j=1}^{m} \int \bar{\zeta}_i \zeta_j \varphi_n(y)\varphi_n(z) f(x_j - x_i + z - y)\, d^\nu y\, d^\nu z \\
&\qquad = \sum_{i,j}^{m} \int \bar{\zeta}_i \zeta_j \varphi_n(y - x_i)\varphi_n(z - x_j) f(z - y)\, d^\nu y\, d^\nu z \\
&\qquad = \int \overline{g(y)}\, g(z) f(z - y)\, d^\nu y\, d^\nu z
\end{aligned} \tag{9.63}
$$

where

$$
g(z) = \sum_{j=1}^{m} \zeta_j \varphi_n(z - x_j)
$$

(9.63) follows by changing the integration variables $y \to y - x_i$, $z \to z - x_j$ in the $i, j$ summand.

To prove (9.62), note first that since $\int \varphi_n(y)\, d^\nu y = 1$, (9.61) implies that

$$
\|\Phi_n(f)\|_\infty \leq \|f\|_\infty \tag{9.64}
$$

On the other hand, if $g \in L^1$, by a change of variables $x \to x - z$,

$$
\begin{aligned}
\int g(x)\Phi_n(f)(x)\, d^\nu x &= \int \varphi_n(y)\varphi_n(z) f(x + z - y) g(x)\, d^\nu y\, d^\nu z\, d^\nu x \\
&= \int f_n(x) g_n(x)\, d^\nu x
\end{aligned}
$$

where $g_n(x) = \int g(x - z)\varphi_n(z)\, d^\nu z$ and $f_n(x) = \int f(x - y)\varphi_n(y)\, d^\nu y$. Now $g_n \to g$ in $L^1$-norm, $\|f_n\|_\infty \leq \|f\|_\infty$, and $f_n \to f$ in $\sigma(L^\infty, L^1)$, so

$$
\lim_n \int g(x)\Phi_n(f)(x)\, d^\nu x = \int g(x) f(x)\, d^\nu x
$$

for all $g$ in $L^1$, that is, $\Phi_n(f) \to f$ in $\sigma(L^\infty, L^1)$. As usual, weak convergence in a dual space implies the norm of the limit can only be smaller, that is,

$$
\liminf \|\Phi_n(f)\|_\infty \geq \|f\|_\infty \tag{9.65}
$$

(9.64) and (9.65) imply (9.62).

(d) By (a), $\|\Phi_n(f)\|_\infty = \Phi_n(f)(0)$. Thus, by (9.62),

$$\|f\|_\infty = \lim_{n\to\infty} \Phi_n(f)(0)$$

Since $f \mapsto \Phi_n(f)(0)$ is additive, so is $\|f\|_\infty$ on the positive definite functions. □

*Proof of Theorem 9.15* (ii) ⇒ (i) is (b) of Proposition 9.18. (iii) ⇒ (ii) since if (9.59) holds, $f$ is continuous by the monotone convergence theorem and

$$\sum_{j=1}^{n} \bar{\zeta}_i \zeta_j f(x_j - x_i) = \int \biggl| \sum_{i,j=1}^{n} \zeta_j e^{ik\cdot x_j} \biggr|^2 d\mu(k) \geq 0 \tag{9.66}$$

Thus, we only need to prove (i) ⇒ (iii). We will do this by using the Strong Krein–Milman theorem.

Let $\mathcal{P} \subset L^\infty$ be the set of positive definite functions on $\mathbb{R}^\nu$ and $\mathcal{P}_1 = \{f \in \mathcal{P} \mid \|f\|_\infty \leq 1\}$. Give $L^\infty$ the $\sigma(L^\infty, L^1)$ (i.e., weak-$*$) topology so $A \equiv \{f \in L^\infty \mid \|f\|_\infty \leq 1\}$ is compact. (9.57) can be rewritten:

$$\int f(x)(\tilde{g} * g)(x)\, d^\nu x \geq 0 \tag{9.67}$$

where

$$\begin{aligned}(\tilde{g} * g)(x) &= \int g(x+y)\,\overline{g(y)}\, d^\nu y \\ &= \int g(x-y)\,\overline{g(-y)}\, d^\nu y\end{aligned} \tag{9.68}$$

written this way since it is the convolution of $g$ and $\tilde{g}(y) = \overline{g(-y)}$. Now $\tilde{g} * g$ is in $L^1$ if $g$ is, so (9.67) is preserved if $g$ is fixed and $f_n \to f$ in $\sigma(L^\infty, L^1)$-topology, that is, $\mathcal{P}_1$ is closed in $A$ and so $\mathcal{P}_1$ is a compact convex set.

By (d) of Proposition 9.18, $\|\cdot\|_\infty$ is additive, so we can apply Proposition 9.13 with $\ell(f) = \|f\|_\infty$. We thus seek extreme points of $f$ of $\mathcal{P}_1$ with $\|f\|_\infty = 1$.

Fix any $f \in \mathcal{P}_1$, $a \in \mathbb{R}$, and $\zeta \in \partial D$. Then by a change of variable,

$$\begin{aligned}0 &\leq \int f(x-y)[\tfrac{1}{2}\, g(x) + \tfrac{1}{2}\, \zeta g(x-a)][\tfrac{1}{2}\, \overline{g(y)} + \tfrac{1}{2}\, \bar{\zeta} g(y-a)]\, d^\nu x\, d^\nu y \\ &= \int f_{a,\zeta}(x-y) g(x)\, \overline{g(y)}\, d^\nu x\, d^\nu y\end{aligned}$$

where

$$f_{a,\zeta}(x) = \tfrac{1}{2}\, f(x) + \tfrac{1}{4}\, \zeta f(x+a) + \tfrac{1}{4}\, \bar{\zeta} f(x-a) \tag{9.69}$$

so each $f_{a,\zeta}$ is also positive definite.

Now suppose $f$ is extreme. Note

$$f_{a,+1} + f_{a,-1} = f$$

so by Proposition 9.18, we conclude that $f_{a,+1} = \theta f$ which means that

$$f(x+a) + f(x-a) = 2\alpha_a f(x)$$

for some real constant $\alpha_a$. Similarly, since $f$ is extreme,

$$f_{a,i} + f_{a,-i} = f$$

implies

$$f(x+a) - f(x-a) = 2i\beta_a f(x)$$

so with $\gamma_a = \alpha_a + i\beta_a$, for each $a$ and a.e. $x$,

$$f(x+a) = \gamma_a f(x) \tag{9.70}$$

Suppose $\|f\|_\infty \neq 0$. Since $\|f(\cdot + a)\|_\infty = \|f\|_\infty$, (9.70) implies $|\gamma_a| = 1$. Since $f(x+a+b) = \gamma_a f(x+b) = \gamma_a \gamma_b f(x)$ for a.e. $x$, we have

$$\gamma_{a+b} = \gamma_a \gamma_b \tag{9.71}$$

Pick $g \in C_0^\infty(\mathbb{R}^\nu)$ so $\int f(x) g(x)\, d^\nu x \neq 0$. Then

$$\gamma_a = \frac{\int f(x) g(x-a) d^\nu x}{\int f(x) g(x)\, d^\nu x}$$

so $\gamma_a$ is a $C^\infty$ function. Differentiating (9.71) at $a = 0$ implies

$$\frac{\partial}{\partial b_j} \gamma_b = \gamma_b \frac{\partial}{\partial a_j} \gamma_a \bigg|_{a=0} \equiv i k_j \gamma_b$$

since $\gamma_a^* \gamma_a = 1$ and $\gamma_{a=0} = 1$ implies $\frac{\partial}{\partial a_j} \gamma_a \Big|_{a=0}$ is pure imaginary. Thus,

$$\gamma_b = e^{ik\cdot b}$$

If

$$f(x+a) = e^{ik\cdot a} f(x) \tag{9.72}$$

for a.e. $x$ and all $a$, it holds for a.e. pairs $x$ and $a$, so picking $x_0$ with $f(x_0) \neq 0$ and so that (9.72) holds for a.e. $a$, we see for a.e. $a$, $f(a) = Ce^{ik\cdot a}$ with $C = e^{-ik\cdot x_0} f(x_0)$. Since $\|f\|_\infty = 1 = \lim_{n\to\infty} \Phi_n(f) = C$, we see that if

$$\varphi_k(x) = e^{ik\cdot x} \tag{9.73}$$

then the $\varphi_k$ are the only candidates for extreme points. Each is in $\mathcal{P}_+$ by a direct calculation (essentially (9.66)) and since $|\varphi_k(x)| = 1$, if $\varphi_k = \theta f + (1-\theta) g$ with $0 < \theta < 1$ and $\|f\|_\infty = \|g\|_\infty = 1$, we must have $f(x) = g(x) = \varphi_k(x)$ for a.e. $x$ showing each $\varphi_k(x)$ is extreme. Thus, we have shown that

$$\mathcal{E}(\mathcal{P}_+) = \{\varphi_k\}_{k\in\mathbb{R}^\nu} \cup \{0\} \tag{9.74}$$

By the Riemann–Lebesgue lemma (see [304, Sect. IX.2]), $\varphi_k \to 0$ in $\sigma(L^\infty, L^1)$ as $k \to \infty$ so $\mathcal{E}(\mathcal{P}_+)$ is closed. Thus, by Theorem 9.2, there is a probability measure $\mu$ on $\mathcal{E}(\mathcal{P}_+) \equiv \mathbb{R}^\nu \cup \{\infty\}$ so that any $f \in \mathcal{P}_+$ is

$$f = \int_{\mathbb{R}^k} \varphi_k \, d\mu + 0\mu(\{\infty\}) \tag{9.75}$$

in the sense that for any $g \in L^1(\mathbb{R}^\nu)$,

$$\int f(x)g(x)\, d^\nu x = \int_{\mathbb{R}^k} \left( \int g(x)e^{ik\cdot x}\, d^\nu x \right) d\mu(k)$$

Since $g \in L^1$ and $\mu$ is a finite measure, we can interchange the $d^\nu x$ and $d\mu$ integral, and then since integrals determine $f$'s in $L^\infty$, we conclude (9.59) holds.

Finally, we must show that the measure is unique. Some care is needed since it is tempting to use the Fourier inversion formula for some convenient class of functions – and that is a bad strategy because some proofs of the Fourier inversion formula use Bochner's theorem (!). We proceed with our bare hands, as follows:

Suppose

$$\int e^{ik\cdot x}\, d\mu(k) = \int e^{ik\cdot x}\, d\nu(k) \tag{9.76}$$

for a.e. $x$. Then with $\hat{f}(k) = (2\pi)^{-\nu/2} \int f(x)e^{-ik\cdot x}\, d^\nu x$, (9.76) implies

$$\int \hat{f}(k)\, d\mu(k) = \int \hat{f}(k)\, d\nu(k) \tag{9.77}$$

for all $f \in L^1$. Since $\hat{\tilde{f}}(k) = \bar{\hat{f}}(k)$, where $\tilde{f}(x) = \overline{f(-x)}$ and $(2\pi)^{-\nu/2}\widehat{f * g}(k) = \hat{f}(k)\hat{g}(k)$, we see that $\{\hat{f} \mid f \in L^1\}$ is a subalgebra of the continuous functions on $\mathbb{R}^\nu$ vanishing at infinity, and it is closed under conjugation. If we show for any $k_1 \neq k_2$, we can find $f \in L^1$ with

$$0 \neq \hat{f}(k_1) \neq \hat{f}(k_2) \tag{9.78}$$

then by the Stone–Weierstrass theorem, $\{\hat{f}\}$ is dense, so (9.77) implies $\mu = \nu$, proving uniqueness. Since $k_1 \neq k_2$, find $x_0 \notin \mathbb{R}^\nu$ with $(k_1 - k_2) \cdot x_0 = \pi$. Let $\varphi_n$ be an approximate identity and let $f_n(x) = \varphi_n(x - x_0)$. Then

$$\lim_{n\to\infty} \hat{f}_n(k_1) = e^{ik_1\cdot x_0}$$

$$\lim_{n\to\infty} \hat{f}_n(k_2) = e^{ik_2\cdot x_0} = -e^{ik_1\cdot x_0}$$

so there exists $f$ for which (9.78) holds. □

*Remark* This proof extends with no real change when $\mathbb{R}^\nu$ is replaced by an arbitrary, locally compact, abelian group. We see that the extreme points are measurable functions on the group that obey (9.72), that is, the group characters.

This completes the discussion of Bochner's theorem. Finally we turn to the proof of the hard part of Loewner's theorem (Theorem 6.5) using these ideas. See Chapter 6 for background and a statement of the theorem. There is also a proof in Chapter 7 and a discussion of other proofs in the Notes. We will need three main inputs from the "preliminaries" in Chapter 6:

(1) Proposition 6.39 which says if $f \in M_\infty(-1,1)$ and $f(t) = 0$, then

$$g_\pm(t) = \left(1 \pm \frac{1}{t}\right) f(t) \in M_\infty(-1,1) \tag{9.79}$$

(2) The result of Dobsch (Theorem 6.26) that if $f \in M_2(-1,1)$ (in particular, if $f \in M_\infty(-1,1) \subset M_2(-1,1)$) and $f$ is nonconstant, then $f'(x)^{-1/2}$ is concave (this is a consequence of the differential inequality $f'f''' - \frac{3}{2}(f')^2 \geq 0$ that $f$ obeys).

(3) $f$ is $C^2$ (true already for $M_3(-1,1)$); see Theorem 6.31.

To get compactness, we have to avoid the fact that if $f \in M_\infty(-1,1)$, so is $a + bf$ for any $a$ and any $b > 0$. To fix this noncompactness, we define

$$\mathcal{L}_1 = \{f \in M_\infty(-1,1) \mid f(0) = 0,\ f'(0) = 1\} \tag{9.80}$$

**Proposition 9.19** *Let $f \in \mathcal{L}_1$ Then*

(a)

$$\tfrac{1}{2} f''(0) \equiv \alpha(f) \in [-1,1] \tag{9.81}$$

(b) *For* $0 \leq \pm x \leq 1$,

$$0 \leq \pm f(x) \leq \frac{|x|}{1-|x|} \tag{9.82}$$

(c) *If* $0 < x < y < 1$,

$$0 \leq f(y) - f(x) \leq |x-y|(1-y)^{-2} \tag{9.83}$$

(d) *If $\alpha(f)$ is given by* (9.81), *then for* $\pm x > 0$,

$$\pm f(x) \geq \pm x(1-\alpha x)^{-1} \tag{9.84}$$

*Proof* Let $c(x) = f'(x)^{-1/2}$, so $c$ is concave and nonnegative on $(-1,1)$, and since $f'(0) = 1$,

$$c(0) = 1 \tag{9.85}$$

and since $f' = c^{-2}$, $f'' = -2c'c^{-3}$ so

$$\alpha(f) = \tfrac{1}{2} f''(0) = -c'(0) \tag{9.86}$$

(a) By concavity of $c$ and (9.86),

$$c(x) \leq 1 - \alpha(f)x \tag{9.87}$$

If $|\alpha(f)| > 1$, $1 - \alpha(f)x$ vanishes on $(-1, 1)$ so (9.87) is incompatible with $c > 0$ on $(-1, 1)$. Thus, $|\alpha(f)| \le 1$.

(b) Since $c(0) = 1$ and $\liminf_{|x|\to\pm 1} c(x) \ge 0$, concavity of $c$ implies

$$c(x) \ge 1 - |x|$$

so

$$f'(x) \le (1 - |x|)^{-2} \tag{9.88}$$

Since $f(0) = 0$ and $\int_0^y dy/(1-y)^2 = y/(1-y)$, (9.88) implies (9.82).

(c) For $0 < x < y$,

$$\int_x^y \frac{du}{(1-u)^2} \le |x - y| \, \frac{1}{(1-y)^2}$$

so (9.88) implies (9.83).

(d) (9.87) says that

$$f'(x) \ge (1 - \alpha x)^{-2} \tag{9.89}$$

Since

$$\int_0^y \frac{dy}{(1-\alpha y)^2} = \frac{y}{1-\alpha y} \tag{9.90}$$

(9.89) implies (9.84). □

*Remark* (9.82) and (9.84) show that if $\alpha = 1$, then for $x > 0$, $f(x) = x/(1-x)$. Once one has analyticity, that is true for all $x$.

We know that the functions $x/(1-\alpha x)$ are special because of what we are seeking to prove, but (9.84) shows us how special they are and is the key to the proof of

**Proposition 9.20** *Let*

$$\varphi_\alpha(x) = \frac{x}{1-\alpha x} \tag{9.91}$$

*Then each $\varphi_\alpha$ is an extreme point of $\mathfrak{L}_1$.*

*Proof* $\varphi'_{\alpha_0}(x) = 1/(1-\alpha_0 x)^2$, $\varphi''_{\alpha_0}(x) = 2\alpha_0/(1-\alpha_0 x)^3$, and $\varphi_{\alpha_0}(0) = 0$, $\varphi'_{\alpha_0}(0) = 1$, so $\varphi_{\alpha_0} \in \mathfrak{L}_1$. Moreover,

$$\alpha(\varphi_{\alpha_0}) \equiv \tfrac{1}{2}\, \varphi''_{\alpha_0}(0) = \alpha_0 \tag{9.92}$$

Suppose

$$\varphi_{\alpha_0} = \theta f + (1-\theta) g \tag{9.93}$$

with $f, g \in \mathfrak{L}_1$ and $\theta \in (0,1)$. Since $\alpha(f) = \frac{1}{2} f''(0)$ is linear in $f$,

$$\theta\alpha(f) + (1-\theta)\alpha(g) = \alpha_0 \tag{9.94}$$

By (9.84),

$$\begin{aligned}\pm\varphi_{\alpha_0}(x) &= \pm[\theta f(x) + (1-\theta)g(x)] \\ &\geq \pm[\theta\varphi_{\alpha(f)}(x) + (1-\theta)\varphi_{\alpha(g)}]\end{aligned} \tag{9.95}$$

Now

$$\frac{\partial^2}{\partial\alpha^2}\left(\frac{|x|}{1-\alpha x}\right) = \frac{2|x|^3}{(1-\alpha x)^3} > 0$$

so $\alpha \mapsto \pm\varphi_\alpha(x)$ is strictly convex in $\alpha$ for each $x \in (-1,1)$, so by (9.94) and (9.95),

$$\pm\varphi_{\alpha_0}(x) > \pm\varphi_{\alpha_0}(x)$$

unless $\alpha(f) = \alpha(g) = \alpha_0$. But if $\alpha(f) = \alpha(g) = \alpha_0$, (9.84) for all $x$ and (9.93) is only possible if $f = g = \varphi_{\alpha_0}(x)$. Thus, $\varphi_{\alpha_0}$ is an extreme point. □

*Proof of Loewner's Theorem (Theorem 6.5)* Consider the vector space, $X$, of all continuous functions on $(-1,1)$ (not necessarily bounded) topologized with the seminorms

$$\rho_n(f) = \sup_{|x|\leq 1-n^{-1}} |f(x)|$$

$\mathfrak{L}_1$ is compact in that topology since, by (9.82), $f$'s in $\mathfrak{L}_1$ are uniformly bounded on each $I_n = [-1+n^{-1}, 1-n^{-1}]$ and, by (9.83), uniformly equicontinuous.

Let $f$ be a point in $\mathfrak{L}_1$ and let $g_\pm$ be given by (9.79). Then

$$g_\pm(0) = \pm 1, \qquad g'_\pm(0) = (1 \pm \alpha(f)) \tag{9.96}$$

Suppose first $|\alpha(f)| < 1$. Then define

$$h_\pm(x) = (g_\pm(x) \mp 1)(1 \pm \alpha(f))^{-1} \tag{9.97}$$

so by (9.79) and (9.96), $h_\pm \in \mathfrak{L}_1$. Moreover, with $\theta = \frac{1}{2}(1+\alpha(f)) \in (0,1)$,

$$f = \theta h_+ + (1-\theta)h_- \tag{9.98}$$

Since $\theta \in (0,1)$, if $f$ is extreme, then $f = h_+$, that is,

$$(1+\alpha)f(x) = \left(1 + \frac{1}{x}\right)f(x) - 1$$

or solving for $f(x)$,

$$f(x) = \frac{1}{\frac{1}{x} - \alpha} = \frac{x}{1-\alpha x} = \varphi_\alpha(x)$$

with $\varphi_\alpha$ given by (9.91).

If $\alpha(f) = 1$, then $g_-'(0) = 0$, so since $g_- \in M_\infty(-1,1)$, $g_-(x) \equiv -1$, that is,

$$f(x) = -\frac{1}{1 - \frac{1}{x}} = \frac{x}{1-x} = \varphi_1(x)$$

Similarly, if $\alpha(f) = -1$, $f(x) = \varphi_{-1}(x)$. This shows that there is a unique $f$ with $\alpha(f) = +1$ or with $\alpha(f) = -1$. Thus, $\{\varphi_\alpha\}_{\alpha \in [-1,1]}$ are the only possible extreme points and, by Proposition 9.20, they are extreme points. Thus,

$$\mathcal{E}(\mathcal{L}_1) = \{\varphi_\alpha\}_{\alpha \in [-1,1]}$$

It is easily seen that $\alpha \mapsto \varphi_\alpha$ is continuous in the topology on $X$, so $\mathcal{E}(\mathcal{L}_1)$ is closed. By the Strong Krein–Milman theorem, for any $f \in \mathcal{L}_1$, we have a probability measure $d\mu$ on $[-1,1]$ so

$$f = \int_{-1}^{1} \varphi_\alpha \, d\mu(\alpha)$$

in the sense of equality when continuous linear functions are applied. Since $f \mapsto f(x)$ is continuous in the topology on $X$,

$$f(x) = \int_{-1}^{1} \frac{x}{1-\alpha x} \, d\mu(\alpha)$$

Putting back $f(0)$, $f'(0)$, and taking $d\mu(\alpha)$ to $d\mu(-\alpha)$, we see any $f \in M_\infty(-1,1)$ has the form

$$f(x) = f(0) + \int_{-1}^{1} \frac{x}{1+\alpha x} \, d\mu(\alpha) \tag{9.99}$$

for a measure $d\mu$ with $\int_{-1}^{1} d\mu = f'(0) < \infty$. □

*Remark* As noted in Chapter 6, $d\mu$ in (9.99) is unique.

**Example 9.21** As a parting example, we want to note that the representation (1.62) for positive, monotone, convex functions on $(0,\infty)$ can be viewed as an example of the Strong Krein–Milman theorem. Let $X$ be the locally convex space of all continuous functions on $[0,\infty)$ with $f(0) = 0$ and the topology of uniform convergence on compact subsets of $[0,\infty)$. Let $C$ be the cone of all positive, monotone, convex functions and $C_1$ the cone of functions obeying

$$\lim_{x\to\infty} \frac{f(x)}{x} < \infty \tag{9.100}$$

$C_1$ is dense in $C$ since, given any $f \in C$, the functions

$$f^{(a)}(x) = \begin{cases} f(x), & x \le a \\ f(a) + (Df^+)(a)(x-a), & x \ge a \end{cases} \tag{9.101}$$

lie in $C_1$ and $f^{(a)} \to f$ as $a \to \infty$.

Let the set $A \subset C_1$ be given by

$$A = \{f \in C \mid 0 \le f(x) \le x \text{ for all } x \in [0,\infty)\} \tag{9.102}$$

We claim $A$ is a compact cap for $C_1$. $A$ is compact because any $f \in A$ has $\lim_{x\to\infty}(Df^-)(x) \le 1$ and thus obeys

$$|f(x) - f(y)| \le |x - y| \tag{9.103}$$

and so $A$ is compact in the topology of $X$ by the Arzelà–Ascoli theorem, and $\ell(f) = \lim_{x\to\infty} f(x)/x = \lim_{x\to\infty} Df^-(x)$ is a linear functional on $C_1$ with $A = \{f \mid \ell(f) \le 1\}$ so $C_1 \backslash A$ is convex.

For $a \in [0,\infty)$, let $g_a$ be the function

$$g_a(x) = (x - a)_+ \tag{9.104}$$

We claim

$$\mathcal{E}(A) = \{g_a\}_{a\in[0,\infty)} \cup \{0\} \tag{9.105}$$

If $f, h \in A$ and $g_a = \theta f + (1-\theta)h$, then $f = h = 0$ on $[0,a]$ and by (9.103), $f(x) \le g_a(x)$ on $[a,\infty]$. Thus, $f \le g_a$ and similarly, $h \le g_a$ so $f = h = g_a$. Thus, each $g_a$ is an extreme point.

Conversely, if $f \in A$ and $f^{(a)}$ is given by (9.101), then $f^{(a)}$ and $f - f^{(a)}$ lie in $A$. If $f$ is extreme, it follows that for each $a$, $f^{(a)} = f$ or $f^{(a)} = 0$. If $f^{(a)} = 0$ for all $a$, then $f = 0$. So if $f \not\equiv 0$, there is an $a$ with $f^{(a)} = f$. Let $a_0 \equiv \inf\{a \mid f^{(a)} = f\}$. Then by continuity of $Df^+$ from the right, $f^{(a_0)} = f$. Since $a_0$ is the inf, if $b < a_0$, $f^{(b)} = 0$, that is, $f(x) = 0$ on $[0, a_0]$. Thus, $f = \theta g^{(a_0)}$ for some $\theta \in [0,1]$. For $f$ to be extreme, $\theta$ must be 1. This proves (9.105).

$\mathcal{E}(A)$ is closed, so any $f \in C_1$ has a representation of the form (1.62) with $\int d\gamma + F'(0) \le 1$ (the $xF'(0)$ term should be thought of as a contribution from a point mass at 0). Using the fact that $C_1$ is dense in $C$, one can find a representation of the form (1.62) for all $f \in C$. □

# 10

# Choquet theory: existence

Throughout the next two chapters, $A$ is a compact convex subset of a locally convex vector space, $X$. For any $x \in A$, we are interested in finding a probability measure $\mu$ on $A$ with barycenter, $x$, and $\mu(\mathcal{E}(A)) = 1$. We will succeed in doing that in case $A$ is metrizable (equivalently, since $A$ is compact, if $A$ is separable). Every example in the last chapter is separable, and in practical applications, that is invariably true. There are some fascinating subtleties in the nonseparable case that have evoked considerable literature – for example, $\mathcal{E}(A)$ may not be a Borel measurable set (see Choquet [83]) in that case, so $\mu(\mathcal{E}(A)) = 1$ does not make sense. Since the separable case covers applications and avoids some unpleasant pathologies, we will settle for constructing *in general* measures $\mu$ with barycenter $x$ which morally want to be supported on $\mathcal{E}(A)$ and prove that in the separable case, $\mu(\mathcal{E}(A)) = 1$ without any detailed discussion on the sense in which $\mu$ is "concentrated" on $\mathcal{E}(A)$ if $A$ is not separable.

Given this setup, we define many subsets of $C_{\mathbb{R}}(A)$, the real functions on $A$: Let $\mathcal{A}(A)$ be the continuous affine functions on $A$ (i.e., $f(\theta x + (1-\theta)y) = \theta f(x) + (1-\theta)f(y)$), $\mathfrak{L}(A) \subset \mathcal{A}(A)$, the restrictions to $A$ of elements of $X^*$, $\mathcal{C}(A)$ as the set of continuous (and so bounded) convex functions on $A$. Since we will focus here on $C_{\mathbb{R}}(A)$, we drop the $\mathbb{R}$ and just write $C(A)$. $R(\cdot)$ will be the barycenter map from $\mathcal{M}_{+,1}(A)$ to $A$.

**Definition** Let $\mu, \nu \in \mathcal{M}_{+,1}(A)$. We say $\mu$ is larger than $\nu$ in the *Choquet order* and write $\mu \succ \nu$ if and only if $\mu(f) \geq \nu(f)$ for all $f \in \mathcal{C}$.

We will see momentarily that if $\mu \prec \nu$, then $R(\mu) = R(\nu)$. Since convex functions take larger values as one gets close to the boundary, one can hope that maximal measures in the Choquet order live very near $\mathcal{E}(A)$. For example, if $A = [0,1]$ and $f_0(x) = x^2$, it is not hard to show that among all $\mu$ with $\int x\, d\mu(x) = \theta$, then $\int x^2\, d\mu(x)$ is maximized precisely when $\mu = \theta\delta_1 + (1-\theta)\delta_0$.

**Definition** A measure $\mu \in \mathcal{M}_{+,1}(A)$ is called a *maximal measure* if and only if it is maximal in the Choquet order, that is, $\mu \prec \nu$ implies $\mu = \nu$.

**Proposition 10.1** (i) $\mathcal{C}(A) \cap [-\mathcal{C}(A)] = \mathcal{A}(A)$
(ii) $\mathcal{C}(A) - \mathcal{C}(A)$ *is dense in* $C(A)$.
(iii) $\prec$ *is antisymmetric, that is,* $\mu \prec \nu$ *and* $\nu \prec \mu$ *implies* $\mu = \nu$.
(iv) *If* $\mu \prec \nu$, *then* $R(\mu) = R(\nu)$.
(v) $\delta_x$ *is a maximal measure if and only if* $x \in \mathcal{E}(X)$.

*Proof* (i) is obvious.

(ii) Consider $C(A)$'s lattice structure, that is,

$$(f \vee g)(x) = \max(f(x), g(x))$$
$$(f \wedge g)(x) = \min(f(x), g(x))$$

If $f, g \in \mathcal{C}(A)$, then $f + g$ and $f \vee g \in \mathcal{C}(A)$, so $f \wedge g = f + g - (f \vee g) \in \mathcal{C}(A) - \mathcal{C}(A)$. It follows that $\mathcal{C}(A) - \mathcal{C}(A)$ is a lattice. Since $\mathcal{L}(A) \subset \mathcal{A}(A) \subset \mathcal{C}(A)$ and $\mathcal{L}$ separates points, so does $\mathcal{C}(A) - \mathcal{C}(A)$. The Kakutani–Krein theorem (see [303, Thm. IV.12]) then says this difference is dense.

(iii) If $\mu \prec \nu$ and $\nu \prec \mu$, then $\mu(f) = \nu(f)$ for all $f \in \mathcal{C}(A)$, so by linearity for all $f$ in $\mathcal{C}(A) - \mathcal{C}(A)$, so by (ii) and continuity for all $f$, that is, $\mu = \nu$.

(iv) Since $\mathcal{L}(A) \subset \mathcal{C}(A) \cap [-\mathcal{C}(A)]$, if $\mu \prec \nu$, then $\mu(\pm\ell) \leq \nu(\pm\ell)$ for every linear functional $\ell$, that is, $\mu(\ell) = \nu(\ell)$, so $R(\mu) = R(\nu)$.

(v) By Bauer's theorem (Theorem 9.3), if $x \in \mathcal{E}(A)$, then $\delta_x$ is the unique measure, $\mu$, with $R(\mu) = x$ so, by (iv), it is maximal. If $x \notin \mathcal{E}(A)$, $x = \frac{1}{2}y + \frac{1}{2}z$ for $y \neq z$. By the definition of convexity of $f$,

$$\tfrac{1}{2}\,\delta_y + \tfrac{1}{2}\,\delta_z \succ \delta_x \tag{10.1}$$

□

**Theorem 10.2** *Given any* $x \in A$, *there is a maximal measure* $\mu$ *with barycenter* $x$.

*Proof* This is a simple Zornification. $\prec$ is clearly transitive and reflexive, and we have proven it to be antisymmetric, so $\prec$ is a partial order.

Let $\{\mu_\alpha\}_{\alpha \in I}$ be a family in $\mathcal{M}_{+,1}(A)$ which is linearly ordered in the Choquet order. Turn it into a net by saying $\alpha > \beta$ if $\mu_\alpha \succ \mu_\beta$. Since $\mathcal{M}_{+,1}(A)$ is compact, let $\mu_\infty$ be a limit point of $\{\mu_\alpha\}$. Since the net is linearly ordered for $\alpha < \beta$, and $f \in \mathcal{C}(A)$, $\mu_\alpha(f) \leq \mu_\beta(f)$, so $\mu_\alpha(f) \leq \mu_\infty(f)$ since $\mu_\infty(f)$ is a limit point of a bounded net of linearly ordered reals, hence its sup. It follows that every linearly ordered subfamily in $\mathcal{M}_{+,1}(A)$ has an upper bound.

By Zorn's lemma, any measure, in particular, $\delta_x$ is dominated by a maximal measure. If $\delta_x \prec \mu$, $R(\mu) = x$ by Proposition 10.1. □

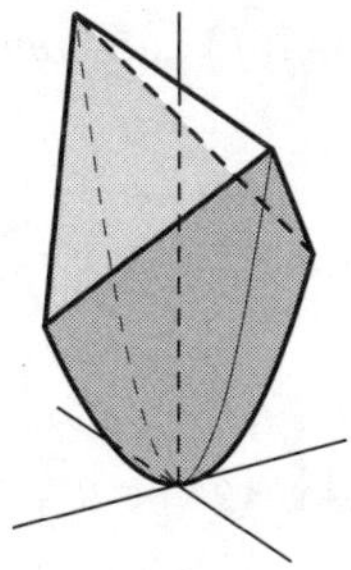

Figure 10.1 A concave envelope only piecewise affine

The remainder of this chapter will show that if $A$ is metrizable, then $\mathcal{E}(A)$ is a $G_\delta$ that is also a Baire set, and also, any maximal measure has $\mu(\mathcal{E}(A)) = 1$. The machinery to do this will be useful in the next chapter. We will only use metrizability late in the game, so we will even provide insight into the general case.

A key role will be provided by the interplay of Choquet order and something very close to the convex envelope used in Chapter 5 to discuss Legendre transforms (see the discussion following Example 5.14), except that we will need essentially $-(-f)_*$.

**Definition** Let $f$ be an arbitrary element in $C(A)$; we define $\hat{f}$, the *concave envelope* of $f$, by

$$\hat{f}(x) = \inf\{g(x) \mid -g \in \mathcal{C}(A),\ g \geq f\} \tag{10.2}$$

*Remark* For any continuous concave function, $g$, we have

$$g(x) = \inf\{\ell(x) \mid \ell \in \mathcal{A}(A),\ \ell \geq g\}$$

essentially by the argument in Theorem 5.15. Thus, one can also write

$$\hat{f}(x) = \inf\{\ell(x) \mid \ell \in \mathcal{A}(A),\ \ell \geq f\} \tag{10.3}$$

but we will mainly use the definition (10.2).

If $\Delta_n$ is the $n$-dimensional simplex, it is easy to see that $\hat{f}$ is the affine function that agrees with $f$ at the extreme points. As Figure 10.1 shows, if $A$ is a square, $\hat{f}$ may only be piecewise affine and concave. In both cases, one can see that if $f$ is strictly convex, $\{x \mid f(x) = \hat{f}(x)\} = \mathcal{E}(A_n)$. We will prove this in great generality and, in particular, for any finite-dimensional, compact, convex subset. Thus, to measure the fact that a maximal $\mu$ is concentrated near the extreme points, we will show $\mu(\hat{f}) = \mu(f)$ for any $f$!

**Proposition 10.3** (i) *$\hat{f}$ is concave and usc.*
(ii) *If $f \in C(A)$ is concave, then $\hat{f} = f$.*

(iii) *For any* $\mu \in \mathcal{M}_{+,1}(A)$, $f \mapsto \mu(\hat{f})$ *is a convex function and a homogeneous function of degree* 1.

(iv)

$$\|\hat{f} - \hat{g}\|_\infty \leq \|f - g\|_\infty \tag{10.4}$$

(v) *For any* $\mu \in \mathcal{M}_{+,1}(X)$,

$$\mu(\hat{f}) = \inf\{\mu(g) \mid -g \in \mathcal{C}(A),\, g > f\} \tag{10.5}$$

*Proof* (i) Immediate since $\hat{f}$ is an inf of concave continuous functions.

(ii) Obvious.

(iii) If $\lambda > 0$, $f \leq g$ if and only if $\lambda f \leq \lambda g$ and $g \in -\mathcal{C}(A)$ if and only if $-\lambda g \in \mathcal{C}(A)$. Thus, $\widehat{\lambda f} = \lambda \hat{f}$ so $f \mapsto \mu(\hat{f})$ is homogeneous of degree 1. If $f_1 \leq g_1$ and $f_2 \leq g_2$ with $-g_i \in \mathcal{C}(A)$, then $f_1 + f_2 \leq g_1 + g_2$ with $-(g_1 + g_2) \in \mathcal{C}(A)$ so

$$\widehat{f_1 + f_2} \leq \hat{f}_1 + \hat{f}_2 \tag{10.6}$$

which implies for $\theta \in [0, 1]$,

$$\begin{aligned} \mu(\widehat{\theta f_1 + (1-\theta) f_2}) &\leq \mu(\widehat{\theta f_1}) + \mu(\widehat{(1-\theta) f_2}) \\ &= \theta\mu(\hat{f}_1) + (1-\theta)\mu(\hat{f}_2) \end{aligned}$$

(iv) By (10.6) with $f_1 = g$ and $f_2 = f - g$,

$$\hat{f} - \hat{g} \leq \widehat{f - g}$$

so by symmetry,

$$\|\hat{f} - \hat{g}\|_\infty \leq \|\widehat{f - g}\|_\infty$$

and thus, it suffices to prove (10.4) when $g = 0$. Since $\|f\|_\infty 1 \in -\mathcal{C}(A)$,

$$-\|f\|_\infty \leq f \leq \hat{f} \leq \|f\|_\infty$$

from which $\|\hat{f}\|_\infty \leq \|f\|_\infty$ is immediate.

(v) The monotone convergence theorem for nets says if $g_\alpha$ is a net of continuous functions decreasing to the function $f$ (which is then, automatically, Baire), then $\mu(g_\alpha)$ converges to $\mu(f)$. This implies (10.5). □

Half of the technical core of the argument is:

**Lemma 10.4** *Let* $\mu \in \mathcal{M}_{+,1}(A)$. *Let* $\nu$ *be a (not a priori continuous) linear functional on* $C(A)$. *Then the following are equivalent:*

(i) $\nu \in \mathcal{M}_{+,1}(A)$ *and* $\nu \succ \mu$

(ii)

$$\nu(f) \leq \mu(\hat{f}) \qquad \textit{for all } f \tag{10.7}$$

*Proof* <u>(i)⇒ (ii)</u> If $-g \in \mathcal{C}(A)$ and $\nu \succ \mu$, then $\nu(g) \le \mu(g)$. Thus, by (10.5),

$$\begin{aligned}\nu(f) \le \nu(\hat{f}) &= \inf\{\nu(g) \mid -g \in \mathcal{C}(A),\, f \le g\} \\ &\le \inf\{\mu(g) \mid -g \in \mathcal{C}(A),\, f \le g\} = \mu(\hat{f})\end{aligned}$$

<u>(ii) ⇒ (i)</u> Let $f \le 0$. Then, since $-0 \in \mathcal{C}(A)$, $\hat{f} \le 0$ so (10.7) implies $\nu(f) \le \mu(0) = 0$. Since $\nu$ is linear, $\nu(f) \ge 0$ if $f \ge 0$, and thus, $\nu \in \mathcal{M}_+(A)$. Since $\pm 1$ are concave functions, $\pm\hat{1} = \pm 1$ so (10.7) implies

$$\nu(\pm 1) \le \mu(\pm 1) = \pm 1$$

which means $\nu(A) = 1$, that is, $\nu \in \mathcal{M}_{+,1}(A)$. If $g \in \mathcal{C}(A)$, then (10.7) implies $\nu(-g) \le \mu(\widehat{-g}) = \mu(-g)$, that is, $\mu(g) \le \nu(g)$, so $\mu \prec \nu$. □

The main theorem that relates maximal measures and concave envelopes is

**Theorem 10.5** *Let $\mu \in \mathcal{M}_{+,1}(A)$. The following are equivalent:*
(i) *$\mu$ is a maximal measure (in Choquet order).*
(ii)

$$\mu(f) = \mu(\hat{f}) \qquad \text{for all } f \in C(A) \tag{10.8}$$

*Proof* <u>(i) ⇒ (ii)</u> We begin with the second half of the technical core of the argument (the first half was (ii) ⇒ (i) in Lemma 10.4). By Proposition 10.3 (iii), $f \mapsto \mu(\hat{f})$ is convex, so by Theorem 1.37, for any $f_0 \in C(A)$, there exists a linear function $\nu$ on $C(A)$ so that for all $f \in C(A)$,

$$\nu(f) - \nu(f_0) \le \mu(\hat{f}) - \mu(\hat{f}_0) \tag{10.9}$$

Taking $f = \frac{1}{2} f_0$ and then $f = \frac{3}{2} f$ and using $\mu(\widehat{\lambda f}) = \lambda \mu(\hat{f})$ for $\lambda$ positive, (10.9) implies $\pm\frac{1}{2}\nu(f_0) \le \pm\frac{1}{2}\mu(\hat{f})$, so

$$\nu(f_0) = \mu(\hat{f}_0) \tag{10.10}$$

so that (10.9) becomes (10.7). Thus, by Lemma 10.4, $\nu \in \mathcal{M}_{+,1}(A)$ and $\nu \succ \mu$. Since $\mu$ is assumed maximal, $\nu = \mu$. But then $\mu(f_0) = \nu(f_0) = \mu(\hat{f}_0)$ by (10.10). Since $f_0$ is arbitrary, (ii) holds.

<u>(ii) ⇒ (i)</u> Suppose (ii) holds and $\nu \succ \mu$. Then for $f \in C(A)$,

$$\begin{aligned}\nu(f) \le \nu(\hat{f}) &= \inf\{\nu(g) \mid -g \in \mathcal{C}(A),\, g \ge f\} && \text{(by (10.6))} \\ &\le \inf\{\mu(g) \mid -g \in \mathcal{C}(A),\, g \ge f\} && \text{(since } \nu \succ \mu) \\ &= \mu(\hat{f}) && \text{(by (10.6))} \\ &= \mu(f) && \text{(by hypothesis (ii))}\end{aligned}$$

Thus, $\nu(\pm f) \le \mu(\pm f)$, so $\nu = \mu$. We conclude $\mu$ is maximal. □

Here is a sense in which, morally, measures that obey (10.8) want to be concentrated on $\mathcal{E}(A)$.

**Corollary 10.6**

$$\bigcap_{f\in C(A)} \{x \mid \hat{f}(x) = f(x)\} = \mathcal{E}(A) \tag{10.11}$$

*In particular, if* $x \in \mathcal{E}(A)$, $\hat{f}(x) = f(x)$ *for all* $f$.

*Proof* By Proposition 10.1, if $x_0 \in \mathcal{E}(A)$, $\delta_{x_0}$ is maximal, so by Theorem 10.5, $\delta_{x_0}(\hat{f}) = \delta_{x_0}(f)$, that is, $\hat{f}(x_0) = f(x_0)$, and thus, $\mathcal{E}(A) \subset \cap_{f\in C(A)}\{x \mid \hat{f}(x) = f(x)\}$. On the other hand, if $x_0 \notin \mathcal{E}(A)$, we can find $y \neq z$ so $x_0 = \frac{1}{2}y_0 + \frac{1}{2}z_0$. Pick $\ell \in X^*$ so $\ell(y_0 - z_0) = 2$ and let

$$f(x) = (\ell(x) - \ell(x_0))^2$$

so

$$f(x_0) = 0, \qquad f(y_0) = f(z_0) = 1$$

If $g$ is concave and $g \geq f$, then $g(x_0) \geq \frac{1}{2}(g(y_0) + g(z_0)) \geq 1$, so $\hat{f}(x_0) \geq 1$. Thus, $f(x_0) \neq \hat{f}(x_0)$ and we see

$$\bigcap_{f\in C(A)} \{x \mid \hat{f}(x) = f(x)\} \subset \mathcal{E}(A)$$ □

*Remark* The proof actually shows it is also true that

$$\mathcal{E}(A) = \bigcap_{f\in \mathcal{C}(A)} \{x \mid \hat{f}(x) = f(x)\} \tag{10.12}$$

for the $f$'s that showed any $x \notin \mathcal{E}(A)$ was not in the intersection were convex.

The end of this proof shows that if $f$ is strictly convex and $x \notin \mathcal{E}(A)$, then $\hat{f}(x) \geq \frac{1}{2}[f(y) + f(z)] > f(x)$, that is,

$$f \text{ strictly convex} \Rightarrow \mathcal{E}(A) = \{x \mid \hat{f}(x) = f(x)\} \tag{10.13}$$

$\mu(\{x \mid \hat{f}(x) = f(x)\}) = 1$ for any maximal measure by Theorem 10.5. There are infinitely many $f$'s so we cannot directly conclude from (10.11) that $\mu(\mathcal{E}(A)) = 1$. Of course, (10.13) says we can, if we can construct a strictly convex $f$ on $A$, and we will do that in case $A$ is separable. It is known (see Hervé [160]) that if $A$ is not separable, then no strictly convex $f$ exists.

**Theorem 10.7** (Choquet's Theorem) *Let $A$ be a metrizable, compact, convex subset of a locally convex space, $X$. Then*

(i) $\mathcal{E}(A)$ *is a Baire* $G_\delta$.

(ii) *For any* $x \in A$, *there is a probability measure* $\mu$ *on* $A$ *whose barycenter is* $x$ *and with* $\mu(\mathcal{E}(A)) = 1$.

*Proof* Suppose we can find a strictly convex, continuous function $f$ on $A$. Let $g = \hat{f} - f \geq 0$. Then by (10.13), $\mathcal{E}(A) = \{x \mid g(x) = 0\}$. Moreover, since $f$ is continuous and $\hat{f}$ is usc, $g$ is usc. Therefore, $\{x \mid g(x) \geq \alpha\}$ is a compact set for any $\alpha$. It follows that

$$\mathcal{E}(A) = \{x \mid g(x) = 0\} = \bigcap_{n=1}^{\infty} \{x \mid g(x) < n^{-1}\}$$

is a $G_\delta$.

To see $\mathcal{E}(A)$ is a Baire set, let

$$G_{mn} = \{x \in A \mid x = \tfrac{1}{2}\, y + \tfrac{1}{2}\, z,\ f(x) < \tfrac{1}{2}\,(f(y) + f(z)) - n^{-1} + (n+m)^{-1}\}$$

This is easily seen to be an open set because $f$ is continuous and is clearly disjoint from $\mathcal{E}(A)$. By compactness, one can see

$$\bigcap_{m=1}^{\infty} G_{mn} \equiv F_n = \{x \in A \mid x = \tfrac{1}{2}\, y + \tfrac{1}{2}\, z;\ f(x) \leq \tfrac{1}{2}\,(f(y) + f(z)) - n^{-1}\}$$

which is compact as the continuous image under $(y, z) \mapsto \frac{1}{2}y + \frac{1}{2}z$ of $\{y, z \in A \times A \mid f(\frac{1}{2}y + \frac{1}{2}z) \leq \frac{1}{2}(f(y) + f(z)) - n^{-1}\}$ which is closed and so, compact in $A \times A$. Thus, $F_n$ is a compact $G_\delta$ and so, Baire. By strict convexity, $\cup F_n = A \backslash \mathcal{E}(A)$ so $\mathcal{E}(A)$ is Baire.

By Theorem 10.5, if $f$ exists, $\mu(\mathcal{E}(A)) = 1$. We are thus left with constructing a strictly convex function.

Since $A$ is a compact metric space, $C(A)$ is separable and all the more so, $\mathcal{A}(A)$ is separable, so pick $\{\ell_n\}_{n=1}^{\infty}$ a countable dense subset of $\mathcal{A}(A)$. If for some $x, y \in A$ with $x \neq y$, $\ell_n(x) = \ell_n(y)$ for all $n$, then by density, $\ell(x) = \ell(y)$ for all $\ell \in \mathcal{A}$, which is impossible since $X^*$ separates points. Thus, $\{\ell_n\}$ separates points. Let

$$f(x) = \sum_{n=1}^{\infty} 2^{-n} (\|\ell_n\|_\infty + 1)^{-2} [\ell_n(x)]^2 \tag{10.14}$$

Each $\ell_n^2$ is convex, so $f$ is convex. If $\ell_n(y) \neq \ell_n(z)$, then $\ell_n(\theta y + (1-\theta) z)^2 < \theta \ell_n(y)^2 + (1-\theta)\ell_n(z)^2$ for all $\theta \in (0,1)$ so, since the $\ell_n$'s separate points, $f$ is strictly convex. □

While we used maximal measures to find the $\mu$ with $\mu(\mathcal{E}(A))$ and $r(\mu) = x$, Theorem 10.7 did not mention maximal measures. There is an aspect still missing that we want to fill in – namely, that in the metrizable case, the only measures supported on $\mathcal{E}(A)$ are the maximal measures:

**Theorem 10.8** *Let $A$ be a metrizable, compact convex set. A probability measure $\mu$ on $A$ has $\mu(\mathcal{E}(A)) = 1$ if and only if $\mu$ is maximal in Choquet order.*

*Proof* The existence of a strictly convex function, $f_0$, on $A$ shows that if $\mu$ is maximal, then, by Theorem 10.5, $\mu$ is supported on $\{x \mid \hat{f}_0(x) = f_0(x)\} = \mathcal{E}(A)$, as we have seen above.

Conversely, if $\mu$ is supported on $\mathcal{E}(A)$, by Corollary 10.6, for all $f \in C(A)$, $\int(\hat{f} - f)(x)\, d\mu(x) = 0$ so $\mu(f) = \mu(\hat{f})$. Hence, $\mu$ is maximal by Theorem 10.5. □

# 11

# Choquet theory: uniqueness

In this chapter, the final one of the four on the Krein–Milman theorem and Choquet theory, we discuss the issue of when a point $x \in A$, a compact convex subset of a locally convex space has a unique representation as an integral of extreme points. We have an elegant solution if $A$ is metrizable and, more generally, if one asks instead that for each $x \in A$, there exists a unique maximal measure with barycenter $x$. Let us begin with an analysis of the finite-dimensional case.

**Theorem 11.1** *Let $A$ be a compact convex subset of finite-dimensional space. Let $\dim(A) = n$. Then $A$ has the property that every point has a unique representation as a convex combination of extreme points if and only if $\mathcal{E}(A)$ has precisely $n+1$ points, in which case $A$ is affinely isomorphic to the $n$-simplex, $\Delta_n$. If $A$ is not a simplex and $x \in A^{\text{iint}}$, the intrinsic interior of $A$, then $x$ has multiple representations as a convex combination of extreme points.*

*Proof* Since $\text{cch}(x_1, \dots, x_\ell)$ for any $\ell$ points has dimension at most $\ell - 1$ (it is the dimension of the span of $\{x_j - x_1\}_{j=2}^{\ell}$ by the proof of Theorem 8.8), any $n$-dimensional set, $A$, must have at least $n+1$ extreme points. If $\mathcal{E}(A)$ has exactly $n+1$ points, they must be affinely independent (given the dimension), in which case $A$ is affinely isomorphic to $\Delta_n$ and representations are unique (by independence of $\{e_j - e_1\}_{j=2}^{n}$ where $\{e_j\}_{j=1}^{n}$ are the extreme points).

Suppose $\mathcal{E}(A)$ has at least $n+2$ points. Pick any $x_0 \in A^{\text{iint}}$. By Theorem 8.11, we can write $x_0$ as a convex combination of at most $n+1$ extreme points. Suppose $c_0$ is an extreme point not among those used in the first representation. By Theorem 8.11 again, we can find a representation $x = \sum_{j=0}^{n} \theta_j c_j$ with $\theta_0 > 0$. It follows this is distinct from the first one. □

Therefore, unique representation by maximal measures in the finite-dimensional case is restricted to simplexes. For each $n$, there is up to affine homeomorphism a unique $n$-dimensional compact convex set with the unique representation property. To get the infinite-dimensional analog, we have to figure out a geometric property that picks out simplexes. Figure 11.1 shows four convex subsets in the plane. The triangle has the property that if two of its translates overlap, then the intersection is similar to the original triangle (i.e., related to the original by a scaling and translation). For the square, the intersection is always affinely the same, but may not be similar; and for the trapezoid and disk, even that fails. It is easy to see this property is true for $\Delta_n$ and it follows from what we will show below (and Theorem 11.1) that it fails for any other finite-dimensional, compact convex set.

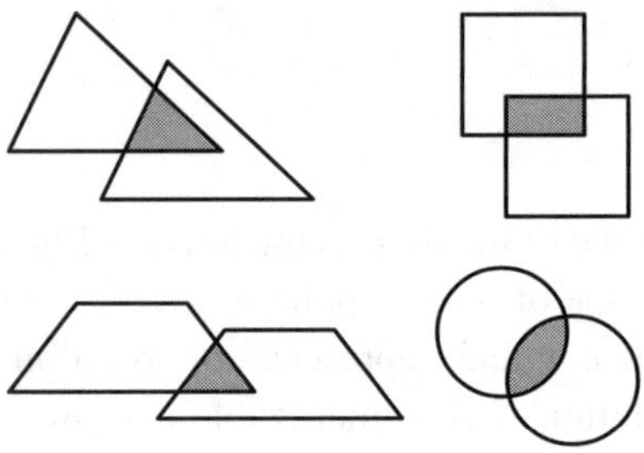

Figure 11.1 Similar and nonsimilar intersections

There is a geometrically even more attractive way to phrase this. Consider the cones with congruent triangular bases and two cones with circular bases. In the first case, their intersection is a translate of the original cone, while not in the second (see Figure 11.1). We will use this idea in the general infinite-dimensional case, so we begin with some preliminaries on cones with a given base and the order they generate. Recall the definition (1.19) of the suspension of a compact set, $A \subset V$, a vector space

$$A_{\text{sus}} = \{(\lambda x, \lambda) \in V \times \mathbb{R} \mid x \in A,\ \lambda \geq 0\} \tag{11.1}$$

**Definition** A cone, $C \subset V$, is called *proper* if and only if $C \cap (-C) = \{0\}$. It is called *generating* if $C - C = V$.

**Definition** Let $C$ be a convex cone in a vector space $V$. A *base* for $C$ is a subset $A \subset C$ so that

(i) $0 \notin A$
(ii) $A$ is convex.
(iii) For each nonzero $x \in C$, there is a unique $\lambda \in (0, \infty)$ and $y \in A$ so

$$x = \lambda y \tag{11.2}$$

This is closely related to Proposition 9.13. In that setting $H_1$ is a base for $C$ and, given a base $C$, if we define $\ell(x)$ to be the $\lambda$ of (11.2), then $\ell$ is additive (because

$C$ is convex). $\{\lambda y \mid y \in A,\, 0 \leq \lambda \leq 1\}$ is a cap for $C$ if it is compact. Notice that if $C$ has a base, it must be proper, since if $x, -x \in C$, $\lambda_1 x, -\mu_1 x \in A$ for $\lambda_1, \mu_1 > 0$, and then $\theta(\lambda_1 x) + (1-\theta)(-\mu_1 x) = 0$ with $\theta = \mu_1/(\lambda_1 + \mu_1)$.

**Proposition 11.2** *Let $A \subset V$ be a convex subset of $V$. Then $\{(x,1) \mid x \in A\}$ is a base for $A_{\text{sus}}$ and it is affinely isomorphic to $A$. If $C$ is any other cone with base $B$ affinely isomorphic to $A$, then the homomorphism can be extended to a linear isomorphism from $C$ onto $A_{\text{sus}}$. If $V$ has a topology and $A$ is compact, all maps can be taken continuous.*

*Proof* Elementary. □

Since all cones with base $A$ are "the same," we talk about the cone with base $A$. In many cases, (e.g., $\mathcal{M}_{+,1}(X)$ in $\mathcal{M}_+(X)$), the cone can be taken in the original space $V$ as the minimal cone containing $A$ (i.e., $0 \notin A$ and for each $\lambda \neq 1$, $x \in A$, $\lambda x \notin A$).

Cones are associated with orders compatible with the vector structure:

**Definition** Let $V$ be a vector space. A partial order, $\lhd$, on $V$ is called a *vector order* if and only if

(i)

$$x \lhd y \Rightarrow x + z \lhd y + z \tag{11.3}$$

(ii) $x \lhd y$ and $\lambda \geq 0$ implies $\lambda x \lhd \lambda y$.

(iii) Any pair, $x, y$, of elements in $V$ has an upper bound, that is, $z$ so $x \lhd z$ and $y \lhd z$.

A vector space with a vector order is called an *ordered vector space*.

**Theorem 11.3** *Let $V$ be an ordered vector space with order $\lhd$. Then*

$$P_\lhd = \{x \in V \mid 0 \lhd x\} \tag{11.4}$$

*(the* positive *elements of $V$) is a proper, generating, convex cone. Conversely, if $P$ is any proper, generating, convex cone, then it is $P_\lhd$ for a unique vector order on $V$.*

*Proof* Any order that obeys (i) is clearly determined by (and determines) $P_\lhd$ given by (11.4). The dictionary below relates properties of $\lhd$ to properties of $P_\lhd$ (if and only if).

Except for the last, these are obvious. Notice that if property (iii) holds and $x \in V$, and $w$ is an upper bound of $x$ and $0$, then $w \in P$ and $w - x \in P$, so $x = w - (w - x) \in P - P$. Conversely, if $P$ is generating, $x, y \in V$ and $x = w_1 - w_2$, $y = w_3 - w_4$ with $w_i \in P$, then $w_1 + w_3$ is an upper bound for $x$ and $y$. □

*Remark* If $V$ is a topologized vector space, then $\lhd$ obeys the condition $x_\alpha \to x$, $y_\alpha \to y$ and $x_\alpha \lhd y_\alpha \Rightarrow x \lhd y$ if and only if $P_\lhd$ is closed.

| $\lhd$ is | $P_\lhd$ obeys |
|---|---|
| transitive | $x, y \in P \Rightarrow x + y \in P$ |
| reflexive | $0 \in P$ |
| antisymmetric | $P \cap (-P) = \{0\}$ |
| condition (ii) | $P$ is a cone |
| condition (iii) | $P$ is generating |

Given a proper, generating, convex cone $C$, we call the order $\lhd$ with $P_\lhd = C$ the *order defined by* $C$. If $A$ is a convex subset of a vector $V$, $A_{\text{sus}}$ in the subspace $W = A_{\text{sus}} - A_{\text{sus}}$ of $V \times \mathbb{R}$ is a proper, generating, convex cone, and so $A_{\text{sus}}$ defines an order on $W$. If $V$ is any ordered vector space so that $P_\lhd$ has base affinely isomorphic to $A$, then there is a linear order equivalent bijection of $V$ to the $W$ just constructed. We will call $W$ with the order defined by $A_{\text{sus}}$, the *order induced by* $A$. When convenient (see Example 11.4 below), we will use an isomorphic $V$ rather than $A_{\text{sus}} - A_{\text{sus}}$.

**Definition** A partially ordered set where each pair of elements has a greatest lower bound, denoted $x \wedge y$, and least upper bound, denoted $x \vee y$, is called a *lattice*. An ordered vector space whose order is a lattice is called a *vector lattice*. A convex subset, $A$, of a vector space $V$ for which the induced order is a vector lattice is called an *algebraic simplex*. If $A$ is compact in a locally convex topology on $V$, $A$ (with its topology) is called a *Choquet simplex* or a *simplex*.

*Remark* Since $x \lhd y$ if and only if $y - x \in P$ and $y - x = -x - (-y)$, we see that $x \mapsto -x$ is order reversing. Thus,

$$x \wedge y = -[(-x) \vee (-y)] \tag{11.5}$$

and to be a vector lattice, it suffices to show any two elements have a least upper bound. (11.5) can be rewritten as

$$x \vee y + x \wedge y = x + y \tag{11.6}$$

for $x + y - y \wedge x = -(-x - y + y \wedge x) = -((-x) \wedge (-y))$ by (11.3).

**Example 11.4** (Measures on a compact Hausdorff space) The space, $\mathcal{M}(X)$, of signed measures on a compact Hausdorff space $X$ (i.e., $C(X)^*$) has a natural order, $\mu \lhd \nu$, if and only if $\int f\, d\mu \geq \int f\, d\nu$ for all positive $f \in C(X)$. As usual, we will write $\mu \geq \nu$. The cone of positive elements is $\mathcal{M}_+(X)$ and it has $\mathcal{M}_{+,1}(X)$ as a base.

If $\mu, \nu \in \mathcal{M}(X)$ and $\gamma = |\mu| + |\nu|$, then $\mu$ and $\nu$ are both $\gamma$-ac and so $d\mu = f\, d\gamma$; $d\nu = g\, d\gamma$ for $f, g \in L^1(X, d\gamma)$. If $d\eta = \max(f, g)\, d\gamma$ (pointwise max), it is easy to see that $\eta$ is a least upper bound of $\mu$ and $\nu$, so $\mathcal{M}(X)$ is a lattice and $\mathcal{M}_{+,1}(X)$ is a simplex.

Note that, by construction, $\mu \vee \nu$ is absolutely continuous with respect to $|\mu| + |\nu|$. □

We can now state the main result of this chapter.

**Theorem 11.5** (Choquet–Meyer Theorem) *Let $A$ be a compact convex subset of a locally convex topological vector space. Then the following are equivalent:*

(i) *$A$ is a simplex.*

(ii) *For each $f \in \mathcal{C}(A)$, its concave envelope, $\hat{f}$, is affine.*

(iii) *If $f \in \mathcal{C}(A)$ and $\mu$ is a maximal measure,*

$$\mu(f) = \hat{f}(r(\mu)) \tag{11.7}$$

(iv) *For each $x \in A$, there is a unique maximal measure, $\mu$, with barycenter $x$.*

*Remark* This theorem does *not* require that $A$ be separable.

The proof will go (i) ⇒ (ii) ⇒ (iii) ⇒ (iv) ⇒ (i) and will require quite a few preliminaries. (iv) ⇒ (i) will come from the fact that the set of maximal measures will always be an algebraic simplex, so if $r$ is one-one from maximal measures to $A$, $A$ will be an algebraic simplex. (i) ⇒ (ii) will need a critical decomposition lemma for lattices. (ii) ⇒ (iii) and (iii) ⇒ (iv) will be easy. We need to begin with restricting order to certain subspaces.

**Proposition 11.6** *Let $V$ be a vector lattice with $P$ its cone of positive elements. Suppose $W$ is a subspace with the property that $x, y \in W$ implies $x \vee y \in W$ ($x \vee y$ in the $V$ order). Then $P \cap W$ is a proper, generating, convex cone for $W$ and $W$ is a lattice in the order it defines. Moreover, if $x, y \in W$, $x \vee y$ (order in $V$) is the least upper bound of $x$ and $y$ in the order defined by $P \cap W$.*

*Remark* Hidden in this statement is a subtlety. Let $C$ be the "standard" cone in $\mathbb{R}^3$ (the one you put ice cream in), $C = \{(x, y, z) \mid x^2 + y^2 \le z^2; z \ge 0\}$. Let $W$ be the two-dimensional plane $\alpha x + \beta y + \gamma z = 0$ where $\alpha^2 + \beta^2 < \gamma^2$. Then $C \cap W = \{0\}$ and this is not generating. In general, because $P \cap W$ is not generating, one cannot restrict order to subspaces. But one can sometimes, and that is the more subtle part of this proposition.

*Proof* $P \cap W$ is clearly a proper convex cone. To show it is generating, we note that by (11.6),

$$x = x \vee 0 + x \wedge 0 \tag{11.8}$$

and that $x \vee 0 \in P \cap W$ and $-(x \wedge 0) = (-x) \vee 0 \in P \cap W$, so (11.8) shows $P \cap W - P \cap W = W$.

Let $x, y \in W$. Since $x \vee y - x$ and $x \vee y - y \in P \cap W$, $x \vee y$ is an upper bound in the $W$ order for $x$ and $y$. If $z \in W$ and $z - x, z - y \in P \cap W$, they are in $P$ and so $z - x \vee y \in P$ and so in $P \cap W$. It follows that $x \vee y$ is an upper bound in the order defined by $P \cap W$, so $W$ is a lattice in the order. □

**Corollary 11.7** *Let $A$ be a compact convex subset of a locally convex space. Let $\mathcal{M}_{+,1}^{\max}(A)$ be those sets of elements in $\mathcal{M}_{+,1}(A)$ which are maximal in the Choquet order. Then $\mathcal{M}_{+,1}^{\max}(A)$ is convex and an algebraic simplex.*

*Proof* By Theorem 10.5, $\mu \in \mathcal{M}_{+,1}(A)$ lies in $\mathcal{M}_{+,1}^{\max}(A)$ if and only if

$$\mu(\hat{f}) = \mu(f) \qquad \text{for all } f \in C(A) \tag{11.9}$$

Let $W$ be the set of all $\mu \in \mathcal{M}(X)$ for which (11.9) holds. If $\mu, \nu \in W$ with $\mu \geq 0$, $\nu \geq 0$, then $\mu + \nu \in W$ and so for any $f \in C(A)$, $\mu + \nu$ is supported on $\{x \mid \hat{f}(x) = f(x)\}$. But $\mu \vee \nu$ is absolutely continuous with respect to $\mu + \nu$ by the construction in Example 11.4, so $\mu \vee \nu$ obeys (11.9) for all $f \in C(A)$, that is, $\mu \vee \nu \in W$. For general $\mu, \nu \in W$,

$$\mu \vee \nu = (|\mu| + |\nu| + \mu) \vee (|\mu| + |\nu| + \nu) - (|\mu| + |\nu|) \tag{11.10}$$

is also in $W$. In (11.10), $|\mu| = \mu \vee 0 + (-\mu) \vee 0$ lies in $W$ by the special case of positive measures in $W$.

Since $\mathcal{M}_{+,1}^{\max}(A)$ is a base of $\mathcal{M}_{+,1}(A) \cap W$, it is an algebraic simplex. □

*Remarks* 1. As we will see, $\mathcal{M}_{+,1}^{\max}$ may not be closed, so it may not be a Choquet simplex.

2. The reason $\mathcal{M}_{+,1}^{\max}(A)$ may not be closed is that $\hat{f}$ may not be continuous. We will see below that, in general, $\hat{f}$ is not continuous if $\mathcal{E}(A)$ is not closed.

*Proof of* (iv) $\Rightarrow$ (i) *in Theorem 11.5* If each $x$ is the barycenter of a unique maximal measure, $r\colon \mathcal{M}_{+,1}^{\max}(A) \to A$ is one-one and it is onto by Theorem 10.2. It is clearly affine, so $A$ and $\mathcal{M}_{+,1}^{\max}(A)$ are affinely (algebraically) homomorphic. It follows that $A$ is an algebraic simplex. Since it is compact, it is a Choquet simplex. □

The next step requires an interesting general result about $\hat{f}$:

**Proposition 11.8** *For any $f \in C(A)$ and $x \in A$,*

$$\hat{f}(x) = \sup\{\mu(f) \mid r(\mu) = x\} \tag{11.11}$$

*Proof* We first show that for any $\mu$ and $f \in C(A)$,

$$\hat{f}(r(\mu)) \geq \mu(f) \tag{11.12}$$

If $\mu = \sum_{i=1}^m \theta_1 \delta_{x_i}$, (11.12) is just the assertion that $\hat{f}$ is concave and $\hat{f} \geq f$. Given any $\mu$, we can find a net $\mu_\alpha \to \mu$ in the weak-$*$ (i.e., $\sigma(\mathcal{M}(A), C(A))$) topology (e.g., by the Krein–Milman theorem). Since $r$ is continuous, $r(\mu_\alpha) \to r(\mu)$ in $A$ and, since $\hat{f}$ is usc,

$$\begin{aligned} \hat{f}(r(\mu)) &\geq \limsup \hat{f}(r(\mu_\alpha)) \\ &\geq \limsup \mu_\alpha(f) \qquad \text{(by (11.12) for point measures)} \\ &= \mu(f) \end{aligned}$$

so (11.12) is proven.

On the other hand, by arguments similar to those proving Theorem 5.17,

$$\{(x, \lambda) \in A \times \mathbb{R} \mid \lambda \leq \hat{f}(x)\} = \text{cch}\{(x, \lambda) \in A \times \mathbb{R} \mid \lambda \leq f(x)\} \tag{11.13}$$

Thus, given $x_0 \in A$, we can find a net $p_\alpha \in \text{cch}\{(x, \lambda) \mid \lambda \leq f(x)\}$ so $p_\alpha \to (x_0, \hat{f}(x_0))$, that is,

$$p_\alpha = \sum_{j=1}^{N(\alpha)} \theta_{\alpha,j}(x_{\alpha,j}, \lambda_{\alpha,j}) \tag{11.14}$$

with

$$\lambda_{\alpha,j} \leq f(x_{\alpha,j}) \tag{11.15}$$

$$\sum_{j=1}^{N(\alpha)} \theta_{\alpha,j} x_{\alpha,j} \to x_0 \tag{11.16}$$

$$\sum_{j=1}^{N(\alpha)} \theta_{\alpha,j} \lambda_{\alpha,j} \to \hat{f}(x) \tag{11.17}$$

Consider the measures

$$\mu_\alpha = \sum_{j=1}^{N(\alpha)} \theta_{\alpha,j} \delta_{x_{\alpha,j}} \tag{11.18}$$

By compactness, we can pass to a subnet so $\mu_\alpha \to \mu$. By (11.16), $r(\mu_\alpha) \to x_0$, so

$$r(\mu) = x_0 \tag{11.19}$$

Since $\mu_\alpha \to \mu$ (after passing to the subnet),

$$\begin{aligned} \mu(f) &= \lim \mu_\alpha(f) \\ &\geq \limsup \sum_{j=1}^{N(\alpha)} \theta_{\alpha,j} \lambda_{\alpha,j} \qquad \text{(by (11.15))} \\ &= \hat{f}(x_0) \qquad \text{(by (11.17))} \end{aligned}$$

so by (11.19), for any $x_0$,

$$\hat{f}(x_0) \le \sup\{\mu(f) \mid r(\mu) = x_0\}$$

This and (11.12) prove (11.11). □

We know the point measures in $\mathcal{M}_{+,1}(X)$ are weak-$*$ dense in $\mathcal{M}_{+,1}(X)$. We need to show this remains true if we require the approximations to have the same barycenter as the limit.

**Proposition 11.9** *Let $x_0 \in A$ and define*

$$\mathcal{M}_{+,1}^{x_0} = \{\mu \in \mathcal{M}_{+,1} \mid r(\mu) = x_0\}$$

*Then the finite convex combinations of point masses in $\mathcal{M}_{+,1}^{x_0}$ are dense (in the $\sigma(\mathcal{M}(A), C(A))$-topology) in $\mathcal{M}_{+,1}^{x_0}$.*

*Proof* Fix $\mu \in \mathcal{M}_{+,1}^{x_0}$. Let $U$ be a convex, balanced, open neighborhood of 0 in the underlying space. By compactness of $A$, find $y_1, \dots, y_\ell$ so that

$$\bigcup_{j=1}^{\ell} (y_j + U) = A \tag{11.20}$$

Let $\chi_j$ be the characteristic function of $y_j + U$ so $\sum_{j=1}^{\ell} \chi_j > 0$ on $A$ and define

$$g_k = \frac{\chi_k}{\sum_{j=1}^{\ell} \chi_j}$$

so

$$\sum_{j=1}^{\ell} g_k = 1 \tag{11.21}$$

Let $\theta_j = \mu(g_k)$, $\mu_j(f) = \mu(g_j f)/\mu(g_j)$, and $x_j = r(\mu_j)$.

Since $y_j + U$ is convex and $\mu_j$ is a measure on $\overline{y_j + U}$, we have

$$x_j \in y_j + U$$

and, in particular, for any $f \in C(A)$,

$$|f(x_j) - \mu_j(f_j)| \le \sup_{\substack{x-y \in \bar{U} \\ x,y \in A}} |f(x) - f(y)| \tag{11.22}$$

For any $L \in X^*$,

$$\begin{aligned} L\biggl(\sum_{j=1}^{\ell} \theta_j x_j\biggr) &= \sum_{j=1}^{\ell} \theta_j L(r(\mu_j)) \\ &= \sum_{j=1}^{\ell} \theta_j \mu_j(L) \qquad \text{(by definition of } r) \end{aligned}$$

$$
\begin{aligned}
&= \sum_{j=1}^{\ell} \mu(g_j L) && \text{(by definition of } \theta_j \text{ and } \mu_j\text{)} \\
&= \mu(L) && \text{(by (11.21))} \\
&= L(x_0)
\end{aligned}
$$

since $\mu \in \mathcal{M}_{+,1}^{x_0}$, so since $X^*$ separates points,

$$\sum_{j=1}^{\ell} \theta_j x_j = x_0 \tag{11.23}$$

We have suppressed $U$ so far but make it explicit in the notation for

$$\nu_U = \sum_{j=1}^{\ell} \theta_j \delta_{x_j}$$

By (11.23), $\nu_U \in \mathcal{M}_{+,1}^{x_0}$. By (11.22) and $\mu = \sum_{j=1}^{\ell} \theta_j \mu_j$,

$$|\nu_U(f) - \mu(f)| \le \sup_{\substack{x-y\in\bar{U} \\ x,y\in A}} |f(x) - f(y)| \tag{11.24}$$

Think of $\nu_U$ as a net of measures in $\mathcal{M}_{+,1}^{x_0}$ ordered by taking $U_1 > U_2$ if $U_1 \subset U_2$. By compactness of $A$, the right side of (11.24) converges to zero in the net index $U$ so $\nu_U \to \mu$ weakly. □

*Remark* The above proof is correct if $A$ is separable, but if not, there are issues of Baire vs. Borel sets we have ignored. We make similar fudges throughout. In this case, we need only take $U$ so $(U + y) \cap A$ is Baire. Readers should either assume $A$ is always separable or else provide the Baire vs. Borel details themselves.

**Corollary 11.10** *Extend $f$ and $\hat{f}$ from functions on $A$ to functions on $A_{\text{sus}}$ by*

$$g((\lambda x, \lambda)) = \lambda g(x) \tag{11.25}$$

*for $g = f$ or $\hat{f}$. Then for any $z \in A_{\text{sus}}$,*

$$\hat{f}(z) = \lim_{n\to\infty} \sup\{f(z_1) + \cdots + f(z_n) \mid z_i \in A_{\text{sus}},\ z_1 + \cdots + z_n = z\} \tag{11.26}$$

*Proof* By scaling (i.e., changing $\lambda$), we can suppose $z = (x_0, 1)$ with $x_0 \in A$, in which case $z_i = (\theta_i x_i, \theta_i)$ with $\theta_i \ge 0$, $\sum_{i=1}^{n} \theta_i = 1$, $\sum_{i=1}^{n} \theta_i x_i = x_0$, and

$$f(z_i) + \cdots + f(z_n) = \sum_{i=n}^{n} \theta_i f(x_i)$$

so (11.26) is equivalent to

$$\hat{f}(x_0) = \sup\{\mu(f) \mid \mu \in \mathcal{M}_{+,1}^{x_0},\ \mu \text{ a finite point measure}\} \tag{11.27}$$

By Proposition 11.9, this sup is the same as $\sup\{\mu(f) \mid \mu \in \mathcal{M}_{+,1}^{x_0}\}$ and then (11.27) is just (11.11). □

Simplexes enter because of the following decomposition lemma:

**Proposition 11.11** *Let $V$ be a vector lattice. Suppose $\{x_i\}_{i=1}^n$ and $\{y_j\}_{j=1}^m$ are positive elements with*

$$\sum_{i=1}^n x_i = \sum_{j=1}^m y_j \tag{11.28}$$

*Then there exist positive elements $\{w_{ij}\}_{1\leq i\leq n;\,1\leq j\leq m}$ so*

$$\sum_{i=1}^n w_{ij} = y_j \tag{11.29}$$

*and*

$$\sum_{j=1}^m w_{ij} = x_i \tag{11.30}$$

*Proof* Consider first the case $n = m = 2$. Let

$$w_{11} = x_1 \wedge y_1, \qquad w_{12} = x_1 - w_{11}$$
$$w_{21} = y_1 - w_{11}, \qquad w_{22} = x_2 - y_1 + w_{11}$$

Since (11.28) holds, $w_{22} = y_2 - x_1 + w_{11}$, so (11.29) and (11.30) hold. Since 0 is a lower bound for $x_1$ and $y_1$, $w_{11} \rhd 0$ and since $x_1 \rhd w_{11}$, $y_1 \rhd w_{11}$, we have $w_{12} \rhd 0$, $w_{21} \rhd 0$. As for $w_{22}$, since the order is compatible with addition,

$$\begin{aligned} w_{22} &= x_2 - y_1 + w_{11} \\ &= (x_2 - y_1 + x_1) \wedge (x_2 - y_1 + y_1) \\ &= (y_1 + y_2 - y_1) \wedge (x_2) = y_2 \wedge x_2 \end{aligned}$$

which is also positive.

Now we use this special case and a double induction. Suppose we have the result for $n = 2$ and some $m_0 \geq 2$. Given $y_1, \dots, y_{m_0+1}$ and $x_1, x_2$ all positive, let $\tilde{y}_1 = \sum_{j=1}^{m_0} y_j$, $\tilde{y}_2 = y_{m_0+1}$. By the special case, find $\{\tilde{w}_{ij}\}_{i\leq 1,\, j\leq 2}$ so (11.29) and (11.30) hold for $\tilde{w}, \tilde{y}_j, x_i$ (and $n = m = 2$). Then

$$\tilde{w}_{11} + \tilde{w}_{21} = \sum_{j=1}^{m_0} y_j$$

which allows, by the induction assumption, a further decomposition, proving the result for $n = 2$ and $m = m_0 + 1$. After that, do a similar induction in $n$. □

*Completion of the proof of Theorem 11.5* (i) ⇒ (ii) Suppose $A$ is a simplex and $f \in \mathcal{C}(A)$. Use (11.25) to extend $f$ and $\hat{f}$ to $A_{\text{sus}}$. Since $\hat{f}$ is concave and homogeneous of degree 1 on $A_{\text{sus}}$, we have for $x_1, x_2 \in A_{\text{sus}}$,

$$\hat{f}(x_1 + x_2) \geq \hat{f}(x_1) + \hat{f}(x_2) \tag{11.31}$$

and since $f$ is convex, we similarly have

$$f(x_1 + x_2) \leq f(x_1) + f(x_2) \tag{11.32}$$

On the other hand, by (11.26) (all $z$'s and $x$'s in $A_{\text{sus}}$)

$$\begin{aligned}
&\hat{f}(x_1 + x_2) \\
&\quad = \lim_{n\to\infty} \sup\{f(z_1) + \cdots + f(z_n) \mid z_1 + \cdots + z_n = x_1 + x_2\} \\
&\quad = \lim_{n\to\infty} \sup\left\{ f(w_{11} + w_{21}) + \cdots + f(w_{1n} + w_{2n}) \,\middle|\, \sum_{j=1}^{n} w_{ij} = x_i \right\} \\
&\qquad\qquad\qquad\qquad\qquad\qquad\qquad\qquad \text{(by Proposition 11.11)} \\
&\quad \leq \lim_{n\to\infty} \sup\left\{ \sum_{j=1}^{n} f(w_{1j}) + \sum_{j=1}^{n} f(w_{2j}) \,\middle|\, \sum_{j=1}^{n} w_{ij} = x_i \right\} \qquad \text{(by (11.32))} \\
&\quad \leq \lim_{n\to\infty} \sum_{i=1}^{2} \sup\left\{ \sum_{j=1}^{n} f(w_{ij}) \,\middle|\, \sum w_{ij} = x_i \right\} \\
&\quad = \hat{f}(x_1) + \hat{f}(x_2) \quad \text{(by (11.26))}
\end{aligned}$$

This result and (11.31) imply that $\hat{f}$ is additive on $A_{\text{sus}}$ and so affine on $A$.

(ii) ⇒ (iii) Let $x_0 = r(\mu)$. Suppose $f \in \mathcal{C}(A)$. By definition,

$$\hat{f}(x_0) \equiv \inf\{g(x_0) \mid g \in -\mathcal{C}(A),\, g \geq f\}$$

Let $\nu \in \mathcal{M}_{+,1}^{x_0}$ be a finite convex combination of point measures. Since $\hat{f}$ is affine by (ii), the combination is finite and $r(\nu) = x_0$, so

$$\hat{f}(x_0) = \nu(\hat{f})$$

Suppose $g \in -\mathcal{C}(A)$ and $g \geq f$. Since $g \geq \hat{f}$,

$$\nu(\hat{f}) \leq \nu(g)$$

Since $g$ is concave, the combination is finite, and $r(\nu)$ is $x_0$,

$$\nu(g) \leq g(x_0)$$

Thus, we have proven

$$\hat{f}(x_0) \leq \nu(g) \leq g(x_0) \tag{11.33}$$

By Proposition 11.9, $\mu$ is a $\sigma(\mathcal{M}, C)$ limit of such $\nu$, so since $g \in C(A)$, (11.33) implies

$$\hat{f}(x_0) \leq \mu(g) \leq g(x_0) \tag{11.34}$$

Now think of the $g$'s as a decreasing net whose monotone limit is $\hat{f}$. By the monotone convergence theorem for nets, $\mu(g) \to \mu(\hat{f})$ and by definition of $\hat{f}$, $g(x_0) \to \hat{f}(x_0)$. Thus, for any $\mu \in \mathcal{M}_{+,1}(X)$ and $f \in \mathcal{C}(A)$,

$$\mu(\hat{f}) = \hat{f}(r(\mu))$$

By Theorem 10.5, if $\mu$ is maximal, $\mu(\hat{f}) = \mu(f)$ so (11.7) holds.

(iii) $\Rightarrow$ (iv) Let $\mu, \nu$ be two measures with some barycenter $x_0$. By (11.7),

$$\mu(f) = \nu(f)$$

for $f \in \mathcal{C}(A)$. By linearity, that is true for $f \in \mathcal{C}(A) - \mathcal{C}(A)$. Since that is dense in $C(A)$ (Proposition 10.1), $\mu = \nu$. □

Combining Theorems 11.5 and 10.8, we have

**Theorem 11.12** *Let $A$ be a metrizable, compact convex subset of a locally convex space. Then for each $x \in A$, there is a unique probability measure $\mu$ with barycenter $x$ obeying $\mu(\mathcal{E}(A)) = 1$ if and only if $A$ is a simplex.*

One can ask if $A$ is a simplex, when is the map from $A$ to $\mathcal{M}_{+,1}(A)$ that takes $x$ into its unique representing maximal measure continuous? Here is the answer:

**Theorem 11.13** *Let $A$ be a Choquet simplex. Then the following are equivalent:*

(i) *$\mathcal{E}(A)$ is closed.*
(ii) *The map $m\colon A \to \mathcal{M}_{+,1}(A)$ that takes $x$ into the unique maximal measure $\mu_x$ with $r(\mu_x) = x$ is continuous.*
(iii) *The family of maximal measures is closed in the $\sigma(\mathcal{M}(A), C(A))$-topology.*
(iv) *For all $f \in \mathcal{C}(A)$, $\hat{f}$ is continuous.*

*Remark* A simplex with one and, hence all, of these properties is called a *Bauer simplex.*

*Proof* We will show (i) $\Rightarrow$ (ii) $\Leftrightarrow$ (iii) $\Rightarrow$ (iv) $\Rightarrow$ (i).

(i) $\Rightarrow$ (ii) Let $\mathcal{E}(A)$ be closed and $\mu \in \mathcal{M}_{+,1}(\mathcal{E}(A))$. Since $\mu(\mathcal{E}(A)) = 1$, (10.11) and Theorem 10.5 imply that each $\mu$ is maximal. If there is a unique maximal measure for each $x \in A$, $r\colon \mathcal{M}_{+,1}(\mathcal{E}(A)) \to A$ is one-one and, by the Strong Krein–Milman theorem, it is onto. As a continuous bijection between compact sets, its inverse map, $m$, is continuous.

(ii) $\Rightarrow$ (iii) If $m$ is continuous, its image as the continuous image of a compact set is closed.

(iii) $\Rightarrow$ (ii) This repeats the argument at the end of (i) $\Rightarrow$ (ii). If the set of maximal measures is closed, it is compact, and if $A$ is a simplex, the map from maximal measures to $A$ is a bijection, so it has a continuous inverse.

(ii) $\Rightarrow$ (iv) (11.7), which holds for $f \in \mathcal{C}(A)$ and $\mu$ maximal, can be rewritten as

$$\hat{f}(x) = [m(x)](f) \tag{11.35}$$

If $m$ is continuous from $A$ to $\mathcal{M}_{+,1}$ in the weak-$*$ topology, (11.35) shows $\hat{f}$ is continuous.

(iv) $\Rightarrow$ (i) By (10.12), $\mathcal{E}(A) = \cap_{f\in\mathcal{C}(A)}\{x \mid \hat{f}(x) = f(x)\}$. If $\hat{f}$ is continuous, $\{x \mid \hat{f}(x) = f(x)\}$ is closed, so $\mathcal{E}(A)$ is an intersection of closed sets. □

Let $A$ be a finite-dimensional compact convex set which is not isomorphic to a $\Delta_n$. Then, by Theorem 11.1, representations by extreme points are not unique, so $A$ is not a Choquet simplex, that is,

**Theorem 11.14** *Let $A$ be a finite-dimensional, compact convex set. Then the order induced by $A$ is a lattice if and only if $A$ is isomorphic to a standard simplex, $\Delta_n$.*

**Example 11.15** (Three examples of Strong Krein–Milman with $\mathcal{E}(A)$ closed) In Chapter 9, we had three examples of the Strong Krein–Milman theorem with $\mathcal{E}(A)$ closed: the sets we called $\mathcal{B}_1$, $\mathcal{P}_1$, and $\mathcal{L}_1$. In all cases, we proved directly that the representing measures were unique. Thus, all three are Choquet simplexes where $\mathcal{E}(A)$ is closed, so (ii)–(iv) of Theorem 11.13 hold. Neither Example 9.5 nor Example 9.6 are simplexes. □

**Example 11.16** (Example 8.17 continued) Let $X$ be a compact Hausdorff space and $T$ a continuous bijection. The invariant measures $\mathcal{M}^I_{+,1}(T)$ were defined in Example 8.17. We claim they are a Choquet simplex. For let $\mathcal{M}^I(T) \subset \mathcal{M}(X)$ be the invariant signed measures. $\mathcal{M}(X)$ is a lattice. Moreover, we claim, if $\mu, \nu \in \mathcal{M}^I(T)$, their $\mathcal{M}(X) - \sup \mu \vee \nu$ is also in $\mathcal{M}^I(T)$ whence $\mathcal{M}^I_{+,1}(T)$ is a simplex by Proposition 11.6.

To see this, suppose $\mu, \nu \geq 0$. Then $\gamma = \mu + \nu$ is invariant so $d\mu = f\,d\gamma$ and $d\nu = g\,d\nu$ have $f$ and $g$ invariant. Hence, their pointwise sup is invariant, so $\mu \vee \nu$ is invariant. □

**Example 11.17** (Example 9.7, the Poulsen simplex, continued) As a set of invariant measures, the Poulsen simplex is a simplex (as the name suggests). But $\mathcal{E}(A)$ is dense, as we showed in Chapter 9. This simplex has the paradoxical property that an $x \notin \mathcal{E}(A)$ can be represented as the barycenter of only one measure with $\mu(\mathcal{E}(A)) = 1$ (since this $A$ is metrizable), but $\delta_x$ is also a limit of pure point measure $\delta_{x_n}$ with $x_n \in \mathcal{E}(A)$! Since $\mathcal{E}(A)$ is dense, *every* continuous, convex, nonaffine function on $A$ must have $\hat{f}$ discontinuous, because if $f$ is not affine, $f \neq \hat{f}$ (since $f$ is convex and $\hat{f}$ is concave) even though $f = \hat{f}$ on $\mathcal{E}(A)$. We will discuss some of the further interesting properties of the Poulsen simplex in the Notes. □

# 12

# Complex interpolation

We have already seen that convexity is behind a number of important inequalities, including Minkowski's, Hölder's, and Jensen's. In the next five chapters, we explore this notion further and explore a number of themes that relate convexity and concavity to inequalities. This short chapter – the only one where analytic functions are key – provides some convexity inequalities for such functions and applies this idea.

**Theorem 12.1** (Hadamard Three-Circle Theorem) *Let $f$ be a function analytic in the annulus $A_{r_0,r_1} = \{z \mid r_0 < |z| < r_1\}$, continuous on $\bar{A}_{r_0,r_1}$. Let*

$$M_j = \sup_{\theta\in[0,2\pi)} |f(r_j e^{i\theta})| \tag{12.1}$$

*Then*

$$\sup|f(z)| \le M_0^{1-\eta} M_1^{\eta} \tag{12.2}$$

$$\eta(z) = \frac{[\log|z| - \log r_0]}{[\log r_1 - \log r_0]} \tag{12.3}$$

*or equivalently, if*

$$M(r; f) = \sup_{\theta\in[0,2\pi]} |f(re^{i\theta})|$$

*then* $\log M(r; f)$ *is a convex function of* $\log r$.

*Remark* This result is most simply proven using subharmonic functions and the maximum principle on the subharmonic function

$$\log|f(z)| - (1 - \eta(z)) \log M_0 - \eta(z) \log M_1$$

Because we only want to use the maximum principle for analytic functions, we use the trick below of proving it first for rational $\alpha$. We need also only require $|f(z)|$ have continuous boundary values if we use subharmonic functions.

*Proof* Define $\alpha \in \mathbb{R}$ by

$$\left(\frac{r_0}{r_1}\right)^{\alpha} = \frac{M_1}{M_0} \tag{12.4}$$

Suppose first $\alpha = m/n$ is rational with $n > 0$ and $m \in \mathbb{Z}$. Define for $r_0^{1/n} \leq |w| \leq r_1^{1/n}$,

$$g(w) = w^n f(w^n)$$

which is analytic in the annulus $A_{r_0^{1/n}, r_1^{1/n}}$ even if $m < 0$. Moreover, (12.4) implies

$$M_0 r_0^{m/n} = M_1 r_1^{m/n} \tag{12.5}$$

so, by the maximum principle,

$$|g(w)| \leq M_0 r_0^{m/n} \tag{12.6}$$

on $A_{r_0^{1/n}, r_1^{1/n}}$ or

$$\begin{aligned} |f(z)| \leq M_0 \left(\frac{r_0}{|z|}\right)^{-\alpha} &= M_1 \left(\frac{r_1}{|z|}\right)^{-\alpha} \\ &= M_0^{1-\eta(z)} M_1^{\eta(z)} \end{aligned} \tag{12.7}$$

since $r_0^{1-\eta} r_1^{\eta} = |z|$.

For general $\alpha$, pick $\alpha_\ell \uparrow \alpha$, with $\alpha_\ell$ rational, and define $M_1^\ell$ by

$$\left(\frac{r_0}{r_1}\right)^{\alpha_\ell} = \frac{M_1^\ell}{M_0}$$

Since $\alpha_\ell < \alpha$ and $r_0 < r_1$, $M_1^\ell > M_1$ so $|f(r_1 e^{i\theta})| \leq M_1^\ell$ and we conclude (12.7) holds for $M_1^\ell$ in place of $M_1$. But $\ell$ is arbitrary and $M_1^\ell \downarrow M_1$ as $\ell \to \infty$, so (12.7) holds for this irrational $\alpha$ case also. □

The following can be viewed as a kind of limit of the three-circle theorem as one circle goes to infinity.

**Theorem 12.2** (Bernstein's Lemma) *Let $P_n(z)$ be a polynomial of degree $n$. Let $R > 1$. Then*

$$\sup_{|z|=R} |P_n(z)| \leq R^n \sup_{|z|=1} |P_n(z)| \tag{12.8}$$

*Proof* Define the reversed polynomial $P_n^*(z) = z^n \overline{P_n(1/\bar{z})}$, so if $P_n(z) = c_n z^n + c_{n-1} z^{n-1} + \cdots + c_0$, then $P_n^*(z) = \bar{c}_0 z^n + \bar{c}_1 z^{n-1} + \cdots + c_n$. By the maximum modulus principle,

$$\sup_{|z|=R^{-1}} |P_n^*(z)| \leq \sup_{|z|=1} |P_n^*(z)| \tag{12.9}$$

Since

$$|P_n(re^{i\theta})| = r^n |P_n^*(r^{-1}e^{i\theta})|$$

(12.9) implies (12.8). □

In extending this to the three-line theorem, we need to deal with the fact that the strip is unbounded. The idea we use can be extended and systematized to the Phragmén–Lindelöf principle.

**Theorem 12.3** (Hadamard Three-Line Theorem) *Let $f$ be a function analytic on*

$$S = \{z \mid 0 < \operatorname{Re} z < 1\} \tag{12.10}$$

*with $f(z)$ continuous on $\bar{S}$. Suppose for each $\varepsilon > 0$, there is a $C_\varepsilon$ so that on $\bar{S}$,*

$$|f(z)| \le C_\varepsilon \exp(\varepsilon |z|^2) \tag{12.11}$$

*Suppose that*

$$M_j = \sup_{y \in \mathbb{R}} |f(j+iy)| \tag{12.12}$$

*is finite for $j = 0, 1$. Then $|f|$ is bounded on $\bar{S}$ and*

$$M(x) = \sup_y |f(x+iy)| \tag{12.13}$$

*obeys*

$$M(x) \le M_0^{1-x} M_1^x \tag{12.14}$$

*that is, $\log M(x)$ is a convex function of $x$.*

*Remark* One can do better than the bound (12.11), but the example $f(z) = \exp(i \exp(i\pi z))$, with $M_0 = M_1 = 1$ but $f$ unbounded, shows some a priori bound on $f$ is needed.

*Proof* Let

$$\tilde{f}(z) = f(z) \left( \frac{M_1}{M_0} \right)^{-z} M_0^{-1}$$

(where $(M_1/M_0)^{-z} = \exp(-z[\log M_1 - \log M_0])$ is entire). Then $M_0(\tilde{f}) = 1 = M_1(\tilde{f})$ and (12.14) for $\tilde{f}$, which says $\tilde{f} \le 1$, implies the result for $f$ since

$$M(x; \tilde{f}) = M(x, f) M_1^{-x} M_0^{x-1}$$

Thus, it suffices to prove the result under the hypothesis $M_0 = M_1 = 1$.

In that case, let

$$g_\varepsilon(z) = e^{\varepsilon z^2} f(z)$$

Since

$$\sup_{0<x<1} e^{\varepsilon(x+iy)^2} = e^{\varepsilon} e^{-\varepsilon y^2} \tag{12.15}$$

goes to zero as $y \to \infty$, by (12.11), for any $\varepsilon > 0$, $|g_\varepsilon(z)| \to \infty$ as $|z| \to \infty$ in the strip, $\tilde{S}$. Thus, surrounding any $x$ in $\tilde{S}$ by a very large rectangle, we can suppose $|g_\varepsilon(z)| \le 1$ on the top and bottom sides. Thus,

$$\begin{aligned} \sup_{z\in\tilde{S}} |g_\varepsilon(z)| &\le \max_{\substack{\operatorname{Re} x=0 \\ \text{or } Rx=1}} |g_\varepsilon(z)| \\ &\le e^{\varepsilon} \end{aligned}$$

by (12.15). Taking $\varepsilon \downarrow 0$, we see that $M(x) \le 1$ and so, in general, (12.14) holds. □

One of the most interesting and powerful applications of the three-line theorem is the Stein Interpolation Theorem. Let $(M, d\mu)$ be a measure space. Let $\mathcal{S}(M, d\mu)$ be the simple functions on $M$, that is, finite linear combinations of characteristic functions of finite measure. $\mathcal{S}$ is a vector space dense in each $L^p(M, d\mu)$; $1 \le p < \infty$.

By a linear map $T$ of $\mathcal{S}$ to $\mathcal{S}^*$, we mean a bilinear form $B_T : \mathcal{S} \times \mathcal{S} \to \mathbb{C}$ which we write as $B_T(f, g) = \langle f, Tg\rangle$. Note that unlike a Hilbert space inner product, $f \mapsto \langle f, Tg\rangle$ is linear, not antilinear.

For $p, q \in [1, \infty]$, we say $T$ is bounded from $L^p$ to $L^q$ if and only if there is a $C$ with

$$|\langle f, Tg\rangle| \le C\|f\|_{q'}\|g\|_p \tag{12.16}$$

with $q' = (1 - q^{-1})^{-1}$ the dual index to $q$. The minimal value of $C$ in (12.17) is written $\|T\|_{p,q}$, that is,

$$\|T\|_{p,q} = \sup\{|\langle f, Tg\rangle| \mid \|f\|_{q'} = \|g\|_p = 1\} \tag{12.17}$$

**Theorem 12.4** (Stein Interpolation Theorem) *Suppose for each $z \in \bar{S}$, we have a map $T(z)$ from $\mathcal{S} \to \mathcal{S}^*$ so that for each $f, g \in \mathcal{S}$, $z \mapsto \langle f, T(z)g\rangle$ is analytic in $S$, continuous on $\bar{S}$. Suppose that for some $p_0$, $q_0$, $p_1$, $q_1$, $M_0$, and $M_1$,*

$$\sup_y \|T(iy)\|_{p_0,q_0} \le M_0 \tag{12.18}$$

$$\sup_y \|T(1+iy)\|_{p_1,q_1} \le M_1 \tag{12.19}$$

*Moreover, suppose for any $A, B \subset M$ with finite measure, we have*

$$\sup_{z\in\bar{S}} |\langle \chi_A, T(z)\chi_B\rangle| < \infty \tag{12.20}$$

*where $\chi_A$ is the characteristic function of $A$.*

*Define*

$$p_t = tp_1^{-1} + (1-t)p_0^{-1} \tag{12.21}$$

$$q_t = tq_1^{-1} + (1-t)q_0^{-1} \tag{12.22}$$

$$M_t = M_1^t M_0^{(1-t)} \tag{12.23}$$

*Then for any* $z = x + iy \in S$, $T(z)$ *is bounded from* $L^{p_x} \to L^{q_x}$ *and*

$$\|T(x+iy)\|_{p_x,q_x} \le M_x \tag{12.24}$$

*Proof* Given $f \in S$, define $u_f$ by

$$u_f(m) = \begin{cases} f(m)/|f(m)|, & \text{if } |f(m)| \ne 0 \\ 0, & \text{if } |f(m)| = 0 \end{cases}$$

Define

$$G(z) = \langle (|f|^{a(z)} u_f), T(z)(|g|^{b(z)} u_g) \rangle \tag{12.25}$$

where $z = x + iy$ and

$$a(z) = z(q_1')^{-1} + (1-z)(q_0')^{-1} \tag{12.26}$$

$$b(z) = zp_1^{-1} + (1-z)p_0^{-1} \tag{12.27}$$

Bearing in mind that *f* and *g* are finite linear combinations of characteristic functions of finite measure and that (12.20) is assumed, we can apply the three-line theorem to $G(z)$.

Since $||f(z)|^{a(x+iy)}| = |f(z)|^{a(x)}$, we see $\| \, |f|^{a(x+iy)} u_f \|_{q_x'} = \|f\|_1^{1/q_x'}$ and similarly, $\| \, |g|^{b(x+iy)} u_g \|_{p_x} = \|g\|_1^{1/p_x}$. The three-line theorem then implies

$$\sup_y |G(x+iy)| \le M_0^{1-x} M_1^x \| \, |f|^{a(x)} \|_{q_x'} \| \, |g|^{b(x)} \|_{p_x}$$

Since $|f|^{a(x)} u_f$ runs through all of $\mathcal{S}$ as $f$ runs through all of $\mathcal{S}$, we see that (12.24) holds. □

The most important special case of this result predates and motivated it.

**Theorem 12.5** (Riesz–Thorin Interpolation Theorem) *Let* $p_0$, $q_0$, $p_1$, $q_1$ *be given and let* $T\colon L^{p_0} \cap L^{p_1} \to L^{q_0} \cap L^{q_1}$ *with*

$$\|Tf\|_{q_0} \le M_0 \|f\|_{p_0}, \qquad \|Tf\|_{q_1} \le M_1 \|f\|_{q_1} \tag{12.28}$$

*Then for* $p_t, q_t, M_t$ *given by* (12.21), (12.22), *and* (12.23),

$$\|Tf\|_{q_t} \le M_t \|f\|_{p_t} \tag{12.29}$$

*Proof* Take $T(z) \equiv T$. □

Here is a classic application of the Riesz–Thorin theorem:

**Theorem 12.6** (Young's Inequality) *Let $f, g \in L^1(\mathbb{R}^\nu) \cap L^\infty(\mathbb{R}^\nu)$. Then for any $p, q, r$ with*

$$r^{-1} = p^{-1} + q^{-1} - 1 \tag{12.30}$$

*we have*

$$\|f * g\|_r \le \|f\|_p \, \|g\|_q \tag{12.31}$$

*Remarks* 1. Once one has this, we can extend $*$ to a bounded bilinear map from $L^p \times L^q$ to $L^r$. One can even use it for $|f|, |g|$ to prove the integral defining convolution converges absolutely for a.e. $x$.

2. The constant (i.e., 1) in (12.31) is not optimal; see the Notes.

3. Below, all functions should be taken a priori in $L^1 \cap L^\infty$ to be sure integrals converge.

*Proof* Define $(T_x g)(y) = g(y-x)$. Let $f \in L^1$. Then by Minkowski's inequality, for $g \in L^p$,

$$\begin{aligned} \|f * g\|_p &= \left\| \int f(x)(T_x g)\, dx \right\| \\ &\le \|f\|_1 \, \|g\|_p \end{aligned} \tag{12.32}$$

On the other hand, if $p'$ is the dual index to $p$, then by Hölder's inequality,

$$\|f * g\|_\infty \le \|f\|_{p'} \, \|g\|_p \tag{12.33}$$

Applying the Riesz–Thorin theorem to $T(f) = f * g$, noting that (12.30) is an affine map, $A(q^{-1})$ of $q^{-1}$ to $r^{-1}$ and $A(q^{-1} = 1) = p^{-1}$, $A(q^{-1} = p^{-1}) = \infty^{-1}$, we obtain (12.31). □

*Remark* (12.32) for $p = 1$ is just Fubini's theorem. (12.32) then holds for all $p$ by interpolation between $p = 1$ and $p = \infty$, that is, one can avoid Minkowski's theorem.

Interestingly enough, so long as one avoids endpoints, one can improve one factor for $L^p$ to $L^p_w$ by using some additional real variable interpolation theorem. Here, $L^p_w$ is the weak $L^p$ space of all functions, with $\|f\|^*_{p,w} < \infty$ finite, where

$$\|f\|^*_{p,w} = \sup_{0<t<\infty} t^{1/p} |\{x \mid |f(x)| < t\}| \tag{12.34}$$

Despite the symbol, $\| \; \|^*_{p,w}$ is not a norm, although for $1 < p < \infty$, it is equivalent to one.

**Theorem 12.7** (Generalized Young Inequalities) *Let $f \in L^p_w(\mathbb{R}^\nu)$ with $1 < p < \infty$. Then for $1 < q < p' = p(p-1)^{-1}$, $f*$ maps $L^q \to L^r$ where $r$ is given by* (12.30) *and*

$$\|f * g\|_r \le C\|f\|^*_{p,w}\,\|g\|_q \tag{12.35}$$

*Proof* First we use Hunt's Interpolation Theorem (Thm. IX.19 of [304]) to interpolate Young's inequality with $g$ fixed in $L^q$ between the pairs $(1, q)$ and $(q', \infty)$ (where $(s, t)$ means as a map of $L^s$ to $L^t$). The result is that (12.35) holds with $\|f * g\|^*_{r,w}$ in place of $\|f * g\|_r$. In this inequality, we have $r^{-1} < q^{-1}$ since $p > 1$ or $q < r$. As a result, the Marcinkiewicz interpolation (Thm. IX.18 of [304]) applies. This implies that (12.35) holds. □

A special case of this is so important that we specify it:

**Theorem 12.8** (Sobolev Inequalities) *Let $\sigma < \nu$ and let $s^{-1} + q^{-1} + \sigma\nu^{-1} = 2$ with $1 < q, s < \infty$. Then for $f \in L^q(\mathbb{R}^\nu)$ and $g \in L^s(\mathbb{R}^\nu)$,*

$$\iint \frac{|f(x)|\,|g(y)|}{|x-y|^\sigma}\, d^\nu x\, d^\nu y \le C\|f\|_q\,\|h\|_s \tag{12.36}$$

*Proof* $|x|^{-\sigma} \in L^p_w$ with $p = \nu/\sigma$. With $s = r'$, (12.36) then follows from (12.35). □

*Remarks* 1. We will see in Chapter 14 (see Theorem 14.23) that this special case actually implies the full Theorem 12.7!

2. These inequalities are important because they imply $L^p$ properties of Sobolev spaces, that is, spaces of functions $f$ with, say, $f$ and $\Delta f$ in some $L^q$. Because $\Delta^{-1}$ is convolution with $C_\nu|x-y|^{-\sigma}$ with $\sigma = \nu - 2$, (12.36) is relevant to knowing what $L^p$ spaces such $f$'s lie in; see the discussion in the Notes.

In a sense, Young's and Hölder's inequalities undo each other for convolution makes the $p$ in $L^p$ larger and multiplication makes it small. Suppose $1 \le p \le s \le \infty$ and $s' = s/(s-1)$ is the dual index to $s$. If $h \in L^p$ and $g \in L^{s'}$, then $g*h \in L^r$, where

$$r^{-1} = p^{-1} + (s')^{-1} - 1 = p^{-1} - s^{-1}$$

Thus, if $f \in L^s$, by Hölder's inequality, $f(g * h)$ is back in $L^p$, that is, if $p \le s$,

$$\|f(g * h)\|_p \le \|f\|_s\,\|g\|_{s'}\,\|h\|_p \tag{12.37}$$

The following is a strengthening of both this and the generalized Young inequality:

**Theorem 12.9** (Strichartz Inequality) *Let $1 < p < s < \infty$. Then*

$$\|f(g * h)\|_p \le C\|f\|^*_{s,w}\,\|g\|^*_{s',w}\,\|h\|_p \tag{12.38}$$

*for functions on $\mathbb{R}^\nu$ where $C$ depends only on $p$, $s$, and $\nu$.*

*Proof* We repeat the strategy that led to Theorem 12.7. Hölder's inequality and the generalized Young inequality imply

$$\|f(g*h)\|_{p_t} \leq C\|f\|_t \, \|g\|^*_{s',w} \, \|h\|_p$$

where $p_t^{-1} = t^{-1} + p^{-1} - s^{-1}$. Taking values of $t = s + \varepsilon$ and $s - \varepsilon$ for $\varepsilon$ small, and using Hunt's interpolation theorem (Thm. IX.19 of [304]), we obtain

$$\|f(g*h)\|^*_{p,w} \leq C\|f\|^*_{s,w} \, \|g\|^*_{s',w} \, \|h\|_p$$

For $f, g$ fixed, this holds for all $p < s$, so by the Marcinkiewicz theorem (Thm. IX.18 of [304]), we get (12.38). □

*Remark* We will see in Chapter 14 that the general case is implied by the special case $f(x) = |x|^{-\nu/s}$, $g(x) = |x|^{-[\nu-(\nu/s)]}$. For $p = 2$, we will discuss this special case with optimal constants in the Notes.

Here is an example of the Stein interpolation theorem. The example to keep in mind here is $P_t = e^{-t\Delta}$.

**Theorem 12.10** *Let $P_t$ be a strongly continuous self-adjoint semigroup of contractions on $L^2(M, d\mu)$ so that for all $1 \leq p \leq \infty$ and $f \in L^1 \cap L^\infty$,*

$$\|P_t\|_p \leq \|f\|_p \tag{12.39}$$

*Then as operators on $L^p$, $P_t$ can be analytically continued in t to the sector*

$$S^{(p)} = \left\{ t \;\middle|\; |\arg t| < \frac{\pi}{2}\left(1 - \left|\frac{2}{p} - 1\right|\right)\right\} \tag{12.40}$$

*obeying* (12.39) *there also.*

*Proof* By the spectral theorem, $P_t = e^{-tA}$ for a self-adjoint operator, so one can analytically continue $P_t$ to $S^{(2)}$ as $L^2$ operators by the functional calculus. Fix $\theta \in (-\frac{\pi}{2}, \frac{\pi}{2})$ and let

$$T(z) = \exp(-e^{iz\theta} A) \tag{12.41}$$

Since simple functions lie in $L^2$, we have that (12.20) holds. For $\operatorname{Re} z = 0$, note

$$T(iy) = \exp(-e^{-y\theta} A)$$

is a contraction from $L^1$ to $L^1$ by hypothesis. For $\operatorname{Re} z = 1$,

$$T(1 + iy) = \exp(-e^{i\theta} e^{-y\theta} A)$$

is a contraction on $L^2$. Thus, by the Stein interpolation theorem, $T(x + iy)$ is bounded on $L^{p_x}$ with $p_x = 2(2 - x)^{-1}$. As $\theta$ runs through all of $(-\frac{\pi}{2}, \frac{\pi}{2})$ and $y \in \mathbb{R}$, we obtain that $T(z)$ is uniformly bounded on $L^p$ for $t \in S^{(p)}$. Since $(f, T(z)g)$ is analytic in that sector for $f, g \in L^2$, by a limiting argument, the same is true for $g \in L^p$ and $f \in L^{p'}$. Duality then handles $p > 2$. □

As a parting issue where complex interpolation provides information, let us prove $x^\alpha$ for $0 \le \alpha \le 1$ is operator monotone without appealing to the Herglotz representation theorem and Example 6.8.

**Example 12.11** Let $A$ and $B$ be $n \times n$ strictly positive Hermitian matrices. Then, since $0 < A < B \Leftrightarrow 0 < C^{-1}AC^{-1} < C^{-1}BC^{-1}$ and $\|D^*D\| = \|D\|^2$, we have that

$$0 < A < B \Leftrightarrow \|A^{1/2}B^{-1/2}\| \le 1 \tag{12.42}$$

So suppose $0 < A < B$. Let

$$F(z) = A^{z/2}B^{-z/2}$$

on the strip $0 \le \operatorname{Re} z \le 1$. $F(z)$ is analytic and bounded on the strip. Moreover,

$$\|F(iy)\| = 1$$

since $A^{iy/2}B^{-iy/2}$ is unitary, and by (12.42),

$$\|F(1+iy)\| = \|F(1)\| \le 1$$

Thus, by the three-line theorem, $\|F(z)\| \le 1$ in the entire strip. In particular, for $0 \le \alpha \le 1$, we have $\|A^{\alpha/2}B^{-\alpha/2}\| \le 1$ which, by (12.42), implies

$$0 < A^\alpha < B^\alpha \tag{12.43}$$

□

# 13
# The Brunn–Minkowski inequalities and log concave functions

In this chapter, among other things, we will prove the isoperimetric inequality. The core will be some inequalities of which one of the first historically is

**Theorem 13.1** (Brunn–Minkowski Inequality) *Let $A_0, A_1$ be nonempty Borel sets in $\mathbb{R}^\nu$. Define for $\theta \in [0,1]$,*

$$A_\theta = \{\theta x_1 + (1-\theta)x_0 \mid x_0 \in A_0,\, x_1 \in A_1\} \tag{13.1}$$

*Then (with $|\cdot|$ = Lebesgue measure),*

$$|A_\theta|^{1/\nu} \geq \theta|A_1|^{1/\nu} + (1-\theta)|A_0|^{1/\nu} \tag{13.2}$$

*Remark* We will first prove this when $A_0$ and $A_1$ are open convex sets. We will defer the general proof until the end of the chapter. In the next chapter, we will only use it in this convex case, which was the first result. This theorem can be stated in terms of sums of sets, as we will note below (see Proposition 13.3).

We will defer the proof of this theorem, even in the convex case, not because it is difficult, but because it will follow from a more general result. We want to first show it implies the isoperimetric inequalities and then explain where the power $1/\nu$ comes from, at the same time reducing (13.2) to the apparently weaker

$$|A_\theta| \geq |A_1|^\theta\, |A_0|^{1-\theta} \tag{13.3}$$

This is weaker since

$$a^\theta b^{1-\theta} \leq \theta a + (1-\theta)b \tag{13.4}$$

on account of convexity of $e^x$ which says $\exp(\theta \log a + (1-\theta)\log b) \leq \theta a + (1-\theta)b$. Moreover, it has no explicit $\nu$, but still we will reduce (13.2) to (13.3).

Given an arbitrary measurable set $A$, we define its surface area by

$$s(A) = \liminf \big[|\{x \mid d(x,A) < \varepsilon\}| - |A|\big]/\varepsilon \tag{13.5}$$

If $A$ has a smooth boundary or is a polyhedron or in many other cases, the limit exists and agrees with an intuitive notion. One cannot argue that "lim inf" is better than "lim sup" – the isoperimetric inequality is stronger if we take lim inf, so we do. One might want to take $\frac{1}{2}|\{x \mid d(x, \partial A) < \varepsilon\}|$ where we take $|\{x \mid d(x, A) < \varepsilon\}| - |A|$ and that gives the same answer as $s(A)$ in reasonable cases like $\partial A$ smooth.

**Theorem 13.2** (Isoperimetric Inequality) *Let $A$ be a bounded measurable set in $\mathbb{R}^\nu$ and let $B$ be the open ball with the same volume as $A$. Then $s(A) \geq s(B)$.*

*Proof* Without loss (by scaling), suppose $A$ has the same volume as the unit ball, $B$. If $\lambda A = \{\lambda x \mid x \in A\}$,

$$|\lambda A| = \lambda^\nu |A| \tag{13.6}$$

and if

$$A^{(\varepsilon)} = \{x \mid d(x, A) < \varepsilon\} \tag{13.7}$$

then if $B$ is the unit ball,

$$\begin{aligned} A^{(\varepsilon)} &= A + \varepsilon B = (1+\varepsilon)[(1+\varepsilon)^{-1}A + \varepsilon(1+\varepsilon)^{-1}B] \\ &= (1+\varepsilon)A_{\theta(\varepsilon)} \end{aligned}$$

with $\theta(\varepsilon) = \varepsilon(1+\varepsilon)^{-1}$, $A_0 = A$ and $A_1 = B$, and $A_\theta$ is given by (13.1). Then by the assumption that $|A| = |B|$, the Brunn–Minkowski inequality, and (13.6),

$$|A^{(\varepsilon)}| \geq (1+\varepsilon)^\nu |B|$$

so that

$$s(A) \geq \nu |B| = s(B)$$

since in terms of polar coordinates, $s(B) = \int d\Omega$ and

$$|B| = \int_0^1 \left(\int d\Omega\right) r^{\nu-1}\, dr = \nu \int d\Omega = \nu s(B) \qquad \square$$

*Note* This application only needed the apparently weaker (13.3).

To understand why (13.2) has the power $\nu^{-1}$, let $A_0$ and $A_1$ be sets of positive measure and suppose $C_0 = \lambda A_0$ and $C_1 = \mu A_1$ where $\lambda$ and $\mu$ may be different. Then

$$\begin{aligned} C_\theta &= \theta C_1 + (1-\theta)C_0 \\ &= \theta\mu A_1 + (1-\theta)\lambda A_0 \\ &= [\theta\mu + (1-\theta)\lambda] A_{\varphi(\theta)} \end{aligned}$$

with

$$\varphi(\theta) = \frac{\theta\mu}{[\theta\mu + (1-\theta)\lambda]} \tag{13.8}$$

so, by (13.6),

$$|C_\theta|^{1/\nu} = [\theta\mu + (1-\theta)\lambda]\,|A_{\varphi(\theta)}|^{1/\nu} \tag{13.9}$$

As a result, (13.2) for $A_\varphi$ is equivalent to (13.2) for $C_\theta$. This shows first of all that (13.2) cannot hold if $1/\nu$ is replaced by any larger power, $\eta$, for take $A_0 = A_1$ in which case (13.2) is automatic for $A$ but only holds for $C$ if for all $\theta$,

$$[\theta\mu + (1-\theta)\lambda]^{\eta\nu} \geq \theta\mu^{\eta\nu} + (1-\theta)\lambda^{\eta\nu} \tag{13.10}$$

This is only true if $\eta\nu \leq 1$ since $x \mapsto x^\beta$ is strictly convex if $\beta > 1$ and the inequality in (13.10) is reversed. This convexity also implies that the inequality becomes stronger as $\eta$ gets larger, so $1/\nu$ is the optimal choice in (13.2).

The equivalence of (13.2) for $A_\varphi$ and $C_\theta$ has another consequence. Since the sets are assumed open, $|A_0| \neq 0 \neq |A_1|$. Picking $\lambda = |A_0|^{-1/\nu}$ and $\mu = |A_1|^{-1/\nu}$, we see $|C_0| = |C_1|$, so (13.2) need only be proven in the special case $|C_0| = |C_1| = 1$, in which case it says $|C_\theta| \geq 1$, that is, (13.2) is implied by

$$|C_0| = |C_1| = 1 \Rightarrow |C_\theta| \geq 1 \tag{13.11}$$

and this is implied by (13.3). Thus, we havc the first part of

**Proposition 13.3** (i) *Brunn–Minkowksi is implied by* (13.3).
(ii) *Brunn–Minkowski is equivalent to*

$$|A+B|^{1/\nu} \geq |A|^{1/\nu} + |B|^{1/\nu} \tag{13.12}$$

*Proof* We have proven (i) by the scaling relation (13.6). This also implies (ii) for

$$A_\theta = [(\theta A_1)] + [(1-\theta)A_\theta] \tag{13.13}$$

so if (13.12) holds, then

$$\begin{aligned} |A_\theta|^{1/\nu} &\geq |\theta A_1|^{1/\nu} + |(1-\theta)A_\theta|^{1/\nu} \\ &= \theta|A_1|^{1/\nu} + (1-\theta)|A_0|^{1/\nu} \end{aligned}$$

by (13.6).

Conversely, if Brunn–Minkowski holds for $\theta = \frac{1}{2}$, then

$$\begin{aligned} |A_0 + A_1|^{1/\nu} &= 2[A_{1/2}]^{1/\nu} && \text{(by (13.6))} \\ &\geq 2[\tfrac{1}{2}\,A_0^{1/\nu} + \tfrac{1}{2}\,A_1^{1/\nu}] && \text{(by (13.2))} \\ &= A_0^{1/\nu} + A_1^{1/\nu} \end{aligned}$$

proving (13.12) □

Before turning to the proof of (13.3) for convex $A$, let us extend this case of Brunn–Minkowski. Let $C$ be a convex set in $\mathbb{R}^\nu \times \mathbb{R}$ and define for $t \in \mathbb{R}$,

$$C(t) = \{x \in \mathbb{R}^\nu \mid (x,t) \in C\}$$

the slices of $C$. Then for $\theta \in (0,1)$, by convexity of $C$,

$$\theta C(1) + (1-\theta)C(0) \subset C(\theta)$$

so (13.3) (one could also take a result of the form (13.2)) implies

**Proposition 13.4**

$$|C(\theta)| \geq |C(0)|^{1-\theta}\, |C(1)|^\theta \tag{13.14}$$

Note that given $C_0, C_1$ in $\mathbb{R}^\nu$, we can define $C \subset \mathbb{R}^{\nu+1}$ by

$$C = \{x, \theta) \mid 0 \leq \theta \leq 1,\, x \in \theta C_1 + (1-\theta)C_0\}$$

and (13.14) for this set is precisely (13.3). Thus, (13.14) implies (13.3) and so (13.2).

Let $\chi_C$ be the characteristic function of $C$ and note that $\chi_C$ obeys

$$\chi_C(\theta x + (1-\theta)y) \geq \chi_C(x)^\theta \chi_C(y)^{1-\theta} \tag{13.15}$$

Indeed, this is equivalent to $C$ being convex. Notice that

$$|C(t)| = \int \chi_C(x,t)\, d^\nu x$$

Thus, (13.14) is implied by (and suggests) an assertion that a function obeying

$$f(\theta x + (1-\theta)y) \geq f(x)^\theta f(y)^{1-\theta} \tag{13.16}$$

obeys the same relation after integrating out some variables. This is where we now head.

**Definition** A function $f\colon \mathbb{R}^\nu \to [0,\infty]$ is called *log concave* if and only if it is lsc, and for any $x, y \in \mathbb{R}^\nu$, and $\theta \in [0,1]$,

$$f(\theta x + (1-\theta)y) \geq f(x)^\theta f(y)^{1-\theta} \tag{13.17}$$

where $0 \cdot \infty$ is interpreted as 0. It is called *log convex* if $f\colon \mathbb{R}^\nu \to [0,\infty)$,

$$f(\theta x + (1-\theta)y) \leq f(x)^\theta f(y)^{1-\theta} \tag{13.18}$$

*Remarks* 1. Many discussions of this subject do not assume the lsc condition, but then blithely assume measurability. It is likely one can prove any $f$ obeying (13.17) is Lebesgue measurable (i.e., in the completion of the Borel sets), although there are examples where it is not Borel measurable. Anyhow, in applications, this certainly holds.

2. lsc is only needed at points where $f(0) = 0$ or rather outside the interior of the set where $f > 0$; see (i) of the next proposition.

3. Since log convex functions appear here for illustrative purposes only, we do not allow them to take value $+\infty$. Once that is done, continuity is automatic, so we do not assume usc (or lsc).

**Proposition 13.5** (i) *If $f$ is log concave, then $\{x \mid f(x) > 0\}$ is an open convex set.*

(ii) *If $f$ is log concave and for some $x_1$, we have $f(x_1) = \infty$, then there is an open convex set, $S$, so $f(x) = \infty$ if $x \in S$ and $f(x) = 0$ if $x \notin S$. In particular, if there is $x_0$ with $0 < f(x_0) < \infty$, then $f$ does not take the value infinity.*

(iii) *If $f$ is log concave and does not take the value $+\infty$, then $f$ is bounded on compact subsets and continuous on $\{x \mid f(x) > 0\}$.*

(iv) *Every log convex function that is not identically zero is everywhere strictly positive, continuous, and convex.*

*Proof* (i) The set is open since $f$ is lsc. This set is convex by (13.17).

(ii) Let $f(x_1) = \infty$ and $f(x_2) > 0$. Since $S \equiv \{y \mid f(y) > 0\}$ is open and convex, the line $[x_1, x_2]$ can be extended past $x_2$, that is, there exist $x_3 \in S$ and $\theta \in (0,1)$ so $\theta x_1 + (1-\theta)x_3 = x_2$. By (13.17), $f(x_2) = \infty$, that is, $f \equiv \infty$ on $S$.

(iii) If $f$ is never $\infty$, then on $\{x \mid f(x) > 0\}$, $-\log f$ is convex and so, by Theorem 1.19, is continuous and bounded on compact subsets of $S$. If $x_n \to x_\infty \in \partial S$ and $f(x_n) \to \infty$, the argument in (ii) shows $f \equiv \infty$ on $S$. Thus, $f$ is bounded on compact subsets of $\mathbb{R}^\nu$.

(iv) If $f(x_0) = 0$, and for a given $x_1, x_2 = x_0 + 2(x_1 - x_0)$, then $f(x_1) \le f(x_0)^{1/2} f(x_2)^{1/2} = 0$ so $f$ is identically zero. Thus, if $f$ is not identically zero, it is strictly positive, so $\log f$ is convex. Then $f = \exp(\log f)$ is convex since exp is a monotone convex function. □

Suppose $F$ is $C^2$ on $\mathbb{R}$ and strictly positive. Then

$$F^{-2}[\log(F)''] = FF'' - (F')^2$$

If this is positive, then a fortiori, $F'' \ge 0$, that is, as noted, $F$ log convex implies $F$ is convex. But it can be negative even if $F'' > 0$. For example,

$$F(x) = \exp(-x^2)$$

is log concave but not concave for large $x$. More generally, if $A$ is a positive matrix on $\mathbb{R}^\nu$, then

$$F(x) = \exp(-\langle x, Ax \rangle)$$

is log concave.

We will need some notions to state some equivalence to log concavity:

**Definition** A nonnegative function on $\mathbb{R}^\nu$ is called *convexly layered* if and only if $\{x \mid F(x) > \alpha\}$ is a balanced convex set for all $\alpha > 0$. It is called *even, radially monotone* if

(i)

$$f(-x) = f(x) \tag{13.19}$$

(ii)

$$0 \le r \le 1 \text{ implies } f(rx) \ge f(x) \tag{13.20}$$

**Proposition 13.6** *Let $f$ be an lsc function on $\mathbb{R}^\nu$ with values on $[0,\infty)$. The following are equivalent:*

(i) *$f$ is log concave.*

(ii) *For each $a \in \mathbb{R}^\nu$, the function*

$$H_a(x; f) = f(a+x)f(a-x) \tag{13.21}$$

*is convexly layered.*

(iii) *For each $a \in \mathbb{R}^\nu$, the function $H_a$ of* (13.21) *is even, radially monotone.*

(iv) *For all $x_0, y_0 \in \mathbb{R}^\nu$,*

$$f(\tfrac{1}{2}\,x_0 + \tfrac{1}{2}\,y_0) \ge f(x_0)^{1/2} f(y_0)^{1/2} \tag{13.22}$$

*Proof* We will show (i) ⇒ (ii) ⇒ (iii) ⇒ (iv) ⇒ (i).

(i) ⇒ (ii) Clearly,

$$H_a(-x) = H_a(x)$$

so $\{x \mid H_a > \alpha\}$ is balanced. Moreover, as a product of log concave functions, $H_a(\cdot)$ is log concave and so, by (13.17), $\{x \mid H_a(x) > \alpha\}$ is convex.

(ii) ⇒ (iii) If $r \in (0,1)$, let $\theta = \frac{1}{2}(1+r) \in (0,1)$, so

$$rx = \theta x + (1-\theta)(-x) \tag{13.23}$$

Let $\alpha < H_a(x)$, $x \in \{y \mid H_a(y) > \alpha\}$. Since this set is balanced, by (13.21), $H_a(rx) > \alpha$. Thus, $H_a(rx) \ge H_a(x)$ so $H_a$ is radially monotone.

(iii) ⇒ (iv) Let $a = \frac{1}{2}x_0 + \frac{1}{2}y_0$ and $x = \frac{1}{2}(x_0 - y_0)$. Since $H_a$ is radially monotone, $H_a(0; f)^{1/2} \ge H_a(x; f)^{1/2}$, which is (13.22).

(iv) ⇒ (i) Since $f$ is lsc, $S \equiv \{x \mid f(x) > 0\}$ is open and, by (13.22), $x_0, y_0 \in S \Rightarrow \frac{1}{2}x_0 + \frac{1}{2}y_0 \in S$. It follows that $S$ is convex and on $S$, $g = \log f$ is midpoint-convex and lsc. By Proposition 1.4, $g$ is convex, so $f$ is log convex. □

The following realization of $f$ will also be especially useful in studying rearrangements in the next chapter:

**Proposition 13.7** (Wedding Cake Representation) *For any function $f$ with $|\{x \mid |f(x)| > \alpha\}| < \infty$ for all $\alpha > 0$,*

$$|f| = \int_0^\infty \chi_{\{|f|>\alpha\}}\, d\alpha \tag{13.24}$$

*in the sense that for any $x$,*

$$|f(x)| = \int_0^\infty \chi_{\{|f|>\alpha\}}(x)\, d\alpha \tag{13.25}$$

*Remark* The colorful name is obvious if you consider the example of where $f$ is spherically symmetric, radially monotone, and takes only finitely many values. Lieb–Loss [231] use "Layer Cake Representation" but since tiered layer cakes are for weddings, I prefer this name.

*Proof* (13.25) says

$$|f(x)| = \int_0^{|f(x)|} d\alpha$$

which is obvious! □

**Lemma 13.8** *Let $f$ be an lsc, layered convex function on $\mathbb{R}^{\nu+1}$ written as $f(x,t)$, $x \in \mathbb{R}^\nu$, $t \in \mathbb{R}$. Suppose $f$ is bounded and has compact support. Let $g$ be the function*

$$g(x) = \int_{\mathbb{R}} f(x,t)\, dt \tag{13.26}$$

*Then $g$ is an even, radially monotone lsc function.*

*Proof* Since sums and integrals of even, radially monotone functions are in the same class, it is enough, by the wedding cake representation, to prove the result when $f$ is the characteristic function of an open, balanced convex set, $S$. In that case, define for $x \in \mathbb{R}^\nu$,

$$I(x) = \{t \in \mathbb{R} \mid (x,t) \in S\}$$

Then $I(x)$ is an open interval

$$I(x) = (c(x), d(x)) \tag{13.27}$$

and

$$g(x) = d(x) - c(x) \tag{13.28}$$

By the fact that $S$ is convex, we have that $c(x)$ is convex and $d(x)$ is concave, so $g(x)$ is concave. Since $S$ is balanced,

$$c(-x) = -d(x), \qquad d(-x) = -c(x) \tag{13.29}$$

so $g(x)$ is even. An even concave function is even, radially monotone. That $g$ is lsc follows by Fatou's lemma. □

*Remark* In essence, this proof used the one-dimensional Brunn–Minkowski theorem (or the special case when $|A_0| = |A_1|$) and this is trivial. Using Brunn–Minkowski, this result extends to functions on $\mathbb{R}^{\nu+\mu}$ where one integrates out $\mu$ variables. But we will use this $\mu = 1$ case to prove Brunn–Minkowski, so we do not state it in this general $\mu$ case.

Here is the main theorem of this chapter:

**Theorem 13.9** (Prékopa's Theorem) *Let $f$ be a log concave function on $\mathbb{R}^{\nu+\mu}$ written $f(x,y)$ with $x \in \mathbb{R}^\nu$ and $y \in \mathbb{R}^\mu$. Then the function $g$ on $\mathbb{R}^\nu$ defined by*

$$g(x) = \int f(x,y)\, d^\mu y \tag{13.30}$$

*is log concave.*

*Proof* Since a product of log concave functions is log concave, $f\chi_{B_n}$ is log concave where $\chi_{B_n}$ is an open ball of large radius. By the monotone convergence theorem, as $n \to \infty$, $g_n$ (associated to $f\chi_{B_n}$) converges to $g$, so we can suppose supp$(f)$ is compact. Moreover, the case where $f$ is infinite on $\{x \mid f(x) > 0\}$ is trivial (and uninteresting), so we can suppose $f$ is bounded. Finally, by repeatedly integrating out one variable at a time, we can suppose $\mu = 1$. Pick $a \in \mathbb{R}^\nu$ and note

$$\begin{aligned} H_a(x;g) &= g(a+x)g(a-x) \\ &= \int f(a+x,z)f(a-x,y)\,dz\,dy \\ &= 2\int f(a+x,u+v)f(a-x,u-v)\,du\,dv \end{aligned}$$

For $a, u$ fixed, the integrand is $H_{a,u}(x,v;f)$ and so convexly layered by Proposition 13.6.

Thus, by Lemma 13.8, the integral over $v$ is an even, radially monotone, lsc function of $x$ for each fixed $a, u$. Since such functions are closed under integrals in an external parameter, we can integrate over $u$ and see that $H_a(x;g)$ is an even, radially monotone function. By Fatou's lemma, it is lsc.

By Proposition 13.6 again, $g$ is log concave. □

*Remark* By (iii) of Proposition 13.5, so long as there is an $x$ with $0 < \int f(x,y)\,d^\nu y < \infty$, the integral is finite for all $x$.

**Example 13.10** Let $S$ be the union of the two triangles with $|x| < 1$, $|y| < 1$, $x > y > 0$ or $x < y < 0$ (see Figure 13.1).

Let $f = \chi_S$, the characteristic function of $S$. For each $y$, $f(x,y)$ is log concave, but

$$g(x) = \begin{cases} |x|, & |x| < 1 \\ 0, & |x| \geq 1 \end{cases}$$

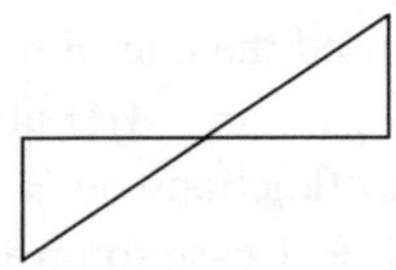

Figure 13.1 Separate log concavity does not imply joint

is definitely not log concave, so joint log concavity is essential. This is in distinction with the log convex case where we have that if $f(x, y)$ is log convex for all $y$, then by Hölder's inequality with $p = \theta^{-1}$,

$$\begin{aligned} g(\theta x + (1-\theta)w) &\leq \int f(x,y)^\theta f(w,y)^{1-\theta}\, dy \\ &\leq \left(\int f(x,y)\, dy\right)^\theta \left(\int fg(w,y)\, dy\right)^{1-\theta} \\ &\leq g(x)^\theta g(w)^{1-\theta} \end{aligned}$$

so $g$ is log concave. (Notice that if $f$ is jointly log convex, the integral defining $g$ is never convergent since

$$f(x,0) \leq f(x,y)^{1/2} f(x,-y)^{1/2} \leq \tfrac{1}{2}\left(f(x,y) + f(x,-y)\right)$$

so $\int_{-R}^R f(x,y)\, dy \geq 2Rf(x,0) \to \infty$ as $R \to \infty$.) This use of Hölder makes Prékopa's theorem surprising (Hölder goes in the wrong direction).

The function $f$ is not radially monotone since $f(0,0) = 0$, but by shifting the triangles so they overlap, one can find an $f$ which is even and radially monotone, but so that $g$ is not. □

*Proof of Theorem 13.1 when $A_0, A_1$ are open convex sets* As noted (Proposition 13.3(i)), it suffices to prove (13.3). In $\mathbb{R}^{\nu+1}$, define the set

$$Q = \{x, \theta) \mid x \in \theta A_1 + (1-\theta)A_0\} \tag{13.31}$$

and let $f$ be the characteristic function of $Q$. Then

$$g(\theta) = \int f(x,\theta)\, d^\nu x = |A_\theta| \tag{13.32}$$

Since $Q$ is convex, $f$ is log concave, and thus, by Prékopa's theorem, $g$ is log concave, that is,

$$|A_\theta| \geq |A_1|^\theta\, |A_0|^{1-\theta}$$

which is (13.3). □

We next give a number of applications of Prékopa's theorem. We will not provide details of the following application which uses path integrals: If $V$ is a convex function on $\mathbb{R}^\nu$ with $V(x) \to \infty$ as $|x| \to \infty$, then the lowest eigenfunction of $-\Delta + V$ is a log concave function.

**Theorem 13.11** *The convolution of two log concave functions is log concave.*

*Proof* If $f, g$ are log concave functions on $\mathbb{R}^\nu$, then $(x, y) \mapsto f(x - y)g(y)$ is log concave on $\mathbb{R}^{2\nu}$, so its integral is log concave. □

**Lemma 13.12** *Let $x = (y, z)$ be the coordinates for $\mathbb{R}^{\mu+\nu}$ with $y \in \mathbb{R}^\mu$, $z \in \mathbb{R}^\nu$. Let $A$ be a strictly positive definite matrix on $\mathbb{R}^{\mu+\nu}$. Then there exist $\nu$ coordinates $w_1, \dots, w_\nu$ so that*

$$w_i(x) = z_i(x) + \sum_{j=1}^{\mu} \rho_{ij} y_j(x) \tag{13.33}$$

*so that*

$$\langle x, Ax\rangle = \langle y, By\rangle + \langle w, Cw\rangle \tag{13.34}$$

*for strictly positive matrices $B$ and $C$ on $\mathbb{R}^\mu$ and $\mathbb{R}^\nu$, respectively.*

*Remark* We think of coordinates $y_j, z_j$ as linear functionals on $\mathbb{R}^{\mu+\nu}$.

*Proof* Let $W \subset \mathbb{R}^{\mu+\nu}$ be the subspace $\{x \mid y_j(x) = 0; j = 1, \dots, \mu\}$ and let $W^\perp$ be the orthogonal complement of $W$ in the inner product defined by $\langle \cdot, A \cdot\rangle$. Any $x \in \mathbb{R}^{\mu+\nu}$ can be uniquely decomposed

$$x = Px + Qx \tag{13.35}$$

with $Px \in W$ and $Qx \in W^\perp$. By the orthogonality in $A$ inner product,

$$\langle x, Ax\rangle = \langle Px, APx\rangle + \langle Qx, AQx\rangle \tag{13.36}$$

Define linear functionals $w_1, \dots, w_\nu$ on $\mathbb{R}^{\mu+\nu}$ by

$$w_i(x) = z_i(Px) \tag{13.37}$$

We claim $\{w_i\}_{i=1}^\nu \cup \{y_j\}_{j=1}^\mu$ are independent as linear functionals and so, a complete coordinate system. For if $y_j(x) = 0$ for $j = 1, \dots, \mu$, then $x \in W$ so $w_i(x) = z_i(Px) = z_i(x)$, and thus, $w_i(x) = 0$ for $i = 1, \dots, \nu$ implies $x = 0$.

Let

$$\eta_i(x) = z_i(Qx) \tag{13.38}$$

Then

$$y_j(x) = 0,\ j = 1, \dots, \mu \Rightarrow x \in W \Rightarrow \eta_i(x) = 0$$

so each $\eta_i$ is a linear combination of the $y$'s, that is,

$$\eta_i(x) = \sum_{j=1}^{\mu} \rho_{ij} y_j(x) \tag{13.39}$$

and (13.33) follows from (13.35), (13.37), (13.38), and (13.39). On the other hand, (13.37) implies (13.34). □

**Theorem 13.13** (Brascamp–Lieb [50]) *Let $A$ be a strictly positive definite matrix on $\mathbb{R}^{\mu+\nu}$. Write $x = (y, z)$ with $x \in \mathbb{R}^{\mu+\nu}$, $y \in \mathbb{R}^\mu$, $z \in \mathbb{R}^\nu$. Let $F(x)$ be jointly log concave (resp. log convex) on $\mathbb{R}^{\mu+\nu}$ and define on $\mathbb{R}^\mu$,*

$$G(y) = \frac{\int \exp(-\langle x, Ax\rangle)F(x)\, d^\nu z}{\int \exp(-\langle x, Ax\rangle\, d^\nu z} \tag{13.40}$$

*Then $G$ is log concave (resp. log convex).*

*Remark* We will use special properties of Gaussians to prove this. It is not true even in the log concave case if $\exp(-\langle x, Ax\rangle)$ is replaced by an arbitrary log concave function, $H$. For example, if $\mu = \nu = 1$ and $F$ is the characteristic function of the set $\{\langle x, y\rangle \mid |x| \leq 1,\ |y| \leq 1\}$ and $H$ of the set $\{\langle x, y\rangle \mid x^2 + y^2 \leq 2\}$, then the numerator is constant on $[-1, 1]$, while the denominator is decreasing. So $G$ is even and monotone increasing in $|y|$ on $[-1, 1]$ so certainly not log concave.

*Proof* Use the lemma and change variables from $z$ to $w$ noting that in either coordinate system, the spaces $\{x \mid y_i(x) = y_i^{(0)}\}$ for $y_i^{(0)}$ fixed are the same (since they are given by the same linear functional values). On that space, we have new coordinates $w_i$ related to $z_i$ by (13.33), that is, a $y_i^{(0)}$-dependent translate. Thus,

$$\begin{aligned} G(y) &= \frac{\int \exp(-\langle y, By\rangle - \langle w, Cw\rangle)F(x)\, d^\nu w}{\int f \exp(-\langle y, By\rangle - \langle w, Cw\rangle)\, d^\nu w} \\ &= N^{-1} \int \exp(-\langle w, Cw\rangle)F(x)\, d^\nu w \end{aligned} \tag{13.41}$$

where $N = \int \exp(-\langle w, Cw\rangle)$. This is because $\exp(-\langle y, By\rangle)$ is $w$-independent and factors out of the integrals and cancels. Since log concavity (resp. log convexity) is a vector space notion, $F$ is log concave (resp. log convex) in the new coordinates.

In the log concave case, $\exp(-\langle w, Cw\rangle)F(x)$ is log concave as a product of log concave functions, so (13.41) implies $G$ is log concave by Prékopa's theorem.

In the log convex case, $\exp(-\langle w, Cw\rangle)F(w, y)$ is log convex in $y$ for each fixed $w$, so $G(y)$ is log convex by Hölder's inequality, as explained in the remark following Theorem 13.9. □

**Corollary 13.14** *Let $\mu_\nu$ be the standard Gaussian measure,*

$$d\mu_\nu(x) = (2\pi)^{-\nu/2} \exp\biggl(-\frac{x^2}{2}\biggr)\, d^\nu x$$

*on $\mathbb{R}^\nu$. Let $C$ be a convex balanced set in $\mathbb{R}^{\kappa+\nu}$. Let $D = \{y \in \mathbb{R}^\nu \mid (0, y) \in C\}$,* *that is, the intersection of $C$ with a coordinate plane and $E \equiv \{x \in \mathbb{R}^\kappa \mid (x, y) \in C$ for some $y\}$, that is, the projection of $C$ onto the other coordinate plane. Then*

$$\mu_{\kappa+\nu}(C) \le \mu_\kappa(E)\mu_\nu(D) \tag{13.42}$$

*Proof* Let $\chi_C$ be the characteristic function of $C$ and

$$G(x) = \int_{\mathbb{R}^\kappa} \chi_C(x, y)\, d\mu_\nu(y)$$

so

$$\mu_\nu(D) = G(0)$$
$$\mu_{\kappa+\nu}(C) = \int_E G(x)\, d\mu_\kappa(x)$$

Since $\chi_C$ and $\exp(-\frac{1}{2}x^2)$ are even and log concave, $G$ is even and log concave, so maximum at $x = 0$, that is,

$$\mu_{\kappa+\nu}(C) \le G(0) \int_E d\mu_\kappa = \mu_\nu(D)\mu_\kappa(E) \qquad \square$$

Finally, we note the following Gaussian version of the Brunn–Minkowski inequality:

**Theorem 13.15** *Define $A_0, A_1, A_\theta$ as in Theorem 13.1 and let $B$ be a strictly positive definite matrix on $\mathbb{R}^\nu$. Define*

$$d\mu_B(x) = \exp(-\langle x, Bx\rangle)\, d^\nu x \tag{13.43}$$

*Then*

$$\mu_B(A_\theta) \ge \mu_B(A_1)^\theta \mu_B(A_0)^{1-\theta} \tag{13.44}$$

*Proof* Define $Q$ by (13.31) and let $f$ be the characteristic function of $Q$ which is log concave since $Q$ is convex. Thus, $x \mapsto f(x)\exp(-\langle x, Bx\rangle)$ is log concave, so by Prékopa's theorem,

$$\mu_B(A_\theta) = \int f(x, \theta) \exp(-\langle x, Bx\rangle)\, d^\nu x$$

is log concave and (13.44) holds. $\square$

Unlike the traditional Brunn–Minkowski inequality, $\nu$ does not enter. That's a virtue! For it lets the inequality extend to various infinite-dimensional contexts, as discussed in the Notes. Finally, we prove the Brunn–Minkowski inequality in the general case:

*Proof of Theorem 13.1* (*General Case*) As noted in (ii) of Proposition 13.5, we need only prove that

$$|A + B|^{1/\nu} \ge |A|^{1/\nu} + |B|^{1/\nu} \tag{13.45}$$

Suppose first $A$ is an open rectangle with sides $a_1, \dots, a_\nu$ and $B$ an open rectangle with sides $b_1, \dots, b_\nu$. Then $A + B$ is a rectangle with sides $a_1 + b_1, \dots, a_\nu + b_\nu$ and (13.45) says

$$\prod_{j=1}^{\nu} c_j^{1/\nu} + \prod_{j=1}^{\nu} d_j^{1/\nu} \le 1 \tag{13.46}$$

with $c_j = a_j/(a_j+b_j)$ and $d_j = b_j/(a_j+b_j)$. By the general arithmetic-geometric mean inequality (1.12),

$$\prod_{j=1}^{\nu} c_j^{1/\nu} + \prod_{j=1}^{\nu} d_j^{1/\nu} \le \frac{1}{\nu}\sum_{j=1}^{\nu}(c_j + d_j) = 1$$

since $c_j + d_j = 1$ and (13.46) holds.

Now suppose $A = \cup_{j=1}^{\ell} A_j$ and $B = \cup_{k=1}^{m} B_k$ are unions of disjoint open rectangles. We will prove (13.45) by induction on $m + \ell$. We have already handled the case $m+\ell = 2$ so suppose $m+\ell \ge 3$. By interchanging $A$ and $B$, we can suppose $\ell \ge 2$. If two open rectangles have intersecting projections onto each coordinate axis, they intersect, so by disjointness, we can find $i \in \{1, \dots, \nu\}$ and $\alpha$ so that $A_1$ and $A_\ell$ are strictly separated by the hyperplane $x_i = \alpha$. Let

$$A_< = A \cap \{x \mid x_i < \alpha\} \tag{13.47}$$
$$A_> = A \cap \{x \mid x_i > \alpha\} \tag{13.48}$$

Then

$$|A| = |A_<| + |A_>| \tag{13.49}$$

and since $A_1$ and $A_\ell$ or $A_\ell$ and $A_1$ are subsets of $A_<$ and $A_>$, respectively, both $A_<$ and $A_>$ are unions of at most $\ell - 1$ rectangles. Let

$$f(t) = |\{x \mid x_i < t\} \cap B|$$

$f(t)$ runs continuous from 0 at very negative $t$ to $|B|$ at large $t$ so we can find $\beta$ with

$$f(\beta)|B|^{-1} = |A_<|\,|A|^{-1} \tag{13.50}$$

Let

$$B_< = B \cap \{x \mid x_i < \beta\} \tag{13.51}$$
$$B_> = B \cap \{x \mid x_i > \beta\} \tag{13.52}$$

Then each of $B_<$ and $B_>$ is a union of at most $m$ rectangles and

$$|B| = |B_<| + |B_>| \tag{13.53}$$

and (13.50) implies

$$|B_<|\,|A_<|^{-1} = |B_>|\,|A_>|^{-1} = |B|\,|A|^{-1} \tag{13.54}$$

In general, two sums of disjoint rectangles may not be disjoint – which is why we have gone through this careful construction – but $A_< + B_<$ is disjoint from $A_> + B_>$ since the hyperplane $\{x \mid x_i = \alpha + \beta\}$ separates them. Thus, using the fact that since $\{A_<, B_<\}$ and $\{A_>, B_>\}$ are unions of at most $\ell - 1 + m$ rectangles, so induction applies,

$$\begin{aligned}
|A+B| &\geq |A_< + B_<| + |A_> + B_>| \qquad \text{(by disjointness)} \\
&\geq (|A_<|^{1/\nu} + |B_<|^{1/\nu})^\nu + (|A_>|^{1/\nu} + |B_>|^{1/\nu})^\nu \\
&\qquad\qquad \text{(by the induction hypothesis)} \\
&= \frac{|A_<|}{|A|}(|A|^{1/\nu} + |B|^{1/\nu})^\nu + \frac{|A_>|}{|A|}(|A|^{1/\nu} + |B|^{1/\nu})^\nu \\
&\qquad\qquad \text{(by (13.54))} \\
&= (|A|^{1/\nu} + |B|^{1/\nu})^\nu
\end{aligned}$$

so (13.45) is proven for $A$ and $B$ unions of disjoint open rectangles.

If $A$ and $B$ are arbitrary compact sets and $A^{(\varepsilon)}$ is the closure of the union of all open rectangles of side $\varepsilon$ and centers in $\varepsilon\mathbb{Z}^\nu$ whose closures intersect $A$ (and similarly for $B^{(\varepsilon)}$), then $|A^{(\varepsilon)}| = |\,\overline{A^{(\varepsilon)}}\,|$, $A^{(\varepsilon)}$ is a finite union of disjoint rectangles, and $|A^{(\varepsilon)}| \to |A|$, $|B^{(\varepsilon)}| \to |B|$, and $|A^{\varepsilon)} + B^{(\varepsilon)}| \to |A+B|$, so (13.45) holds for arbitrary compact sets. Since any Borel set, $A$, is up to a set of Lebesgue measure 0, an increasing union of compact sets, (13.45) holds in general. □

# 14

# Rearrangement inequalities, I. Brascamp–Lieb–Luttinger inequalities

In this chapter and the next, we turn to a beautiful and fascinating issue: decreasing rearrangements and the associated inequalities. To start with a simple example, let $a = (a_1, \dots, a_n)$ be a sequence of nonnegative numbers. Its *decreasing rearrangement* is defined to be that sequence $a^* = (a_1^*, \dots, a_n^*)$ with

$$a_1^* \geq a_2^* \geq \cdots \geq a_n^* \geq 0 \tag{14.1}$$

obtained by permuting the indices. (So if the $a_j$'s are distinct, $a^*$ is determined by (14.1) and the fact that $\{a_j\}_{j=1}^n$ and $\{a_j^*\}_{j=1}^n$ are identical sets. If some $a_i$'s are equal, we need to specify multiplicities.) A little thought will convince you that in forming $\sum_{j=1}^n a_j b_j$ to get the largest result, one should team up the largest $b$'s to the largest $a$'s, that is,

$$\sum_{j=1}^n a_j b_j \leq \sum_{j=1}^n a_j^* b_j^* \tag{14.2}$$

The easiest way to prove this is to use summation by parts or induction to see that

$$\sum_{j=1}^n a_j b_j = b_n \left( \sum_{j=1}^n a_j \right) + (b_{n-1} - b_n) \sum_{j=1}^{n-1} a_j + \cdots + (b_1 - b_2) a_1 \tag{14.3}$$

(14.2) follows if we note first that there is a permutation $\pi$ of $\{1, \dots, n\}$ with

$$\sum_{j=1}^n a_j b_j = \sum_{j=1}^n a_{\pi(j)} b_j^* \tag{14.4}$$

then, that if $b_1 \geq b_2 \geq \cdots \geq b_n$, the right side of (14.3) is increased by replacing $a$ by $a^*$ since

$$\sum_{j=1}^k a_j \leq \sum_{j=1}^k a_j^* \tag{14.5}$$

and finally that $a_\pi^* = a^*$.

This argument also shows if $b_1 > b_2 > \cdots > b_n$, then equality holds in (14.2) only if $a = a^*$.

*Remark* (14.3) also shows the lower bound

$$\sum_{j=1}^{n} a_j b_j \geq \sum_{j=1}^{n} a_j^* b_{n-j+1}^*$$

This also follows from (14.2) by picking $c = \max(b_j)$ and noting that if $d_j = c - b_j$, then $d_j^* = c - b_{n-j+1}^*$.

Two themes will be discussed. The one in this chapter involves generalizations of (14.2) to a lot more than sums of products and to include more than finite sums: explicitly, we want to allow infinite sums and integrals. To illustrate, let us state one of the earliest continuum results. In considering functions on $(-\infty, \infty)$ rather than one-sided $\{1, \dots, n\}$, the natural generalization of $a^*$ is:

**Definition** Let $f$ be a nonnegative function on $\mathbb{R}$ so that

$$|\{x \mid f(x) > t\}| < \infty \qquad \text{for each } t > 0 \tag{14.6}$$

Then $f^*$ is the unique function on $\mathbb{R}$ that obeys:

(i)

$$|\{x \mid f^*(x) > t\}| = |\{x \mid f(x) > t\}| \qquad \text{for all } t \tag{14.7}$$

(ii) $f^*(-x) = f^*(x)$
(iii) $0 \leq x \leq y$ implies $f^*(x) \geq f^*(y) \geq 0$.
(iv) $f^*$ is lsc.

$f$ is called the *symmetric decreasing rearrangement* of $f$.

We defer the questions of existence and uniqueness of $f^*$ until later (when we discuss the $n$-dimensional generalization). That this is a reasonable generalization of the $a \to a^*$ construction for sequences is not only intuitive but illustrated by the following: If $f$ is given by

$$f(x) = \begin{cases} 0, & x < 0 \\ a_j, & 2(j-1) \leq x < 2j, \quad j = 1, \dots, n \\ 0, & x \geq 2n \end{cases}$$

then $f^*$ is given by

$$f^*(x) = \begin{cases} a_j^*, & j - 1 \leq |x| < j \\ 0, & |x| > n \end{cases}$$

Typical of the result we will prove is the following:

**Theorem 14.1** (Riesz's Rearrangement Inequality) *Let $f, g, h$ be three nonnegative functions on $\mathbb{R}$ obeying* (14.6). *Then*

$$\int f(x)g(x-y)h(y)\,dx\,dy \leq \int f^*(x)g^*(x-y)h^*(y)\,dx\,dy \tag{14.8}$$

The second theme, discussed in the next chapter, involves the fact that in proving (14.2), we did not need that the $a_j^*$'s were a rearrangement of the $a_j$'s but only that (14.5) holds. Thus, if $a$ and $c$ are the sequences with

$$\sum_{j=1}^{k} c_j^* \leq \sum_{j=1}^{k} a_j^*, \qquad k = 1, \dots, n \tag{14.9}$$

then

$$\sum_{j=1}^{n} b_j^* c_j^* \leq \sum_{j=1}^{n} b_j^* a_j^*$$

so taking $b_j = c_j$ and then $a_j$, we see that if (14.9) holds, then

$$\sum_{j=1}^{n} c_j^2 \leq \sum_{j=1}^{n} a_j^2 \tag{14.10}$$

More generally, we will prove that

**Theorem 14.2** (Hardy–Littlewood–Pólya Theorem) *If* (14.9) *holds with equality for $k = n$, then for any convex function, $\varphi$,*

$$\sum_{j=1}^{n} \varphi(c_j) \leq \sum_{j=1}^{n} \varphi(a_j) \tag{14.11}$$

We turn now to the first theme. We begin by defining a notion of rearrangement on $\mathbb{R}^\nu$. Throughout the rest of this chapter, we will assume $f$ obeys

$$m_f(t) \equiv |\{x \mid |f(x)| > t\}| < \infty \qquad \text{all } t > 0 \tag{14.12}$$

This is obviously true, for example, if $\lim_{|x|\to\infty}|f(x)| = 0$. For sequences on $\{1, 2, \dots, \}$ or on $\mathbb{Z}$, the analog of (14.12) holds if and only if

$$\lim_{n\to\infty} a_n = 0 \tag{14.13}$$

and we will assume this.

**Definition** Two functions, $f$ and $g$, are called *equimeasurable* if and only if for all $t > 0$,

$$m_f(t) = m_g(t) \tag{14.14}$$

Two sequences, $a$ and $b$, are called *equimeasurable* if and only if

$$\#\{n \mid |a_n| > t\} = \#\{n \mid |b_n| > t\} \qquad \text{for all } t > 0$$

It is easy to see that if (14.13) holds for $a$ and $b$, this is the same as saying $\max_n |a_n| = \max_n |b_n|$, $\max_{n \neq j}(|a_n| + |a_j|) = \max_{n \neq j}(|b_n| + |b_j|)$, etc. Put more precisely, the largest elements are the same, the second largest are the same, $\dots$

**Definition** Let $f$ be a function on $\mathbb{R}^\nu$. The *symmetric decreasing rearrangement* of $f$ is the unique nonnegative function, $f^*$, with

(i) $f^*$ and $f$ are equimeasurable.

(ii) For every rotation (including reflections), $R$,

$$f^*(Rx) = f^*(x) \tag{14.15}$$

(iii) $0 \leq |x| \leq |y|$ implies $f^*(x) \geq f^*(y) \geq 0$.

(iv) $f^*$ is lsc.

If $f$ is a function on $[0, \infty)$, we define its decreasing rearrangement, also denoted $f^*$, as the unique nonnegative function obeying (i), (iii), and (iv).

We note that reflections are only needed in case $\nu = 1$. In all other dimensions, (14.15) for $R \in SO(n)$ implies the result for $R \in O(n)$.

**Definition** $\chi_{\{|f|>\alpha\}}$ will be the symbol for the characteristic function of the set $\{x \mid |f(x)| > \alpha\}$.

**Proposition 14.3** (i) *$f^*$ exists and is uniquely determined. Indeed, if $\tau_\nu$ is the volume of the unit ball in $\nu$ dimensions if $m_f(r)$ is strictly monotone, $f^*$ is determined by*

$$f^*((m_f(\lambda)\tau_\nu^{-1})^{1/\nu}\omega) = \lambda \tag{14.16}$$

*for all unit vectors $\omega \in S^{n-1}$ and, more generally,*

$$f^*(x) = \sup\{\lambda \mid \tau_\nu |x|^\nu < m_f(\lambda)\} \tag{14.17}$$

(ii) *Symmetric decreasing rearrangement is order preserving in the sense that*

$$0 \leq |f| \leq |g| \Rightarrow 0 \leq f^* \leq g^* \tag{14.18}$$

(iii) *For any positive monotone function $F$ (or, more generally, any function $C^1$ $F$ where $G(y) = \int_0^y |F'(s)|\, ds$ has $\int G(f(x))\, d^\nu x < \infty$), we have*

$$\int F(|f(x)|)\, d^\nu x = \int F(f^*(x))\, d^\nu x \tag{14.19}$$

*In particular, $f^* \in L^p$ if and only if $f \in L^p$ and $\|f^*\|_p = \|f\|_p$.*

(iv)

$$\chi^*_{\{|f|>\alpha\}} = \chi_{\{f^*>\alpha\}} \tag{14.20}$$

*More generally, if $G$ is any monotone increasing, lsc function on $[0, \infty)$,*

$$(G \circ |f|)^* = G \circ f^* \tag{14.21}$$

*Proof* (i) Uniqueness is immediate, since any function, $g$, is determined by the sets $S_\alpha(g) = \{x \mid g(x) > \alpha\}$ via

$$g(x) = \sup\{\alpha \mid x \in S_\alpha(g)\} \tag{14.22}$$

and the sets $S_\alpha(f^*)$ are open balls centered at 0 of volume $m_f(\alpha)$. (They are open since $f^*$ is lsc.)

This means that

$$S_\alpha = \{x \mid \tau_\nu |x|^\nu < m_f(\alpha)\}$$

which, given (14.22), implies (14.17) holds for $f^*$ if it exists.

To see existence, it is easy to see the function defined by (14.17) is equimeasurable with $f$ symmetric and decreasing. It is lsc since $\{x \mid f^*(x) > \lambda\}$ is seen to be the open ball of volume $m_f(\lambda)$, and so open. When $m_f$ is strictly monotone, (14.17) implies (14.16).

(ii) $0 \le |f| \le |g|$ implies $m_g(\lambda) \le m_f(\lambda)$, which implies $f^* \le g^*$ by (14.17).

(iii) Since $|f|$ and $f^*$ are equimeasurable, this is immediate.

(iv) (14.20) is immediate from the fact that $\{x \mid f^*(x) > \alpha\}$ is the unique open ball centered at the origin with the same volume as $\{x \mid f(x) > \alpha\}$. To prove (14.21), note that

$$\{x \mid (G \circ |f|)(x) > \alpha\} = \{x \mid |f(x)| > \beta_G(\alpha)\}$$

where

$$\beta_G(\alpha) = \inf\{\gamma \mid G(\gamma) > \alpha\}$$

(which is $G^{-1}$ if $G$ is continuous and strictly monotone), and thus, $(G \circ |f|)^*$ and $G \circ f^*$ are equimeasurable and symmetric decreasing. Since the composition of an lsc function and a monotone increasing lsc function is lsc, they are equal. □

The wedding cake representation (see Proposition 13.7) fits in especially well with rearrangements. By (14.20), we have

$$f^* = \int_0^\infty \chi^*_{\{|f|>\alpha\}}\, d\alpha \tag{14.23}$$

The power of the wedding cake representation is seen by the following proof (compare with our earlier proof of (14.2)):

**Theorem 14.4** *For any functions $f, g$, we have*

$$\int |f(x)g(x)|\, d^\nu x \le \int f^*(x) g^*(x)\, d^\nu x \tag{14.24}$$

*Remark* (14.24) is intended to allow both sides or the right side to be infinite.

*Proof* By the wedding cake representation and (14.23), (14.24) holds if we prove it for the special case where $f = \chi_A$, $g = \chi_B$, the characteristic functions of sets. In this case where $A^*$ is the open ball about 0 with $|A^*| = |A|$, (14.24) says

$$|A \cap B| \le |A^* \cap B^*| \tag{14.25}$$

But, the balls at 0 are nested so

$$\begin{aligned} |A^* \cap B^*| &= \min(|A^*|, |B^*|) \\ &= \min(|A|, |B|) \\ &\ge |A \cap B| \end{aligned}$$

and (14.25) holds. □

Since $\|f - g\|_2^2 \ge \| |f| - |g| \|_2 = \|f\|_2^2 + \|g\|_2^2 - 2\langle |f|, |g|\rangle$ and $\langle |f|, |g|\rangle \le \langle f^*, g^*\rangle$, (14.24) implies that $\|f^* - g^*\|_2 \le \|f - g\|_2$. The following implies that this is true for $L^p$ norms or, more generally, for any Orlicz norm:

**Theorem 14.5** *Let $F$ be a nonnegative convex function on $\mathbb{R}$ with $f(0) = 0$. Then for any nonnegative functions $f, g$,*

$$\int F(f^*(x) - g^*(x))\, d^\nu x \le \int F(f(x) - g(x))\, d^\nu x \tag{14.26}$$

*Proof* Let $F_+$ and $F_-$ be defined by

$$F_\pm(y) = \begin{cases} F(y), & \text{if } \pm y \ge 0 \\ 0, & \text{if } \pm y \le 0 \end{cases}$$

Then $F_\pm$ are convex, so by symmetry, we need only prove the result for $F_+$, that is, without loss we can suppose $F(y) = 0$ for $y \le 0$, that is, the integrals only go over $x$'s with $f(x) \ge g(x)$ (or $f^*(x) \ge g^*(x)$). Let $D^- F$ be the derivative from the left which is nonnegative, monotone, and lsc. Moreover, by (1.57),

$$\begin{aligned} F(f(x) - g(x)) &= \int_{g(x)}^{f(x)} (D^- F)(f(x) - s)\, ds \\ &= \int_0^\infty (D^- F)(f(x) - s)\chi_{\{g \le s\}}(x)\, ds \end{aligned}$$

since $(D^- F)(f(x) - s) = 0$ if $s \ge f(x)$ ($F = 0$ on $(-\infty, 0]$). Thus,

$$\int F(f(x) - g(x))\, d^\nu x = \int_0^\infty \left( \int (D^- F)(f(x) - s)\chi_{\{g \le s\}}(x)\, d^\nu x \right) ds \tag{14.27}$$

and thus, (14.26) is implied by

$$\int (D^- F)(f(x) - s)\chi_{\{g \le s\}}(x)\, d^\nu x \ge \int (D^- F)(f^*(x) - s)\chi_{\{g^* \le s\}}(x)\, d^\nu x \tag{14.28}$$

for each $s \geq 0$. For each such $s$, $y \mapsto (D^- F)(y - s)$ is an lsc monotone function of $y$, so by (14.21),

$$(D^- F)(f(\cdot) - s)^* = (D^- F)(f^*(\cdot) - s)$$

and (14.28) is implied by

$$\int h(x)\chi_{\{g \leq s\}}\, d^\nu x \geq \int h^*(x)\chi_{\{g^* \leq s\}}\, d^\nu x \tag{14.29}$$

for all nonnegative functions, $h$ obeying (14.6).

By the wedding cake representation, (14.29) is implied by

$$\int \chi_{\{h > \alpha\}}\chi_{\{g \leq s\}}\, d^\nu x \geq \int \chi_{\{h^* > \alpha\}}\chi_{\{g^* \leq s\}}\, d^\nu x \tag{14.30}$$

To prove (14.30), subtract

$$\int \chi_{\{h > \alpha\}}\chi_{\{g > s\}}\, d^\nu s \leq \int \chi_{\{h^* > \alpha\}}\chi_{\{g^* > s\}}\, d^\nu x$$

(which is (14.24)) from

$$\int \chi_{\{h > \alpha\}}\, d^\nu x = \int \chi_{\{h^* > \alpha\}}\, d^\nu x$$

This proves (14.30) which implies (14.29) which implies (14.28) which implies (14.26). □

**Corollary 14.6** *For any $f, g \in L^p(\mathbb{R}^\nu)$,*

$$\|f^* - g^*\|_p \leq \|f - g\|_p \tag{14.31}$$

*More generally, for any Orlicz norm,*

$$\|f^* - g^*\|_F \leq \|f - g\|_F \tag{14.32}$$

*Proof* (14.31) follows from (14.26) with $F(y) = y^p$. More generally, (14.26) implies

$$Q_F(f^* - g^*) \leq Q_F(f - g)$$

Since $(\lambda f)^* = \lambda f^*$ for any $\lambda \geq 0$, (14.32) follows. □

**Corollary 14.7** *Let $A_n$ be a sequence of bounded measurable sets and $A_\infty$ a bounded measurable set, so*

$$\lim_{n \to \infty} \mu(A_n \triangle A_\infty) \to 0 \tag{14.33}$$

*Then for any $p < \infty$,*

$$\lim_{n \to \infty} \|\chi_{A_n} - \chi_{A_\infty}\|_p \to 0 \tag{14.34}$$

$$\lim_{n \to \infty} \|\chi^*_{A_n} - \chi^*_{A_\infty}\|_p \to 0 \tag{14.35}$$

*Proof* $|\chi_A - \chi_B| = \chi_{A\triangle B}$ so $\|\chi_A - \chi_B\|_p = \mu(A\triangle B)^{1/p}$, and (14.33) implies (14.34). (14.35) then follows from Corollary 14.6. □

The main result in this chapter is

**Theorem 14.8** (Brascamp–Lieb–Luttinger Inequalities (BLL Inequalities, for short)) *Let* $f_1, \dots, f_\ell$ *be* $\ell$ *nonnegative functions on* $\mathbb{R}^\nu$. *Fix an integer* $n$. *Let* $\{a_{jm}\}_{1\le j\le \ell;\, 1\le m\le n}$ *be an* $\ell \times n$ *real matrix. Then*

$$\int_{(\mathbb{R}^\nu)^n} \prod_{m=1}^n d^\nu x_m \prod_{j=1}^\ell f_j\biggl(\sum_{m=1}^n a_{jm} x_m\biggr) \le \int_{(\mathbb{R}^\nu)^n} \prod_{m=1}^n d^\nu x_m \prod_{j=1}^\ell f_j^*\biggl(\sum_{m=1}^n a_{jm} x_m\biggr) \tag{14.36}$$

The proof will be in several steps given below, starting with the case $\nu = 1$.

**Example 14.9** Take $\ell = 3$, $m = 2$, and

$$A = \begin{pmatrix} 1 & 0 \\ 1 & -1 \\ 0 & 1 \end{pmatrix}$$

and Theorem 14.8 becomes

$$\int d^\nu x\, d^\nu y\, f(x) g(x-y) h(y) \le \int d^\nu x\, d^\nu y\, f^*(x) g^*(x-y) h^*(y)$$

which is a multidimensional generalization of Riesz's rearrangement inequality, so, in particular, Theorem 14.1 is proven once we prove (14.36). □

As a preliminary, we note that if $\chi_r$ is the characteristic function of the ball of radius $r$, then it suffices to prove the theorem when $\prod_{m=1}^n \chi_r(x_m)$ is inserted into the integral (or equivalently, if among the $f_j(\sum_{m=1}^n a_{jm} x_m)$ are the $\chi_r(x_m)$). For $\chi_r^* = \chi_r$ and the integrals converge monotonically as $r \to \infty$ to the integrals with no $\chi_r$ factors. Thus, in the proofs below, we will suppose that the $\chi_r$'s have been inserted.

*Proof of Theorem 14.8 in case* $\nu = 1$ By the wedding cake representation, we can suppose each $f_j$ is the characteristic function of a set $A_j$. Suppose first each $A_j$ is a single interval

$$A_j = (\alpha_j - \beta_j, \alpha_j + \beta_j)$$

For $t$ in $[-1, 1]$, let $I(t)$ be the integral

$$I(t) = \int_{\mathbb{R}^n} dx_1 \dots dx_n \prod_{j=1}^\ell \chi_{A_j(t)}\biggl(\sum_{m=1}^n a_{jm} x_m\biggr) \tag{14.37}$$

where $A_j$ is the interval

$$A_j = (\alpha_j t - \beta_j, \alpha_j t + \beta_j) \tag{14.38}$$

For this case, what (14.36) says is

$$I(0) \geq I(1) \tag{14.39}$$

since $\chi^*_{A_j} = \chi_{A_j(0)}$. In $\mathbb{R}^{n+1}$, let $C$ be the set

$$C = \left\{ (x_1, x_2, \dots, x_n, t) \,\middle|\, \alpha_j t - \beta_j \leq \sum_{m=1}^{n} a_{jm} x_m \leq \alpha_j t + \beta_j \text{ for } j = 1, \dots, \ell \right\}$$

As an intersection of half-spaces, $C$ is a convex set and each set of inequalities preserved by $(x, t) \to (-x, -t)$, so it is balanced. If $C(t) \subset \mathbb{R}^n$ is the fixed $t$ slice, by (14.37),

$$|C(t)| = I(t)$$

By the variant of Brunn–Minkowski that is Proposition 13.4,

$$I(0) \geq I(1)^{1/2} I(-1)^{1/2} \tag{14.40}$$

Since $C$ is balanced, $I(t) = I(-t)$, so (14.40) implies (14.39).

We have actually proven more. (13.14) implies for $0 \leq t \leq s \leq 1$,

$$I(t) \geq I(s)^\theta I(-s)^{1-\theta} = I(s)$$

where $\theta = \frac{1}{2}(1 + ts^{-1})$, that is, $I$ is monotone,

$$0 \leq t \leq s \Rightarrow I(t) \geq I(s) \tag{14.41}$$

Next, suppose each $A_j$ is a finite union of disjoint intervals. Define $A_j(t)$ to be the union of intervals obtained using the transformation (14.38) on each interval. As $t$ decreases, the integral as a sum of single interval integrals increases by what we have just proven. Also, the intervals move towards each other (since at $t = 0$, they all overlap). At some first $t$, two intervals first touch. At that point, merge them and restart the process with one fewer interval (or more, if several collide at once) or alternatively, make an induction on the total number of intervals and appeal to the induction hypothesis once two intervals touch.

Either way, this proves (14.36) when each $f_j$ is the characteristic function of a union of intervals. By regularity of measures and the fact that the intervals are a base for the topology on $\mathbb{R}$, given any bounded Borel set, $A$, we can find a sequence, $A_\ell$, each a finite union of intervals, so $|A_\ell \triangle A| \to 0$. By Corollary 14.7,

$$\int_{S_r} \left| \chi_{A_\ell}\left( \sum_{m=1}^{n} a_{jm} x_m \right) - \chi_A \left( \sum_{m=1}^{n} a_{jn} x_m \right) \right|^p \prod_{m=1}^{n} dx_n \to 0$$

where $S_r = \{(x_1, \ldots, x_n\} \mid |x_i| \leq r\}$ and the same is true for $\chi^*$. It follows, by Hölder's inequality, that when $\prod_{m=1}^{n} \chi_r(x_m)$ are among the $f_j$'s, then we have convergence of the integrals when each $f_j$ is a characteristic function, and we approximate by characteristic functions of finite unions of intervals. Thus, (14.36) holds for arbitrary characteristic functions and so, by the wedding cake representation, for all positive functions. □

We know the integrals in (14.36) go up if we do a symmetrization in one dimension, leaving the orthogonal coordinates unchanged. By repeatedly doing symmetrization in enough different directions, we expect to converge to a spherical rearrangement. (For anyone who has ever packed a snowball, this expectation will be a familiar one.) Since we can use the wedding cake representation to reduce to consideration of sets, we will discuss the notion on sets rather than functions.

**Definition** Let $e$ be a unit vector in $\mathbb{R}^\nu$ and $x_1, \ldots, x_\nu$ an orthonormal coordinate system with $x_1$ the coordinate along $e$. Given a Borel set, $A$, with finite measure for each $x_2^{(0)}, \ldots, x_\nu^{(0)}$, let $d(x_2^{(0)}, \ldots, x_\nu^{(0)})$ be the linear Lebesgue measure of $\{t \mid (t, x_2^{(0)}, \ldots, x_\nu^{(0)}) \in A\}$, the intersection of the line through $(0, x_2^{(0)}, \ldots, x_\nu^{(0)})$ parallel to $e$. Then the *Steiner symmetrization*, $\sigma_e(A)$, is defined by

$$\sigma_e(A) = \{x \mid |x_1| \leq \tfrac{1}{2}\, d(x_2^{(0)}, \ldots, x_\nu^{(0)})\} \tag{14.42}$$

Notice that $\sigma_e$ is only dependent on $e$ and not on the $x_2, \ldots, x_\nu$ coordinates. Since $A$ is Borel, $\sigma_e(A)$ is a Borel set and it is clearly equimeasurable with $A$. Of course, $\chi_{\sigma_e(A)}$ is just the result of applying one-dimensional symmetric rearrangement to $\chi_A$. Let $\mathcal{G}$ be the semigroup of all finite products of $\sigma_e$'s.

The one-dimensional version of (14.36) which we have proven shows integrals like that on the left side with $f_j = \chi_{A_j}$, a characteristic set, only increase if each $A_j$ is replaced by $\sigma_e(A_j)$ so if $A_j$ is replaced by $g(A_j)$ for any $g \in \mathcal{G}$.

Define the Lebesgue distance between Borel sets $A$ and $B$ by

$$\rho(A, B) = |A \triangle B| \tag{14.43}$$

We need a somewhat technical-looking lemma for the key fact proven in the next proposition – that if $\rho(gA, S) = \rho(A, S)$ for all $g \in \mathcal{G}$, where $S$ is the ball with $|A| = |S|$, then $A = S$. This key fact will assure us that if we use properly chosen Steiner symmetrizations, we can keep getting closer to a sphere. In the lemma, we need to discuss absolute values of differences of Lebesgue measures of sets. To avoid expressions like $|\,|A| - |B|\,|$ where one $|\cdot|$ is numeric and two $|\cdot|$'s are Lebesgue measure, we use $\mu_1(\cdot)$ for one-dimensional Lebesgue measure *in this lemma only*.

**Lemma 14.10** *Let $S$ be a balanced interval in $\mathbb{R}$. Let $A$ be a Borel subset of finite measure and $A^*$ the balanced interval with $\mu_1(A^*) = \mu_1(A)$. Then for any $x \in \mathbb{R}$,*

$$\mu_1(A^* \cap S) - \mu_1(A \cap S) \geq \mu_1((A \backslash S) \cap [(S \backslash A) + x]) \tag{14.44}$$

*Remark* This is not restricted to a one-dimensional result. If $A^* \subset S$ or $S \subset A^*$, and $A$ is equimeasurable with $A^*$, then (14.49) below holds in any measure space and (14.44) in any abelian group with Haar measure.

*Proof* For any sets $C, D$, by considering $C \backslash D$, $C \cap D$, and $D \backslash C$,

$$\mu_1(C \triangle D) - (\mu_1(C) - \mu_1(D)) = 2\mu_1(D \backslash C) \tag{14.45}$$

so by symmetry,

$$\mu_1(C \triangle D) - |\mu_1(C) - \mu_1(D)| \geq 2 \min(\mu_1(D \backslash C), \mu_1(C \backslash D)) \tag{14.46}$$

Since either $A^* \subseteq S$ or $S \subseteq A^*$, (14.45) implies

$$\begin{aligned} \mu_1(A^* \triangle S) &= |\mu_1(A^*) - \mu_1(S)| \\ &= |\mu_1(A) - \mu_1(S)| \\ &\leq \mu_1(A \triangle S) - 2 \min(\mu_1(A \backslash S), \mu_1(S \backslash A)) \end{aligned} \tag{14.47}$$

by (14.46).

Since

$$\mu_1(A \triangle B) = \mu_1(A) + \mu_1(B) - 2\mu_1(A \cap B) \tag{14.48}$$

(14.48) implies that

$$\begin{aligned} \mu_1(A^* \cap S) - \mu_1(A \cap S) &\geq \min(\mu_1(A \backslash S), \mu_1(S \backslash A)) \\ &= \min(\mu_1(A \backslash S), \mu_1((S \backslash A) + x)) \\ &\geq \mu_1((A \backslash S) \cap [(S \backslash A) + x]) \end{aligned} \tag{14.49}$$

since $\mu_1(C \cap D) \leq \min(\mu_1(C), \mu_1(D))$. □

**Proposition 14.11** (i) *$\rho$ is a metric on the equivalence classes of Borel set modulo sets of measure zero.*

(ii) *For any $g \in \mathcal{G}$,*

$$|gA \cap gB| \geq |A \cap B| \tag{14.50}$$

(iii) *For any $g \in \mathcal{G}$,*

$$\rho(gA, gB) \leq \rho(A, B) \tag{14.51}$$

(iv) *If $A$ has finite measure, $S$ is the ball with the same volume and $\rho(\sigma_e[A], S) = \rho(A, S)$ for all Steiner symmetrizations, $\sigma_e$, then $A = S$ modulo sets of measure zero.*

(v) *If $A$ has finite measure and $\rho(gA, A) = 0$ for all $g \in \mathcal{G}$, then $A$ is (up to a set of measure zero) a ball centered at the origin.*

*Proof* (i) Clearly, $\rho(A, B) = 0$ if and only if $A$ and $B$ are equal modulo sets of measure zero. $\rho(A, B) = \rho(B, A)$ is obvious and the triangle inequality follows from

$$\rho(A, B) = \int |\chi_A - \chi_B| \, d^\nu x \tag{14.52}$$

or from $A \triangle C \subset (A \triangle B) \cup (B \triangle C)$ (look at the Venn diagrams!).

(ii) This follows for $\sigma_e$ from (14.28) done on the $x_1$ integrals only and then inductively on finite products of $\sigma_e$'s.

(iii) Immediate from (14.50) and (14.48).

(iv) Let $f = \chi_A(1 - \chi_S) = \chi_{A \setminus S}$ and $g = \chi_S(1 - \chi_A) = \chi_{S \setminus A}$. Then since $|A| = |S|$,

$$|A \setminus S| = |S \setminus A| = \tfrac{1}{2}\, \rho(A, S)$$

so $\|f\|_1 = \|g\|_1 = \frac{1}{2}\rho(A, S)$. Thus,

$$\begin{aligned} \int d^\nu x \int d^\nu y \, f(y) g(y + x) &= \int d^\nu y f(y) \int d^\nu x \, g(y + x) \\ &= \tfrac{1}{4}\, \rho(A, S)^2 \end{aligned}$$

It follows that if $\rho(A, S) \neq 0$, then for some $x \neq 0$,

$$q = \int d^\nu y \, f(y) g(y - x) > 0 \tag{14.53}$$

Pick an orthonormal coordinate system where $x = (0, 0, \dots, \alpha)$. For $\tilde{y} \equiv (y_1, \dots, y_{\nu-1})$ fixed, let $A(\tilde{y})$, $S(\tilde{y})$ be the intersection of $A, S$ with the line $x_1 = y_1, \dots, x_{\nu-1} = y_{\nu-1}$. Let $A^*$ be the one-dimensional symmetrization, so

$$[A(\tilde{y})]^* = [\sigma_e(A)](\tilde{y}) \tag{14.54}$$

where $\sigma_e$ is Steiner symmetrization in direction $e = (0, \dots, 1)$. (14.53) says

$$\int d^{\nu-1}\tilde{y} \big| \, [A(\tilde{y}) \setminus S(\tilde{y})] \cap [S(\tilde{y}) \setminus A(\tilde{y}) + x] \, \big| = q$$

By (14.44) (i.e., Lemma 14.10) and (14.54), this implies

$$|\sigma_e(A) \cap S| - |A \cap S| \geq q$$

so by (14.48),

$$\rho(\sigma_e(A), S) \leq \rho(A, S) - 2q$$

We have thus shown that if $\rho(A, S) \neq 0$, there is a $\sigma_e$ with $\rho(\sigma_e(A), S) < \rho(A, S)$.

(v) This follows from (iv) and the following consequence of the triangle inequality,

$$|\rho(gA, S) - \rho(A, S)| \le \rho(gA, A) \qquad \square$$

We are heading towards showing that if $A$ is a bounded Borel set, there is a sequence $g_n \in \mathcal{G}$ so $g_n A \to S$, the ball with $|S| = |A|$, in the $\rho$-metric. A family of compactness results will be critical:

**Proposition 14.12** (i) *Fix $R$ and $C$. Then the family of sets, $A$, obeying*

$$\text{(a)} \qquad A \subset \{x \mid |x| \le R\} \tag{14.55}$$

$$\text{(b)} \qquad |A\triangle(A + x)| \le C|x| \quad \textit{for all } x \tag{14.56}$$

*is compact in the $\rho$-metric.*

(ii) *If, in some coordinate system $(x_1, \dots, x_n)$ on $\mathbb{R}^\nu$, $A$ has the property*

$$x \in A \Rightarrow \{y \mid |y_i| \le |x_i|,\, i = 1, \dots, \nu\} \subset A \tag{14.57}$$

*and $A$ obeys* (14.55), *then* (14.56) *holds with $C = 2^\nu \sqrt{\nu} R^{\nu-1}$.*

(iii) *If $e_1, \dots, e_\nu$ are an orthonormal basis of $\mathbb{R}^\nu$ and $\sigma_i$ is Steiner symmetrization in direction $e_i$, then $A = \sigma_\nu \dots \sigma_2\sigma_1(B)$ obeys* (14.57) *for an arbitrary set $B$.*

*Proof* (i) Since $\rho(A_n, A) \to 0$ if and only if $\|\chi_{A_n} - \chi_A\|_1 \to 0$ and (14.56) is equivalent to $\|\chi_A - \chi_{A+x}\| \le C|x|$, the set of $A$'s obeying (14.56) is closed in the $\rho$-topology and that is obviously true of (14.55). So it suffices to prove the set is precompact – and here we pull a rabbit out of the hat, namely, a theorem of M. Riesz (see [305, Thm. XIII.66]) that a set $C \subset L^1(\mathbb{R}^\nu, d^\nu x)$ is precompact if it obeys:

(a) $\sup_{f\in C} \|f\|_1 < \infty$

(b) For all $\varepsilon$, there is an $R$ so

$$\sup_{f\in C} \int_{|x|>R} |f(y)|\, d^\nu y < \varepsilon$$

(c) For all $\varepsilon$, there is a $\delta$ so that for $|y| < \delta$,

$$\sup_{f\in C} \int |f(x) - f(x + y)|\, d^\nu x < \varepsilon$$

(See the remark after the proof.) This implies the theorem, for any $L^1$ limit of characteristic functions is a characteristic function, since a subsequence converges pointwise.

(ii) (14.57) implies for each fixed $x_2^{(0)}, \dots, x_\nu^{(0)}$ and $\alpha$,

$$\begin{aligned} &|\{x \in A \mid x_i = x_i^{(0)},\, i = 2, \dots, \nu\}\triangle \\ &\qquad \{x - \alpha(1, 0, \dots) \in A \mid x_i = x_i^{(0)},\, i = 2, \dots, \nu\}| \le 2|\alpha| \end{aligned}$$

so

$$|A \triangle A_{(\alpha,0,0,\dots,0)}| \le 2^\nu R^{\nu-1} |\alpha|$$

where $A_x \equiv A + x$, so repeating for each coordinate,

$$\begin{aligned} |A \triangle A_{(\alpha_1,\dots,\alpha_\nu)}| &\le 2^\nu R^{\nu-1}(|\alpha_1| + \cdots + |\alpha_\nu|) \\ &\le \sqrt{\nu}\, 2^\nu R^{\nu-1} |\alpha| \end{aligned}$$

(iii) Let $C_i$ be the condition

$$x \in A,\ y_1 = x_1, \dots, y_{i-1} = x_{i-1}, \dots, y_\nu = x_\nu \text{ and } |y_i| \le |x_i| \Rightarrow y \in A$$

Then after applying $\sigma_i$ to any set, $C_i$ clearly holds. Moreover, if $C_j$ holds for a set $D$, and $i \neq j$, it holds for $\sigma_i(D)$ since $\sigma_i$ merely rearranges the slices orthogonal to the $i$-axis. Thus, $C_1$ holds for $\sigma_1(B)$, $C_2$ and $C_1$ for $\sigma_2(\sigma_1(B)), \dots, C_\nu, \dots, C_1$ for $\sigma_\nu \sigma_{\nu-1} \dots \sigma_1(B)$. If $C_1, \dots, C_\nu$ all hold, then (14.56) holds. □

*Remark* The Riesz theorem says the conditions (a)–(c) stated in the proof are not only sufficient but also necessary. The proof of the direction we need is easy: to cover $C$ with $\varepsilon$-balls, we first use (b) to restrict to functions inside a ball modulo an $\varepsilon/2$ error. Then by (c), we can approximate the functions, $f$, uniformly by suitable $j_\delta * f$ with $j_\delta$ an approximate identity. The $\{j_\delta * f\}$ are uniformly equicontinuous by (a), so Arzelà's theorem says they can be covered by finitely many $\varepsilon/3$ balls in $\|\cdot\|_\infty$, and so in $\|\cdot\|_1$ norm.

Here is the key result which will let us go from (14.36) in case $\nu = 1$ to general $\nu$.

**Theorem 14.13** *Let $A_1, \dots, A_\ell$ be bounded Borel sets in $\mathbb{R}^\nu$ and let $S$ be the ball centered at 0 with $|S| = |A_1|$. Then there exist $g_n \in \mathcal{G}$ and sets $\tilde{A}_2, \dots, \tilde{A}_\ell$, so in the $\rho$-metric,*

$$g_n A_1 \to S, \qquad g_n A_j \to \tilde{A}_j, \qquad j = 2, \dots, \ell \tag{14.58}$$

*Proof* We will pick unit vectors $e_1, e_{\nu+1}, e_{2\nu+1}, \dots$ inductively in a way we will describe momentarily. But once we pick $e_{m\nu+1}$, we will choose $e_{m\nu+2}, \dots, e_{m\nu+\nu}$ to fill out an orthonormal basis in order to use (iii) of the last proposition. Then define $h_0 =$ identity and

$$h_m = \sigma_{e_{m\nu}} \sigma_{e_{m\nu-1}} \dots \sigma_{e_1} \tag{14.59}$$

Pick $e_{m\nu+1}$ so

$$\rho(\sigma_{e_{m\nu+1}}(h_m A_1), S) \le \gamma_m + \frac{1}{m} \tag{14.60}$$

where $\gamma_m = \inf_e \rho(\sigma_e(h_m A_1), S)$.

By Proposition 14.12 and the fact that the $A$'s all lie in some ball and $g(A_i)$ must lie in the same ball, $\{h_m A_i\}_{m=1,\dots,\, i=1,\dots,\ell}$ lies in a compact set, so we can pick a subsequence $g_n$ of the $h_m$'s and $\tilde{A}_1, \dots, \tilde{A}_\ell$ so $g_n A_i \to \tilde{A}_i$ for $i = 1, \dots, \ell$. All that remains is to use (14.60) to show $\tilde{A}_1 = S$.

Fix some unit vector $\eta \in \mathbb{R}^\nu$ and note (14.60) implies

$$\rho(\sigma_{e_{m\nu+1}}(h_m A_1), S) \leq \rho(\sigma_\eta(h_m A_1), S) + \frac{1}{m} \tag{14.61}$$

Pick the subsequence of $m$'s for which the $h_m$'s correspond to the $g_n$'s. Since $g_n A_1 \to \tilde{A}_1$, $\sigma_\eta$ is continuous by Theorem 14.5 and the $m$'s go to infinity, the right side of (14.61) goes to $\rho(\sigma_\eta(\tilde{A}_1), S)$.

On the other hand, by (14.51) and the fact that $\sigma_e(S) = S$ for all $e$, we have that $\alpha_k = \rho(\sigma_{e_k}\sigma_{e_{k-1}} \dots \sigma_{e_1}(A_1), S)$ is monotone decreasing. Since a subsequence of the $\alpha_{m\nu}$ converges to $\rho(\tilde{A}_1, S)$ (since $g_n A_1 \to \tilde{A}_1$), the whole sequence does and, in particular, the left side of (14.61) does. Hence

$$\rho(\tilde{A}_1, S) \leq \rho(\sigma_\eta(\tilde{A}_1), S)$$

By (14.51), the opposite inequality holds. Thus, by (iv) of Proposition 14.11, $\tilde{A}_1 = S$. □

We can now complete

*Proof of Theorem 14.8* By the wedding cake representation, we need only prove the result for each $f_j$, a characteristic function, and as noted, we can insert $\prod_{m=1}^n \chi_r(x_m)$ and so without loss suppose the sets involved $A_1, \dots, A_\ell$ are all bounded.

By the one-dimensional result we have proven, the integral can only increase if all the $A$'s undergo the same Steiner symmetrization. By Theorem 14.13, we can replace $A_1$ by an equimeasurable sphere $S_1$ and $A_2, \dots, A_\ell$ by equimeasurable sets, and, by Hölder's inequality, the integrals converge. Then apply Theorem 14.13 to $A_2$ using the fact that $\sigma_e(S_1) = S_1$ to replace $\tilde{A}_2$ by a sphere. After $\ell$ steps, we obtain an upper bound where each $\chi_{A_i}$ is replaced by $\chi_{A_i}^*$. □

Having proven the BLL inequalities, we turn to some applications. The nice thing is that while we had to work hard to prove the BLL inequalities, the applications are all very easy once one has the right setup. The first involves some general isoperimetric inequalities, starting with the classical one. Rather than measure surface area by points near $A$ but outside $A$ as we did in (13.5), we will measure it with points inside $A$ but near its complement,

$$s_i(A) = \liminf |\{x \in A \mid \text{dist}(x, A^c) < \varepsilon\}|/\varepsilon \tag{14.62}$$

where $A^c = \mathbb{R}^\nu \backslash A$. This agrees with the number $s(A)$ in (13.5) for sets with "reasonable" boundary.

**Theorem 14.14** *For any Borel set, A,*

$$s_i(A) \geq s_i(A^*) \tag{14.63}$$

*where $A^*$ is the ball with the same volume as A.*

*Proof* Let $\chi_\varepsilon$ be the characteristic function of the ball of radius $\varepsilon$ and $V_\varepsilon$ its volume

$$V_\varepsilon^{-1}(\chi_\varepsilon * \chi_A)(x) = \begin{cases} 1, & |\{y \mid |y-x| < \varepsilon\} \cap A| = V_\varepsilon \\ < 1, & \text{otherwise} \end{cases}$$

In particular, if $d(x, A^c) \geq \varepsilon$, this convolution is 1. On the other hand, for $A^*$, the convolution is 1 precisely on the set where $d(x, A) \geq \varepsilon$. Note if $0 \leq f \leq 1$, $|\{x \mid f(x) = 1\}| = \lim_{n\to\infty} \int f(x)^n\, dx$. Thus,

$$\begin{aligned} |\{x \mid d(x, A^c) \geq \varepsilon\}| &\leq |\{x \mid V_\varepsilon^{-1}(\chi_\varepsilon * \chi_A)(x) \leq 1\}| \\ &= \lim_{n\to\infty} \int [V_\varepsilon^{-1}\chi_\varepsilon * \chi_A]^n(x)\, dx \\ &\leq \lim_{n\to\infty} \int V_\varepsilon^{-1}[\chi_\varepsilon * \chi_{A^*}]^n(x)\, dx \\ &= |\{x \mid V_\varepsilon^{-1}(\chi_\varepsilon * \chi_A)(x) = 1\}| \\ &= |\{x \mid d(x, (A^*)^c) \geq \varepsilon\}| \end{aligned} \tag{14.64}$$

We use the BLL inequalities in (14.64). Subtracting this inequality from $|A| = |A^*|$, dividing by $\varepsilon$, and taking limits, we obtain (14.63). □

Here is an inequality related to isoperimetric inequalities. To state it in broad generality, consider a general potential $V$ on $\mathbb{R}^\nu$ and define

$$V_{n,m} = \begin{cases} n, & V(x) \geq n \\ V(x), & -m \leq V(x) \leq n \\ -m, & V(x) \leq -m \end{cases}$$

and

$$H(V) = \operatorname*{f\text{-}lim}_{m\to\infty} \left[ \operatorname*{f\text{-}lim}_{n\to\infty} -\Delta + V_{n,m} \right] \tag{14.65}$$

where f-lim means convergence in the sense of quadratic forms (see Kato [189]). By the monotone convergence theorem for forms, the limit as $n \to \infty$ always exists [189] and there is convergence in strong resolvent sense (srs). So long as $H(V)$ is bounded below (in the sense that inf spec$(-\Delta + V_{\infty,m})$ is bounded below), the limit as $m \to \infty$ also exists [189] in srs. Moreover, if we define

$$E(V) = \inf \operatorname{spec}(H(V)) \tag{14.66}$$

when $H(V)$ is bounded below, then $E$ is always continuous as $m \to \infty$ (since a strong limit of resolvents can at worst decrease norms but the resolvents are increasing). In reasonable cases (e.g., $V \in L^1_{\text{loc}}$), this is also true as $n \to \infty$. We are interested in rearrangement of potentials which are negative:

**Theorem 14.15** *Let $W$ be a nonnegative function on $\mathbb{R}^\nu$ obeying* (14.6) *and let $W^*$ be its symmetric rearrangement. Then if $H(-W^*)$ is semibounded, so is $H(-W)$ and*

$$E(-W^*) \le E(-W) \tag{14.67}$$

*Remark* In many examples, $E(-W)$ is the lowest eigenvalue, the ground state energy.

*Proof* Let $W_m = \min(m, W)$ so $(W^*)_m = (W_m)^*$. Since $E(-W) = \lim_{m\to\infty} E(-W_m)$ with $E(-W_m)$ decreasing, it suffices to prove the inequality (14.67) for bounded $W$. Pick $a > m$. Then

$$\|(H(-W)+a)^{-1}\| = (E(-W)+a)^{-1} \tag{14.68}$$

by the spectral theorem (and $E(-W) \ge m$). Thus, (14.67) follows from

$$\|(H(-W)+a)^{-1}\| \le \|(H(-W^*)+a)^{-1}\| \tag{14.69}$$

Since $\|\varphi\|_2 = \|\,|\varphi|\,\|_2 = \|\,|\varphi|^*\|_2$, this in turn follows from

$$|\langle\varphi, (H(-W)+a)^{-1}\psi\rangle| \le \langle|\varphi|^*, H(-W^*)+a^{-1}|\psi|^*\rangle \tag{14.70}$$

To prove this, let $H_0 = -\Delta$ and note that since $|W| \le m$ and $\|(H_0+a)^{-1}\| = a^{-1}$, $\|W(H_0+a)^{-1}\| < 1$, and similarly for $W^*$. Thus, we have norm convergent series

$$(H(-W)+a)^{-1} = \sum_{n=0}^{\infty}(H_0+a)^{-1}(W(H_0+a)^{-1})^n$$

so (14.70) follows from

$$|\langle\varphi, (H_0+a)^{-1}[W(H_0+a)^{-1}]^n\psi\rangle| \le \langle|\varphi|^*, (H_0+a)^{-1}[W(H_0+a)^{-1}]^n|\psi|^*\rangle \tag{14.71}$$

Once we have the lemma below, (14.71) is a direct consequence of the BLL inequality. □

**Lemma 14.16** *Let $H_0 = -\Delta$ on $L^2(\mathbb{R}^\nu)$. Then $(H_0+a)^{-1}$ is convolution with a symmetric decreasing function.*

*Proof* One can deduce this from a detailed study of the kernel as a Bessel function, but here is a direct proof using the fact that we know the integral kernel of $e^{-tH_0}$, so by writing the resolvent as the Laplace transform of the semigroup, $(H_0+a)^{-1}$ is convolution with the function

$$G_0(x;a) = \int_0^\infty (4\pi t)^{-\nu/2} e^{-x^2/4t} e^{-at}\,dt \tag{14.72}$$

which is clearly symmetric decreasing since each $e^{-x^2/4t}$ is. □

While Theorem 14.15 refers to negative potentials which go to zero at infinity, it can be modified slightly to deal with positive potentials which go to infinity at infinity. First, we need a definition.

**Definition** Let $V$ be a nonnegative function on $\mathbb{R}^\nu$ so that for each $\alpha$, $|\{x \mid V(x) < \alpha\}| < \infty$. Then $V^\star$, the *symmetric increasing rearrangement* of $V$, is the unique function

(i) $|\{x \mid V^\star(x) < \alpha\}| = |\{x \mid V(x) < \alpha\}|$ for all $\alpha$.
(ii) $V^\star$ is spherically symmetric.
(iii) $0 \le x \le y$ implies $V^\star(y) \ge V^\star(x)$.
(iv) $V^\star$ is usc.

**Theorem 14.17** *Let V be positive on $\mathbb{R}^\nu$ and go to infinity at infinity. Then*

$$E(V^\star) \le E(V) \tag{14.73}$$

*If $\{\lambda_n(V)\}_{n=1}^\infty$ are the eigenvalues of $H(V)$ counting multiplicity, then for any $t > 0$,*

$$\sum_{n=1}^\infty e^{-t\lambda_n(V)} \le \sum_{n=1}^\infty e^{-t\lambda_n(V^\star)} \tag{14.74}$$

*Remarks* 1. If $V \to \infty$ at $\infty$, $H(V)$ has only discrete spectrum; see [305, Thm. XIII.67].

2. (14.74) is intended in the sense that if the right side is finite, so is the left.

3. (14.73) is related to (14.74) in that if the sums in (14.74) are finite for any $t > 0$, then as $t \to \infty$, they imply (14.73), since if $\sum_{n=1}^\infty e^{-t_0\lambda_n} < \infty$, then

$$\lim_{t\to\infty} t^{-1} \log\left(\sum_{n=1}^\infty e^{-t\lambda_n}\right) = -\lambda_0$$

*Proof* Let $V_m = \min(m, V)$ and $W_m = -(V_m - m) \ge 0$. Then $W_m^* = m - (V^\star)_m$. Thus,

$$\begin{aligned} E(V_m) &= m + E(-W_m) \\ &\ge m + E(-W_m^*) \\ &= m + E(-m + (V^\star)_m) \\ &= E((V^\star)_m) \end{aligned}$$

Now take $m \to \infty$ and obtain (14.73).

(14.74) will require some tools: the Trotter product formula in a refined form (see [349]) and the theory of trace class and Hilbert–Schmidt operators (see [350]). Basically, (14.74) says

$$\mathrm{tr}(e^{-tH(V)}) \le \mathrm{tr}(e^{-tH(V^\star)}) \tag{14.75}$$

and this is true if

$$\int K_{t/2}(x,y;V)^2\,dx\,dy \le \int K_{t/2}(x,y;V^\star)^2\,dx\,dy \tag{14.76}$$

where $K_s(x,y;V)$ is the integral kernel of $e^{-sH(V)}$. (14.76) comes from $\operatorname{tr}(e^{-tH}) = \|e^{-tH/2}\|_2^2$, where $\|\cdot\|_2$ is the Hilbert–Schmidt norm (see [350]). By the Trotter product formula (in a refined form that holds for integral kernels; see [349]),

$$e^{-t(H_0+V)}(x,y) = \lim_{n\to\infty}\biggl[\int e^{-tH_0/n}(x_1-x_2)e^{-tV(x_2)}\dots \\ \dots e^{-tH_0/n}(x_{n-1}-x_n)e^{-tV(x_n)}\,dx_2\dots dx_{n-1}\Big|_{x_0=x_1-x_n=y}\biggr] \tag{14.77}$$

(14.76) then follows from the BLL inequalities and

$$(e^{-tV})^* = e^{-tV^\star} \tag{14.78}$$

□

For the next isoperimetric inequality we want to consider, we need a fact about symmetric rearrangements of independent interest:

**Theorem 14.18** *Let $H_0 = -\Delta$ on $L^2(\mathbb{R}^\nu)$, and let $q_{H_0}$ be its quadratic form on $Q(H_0)$, the quadratic form domain. Then if $\varphi \in Q(H_0)$, we have $|\varphi|^* \in Q(H_0)$ and*

$$q_{H_0}(|\varphi|^*) \le q_{H_0}(\varphi) \tag{14.79}$$

*Remark* Formally,

$$q_{H_0}(\varphi) = \int |\nabla\varphi|^2\,d^\nu x \tag{14.80}$$

and this holds in classical sense if $\varphi \in C_0^\infty$, and in distributional sense in general since $Q(H_0)$ is exactly those $\varphi \in L^2$ whose distributional derivatives lie in $L^2$ and (14.80) holds. Thus, (14.79) can be written

$$\int |(\nabla|\varphi|^*(x)|^2\,d^\nu x \le \int |(\nabla\varphi)(x)|^2\,d^\nu x \tag{14.81}$$

*Proof* For any semibounded self-adjoint operator, $A$, we have

$$q_A(\varphi) = \lim_{t\downarrow 0} t^{-1}\langle\varphi,(1-e^{-tA})\varphi\rangle \tag{14.82}$$

(both sides may be infinite) by the spectral theorem since

$$t^{-1}(1-e^{-tx}) = x\int_0^1 e^{-t\alpha x}\,d\alpha$$

converges monotonically upwards to $x$. For any $\varphi$,

$$\langle \varphi, \varphi \rangle = \langle |\varphi|^*, |\varphi|^* \rangle$$
$$\langle \varphi, e^{-tH_0} \varphi \rangle \le \langle |\varphi|, e^{-tH_0} |\varphi| \rangle \le \langle |\varphi|^*, e^{-tH_0} |\varphi|^* \rangle \tag{14.83}$$

by the BLL inequality, since $e^{-tH_0}$ is given by convolution with a spherically symmetric decreasing function. Thus, by (14.82), (14.79) holds where one or both sides may be infinite. □

For each open set $\Omega \subset \mathbb{R}^\nu$, we defined the Dirichlet Laplacian, $-\Delta_\Omega^D$, by closing the quadratic form $\varphi \mapsto \|\nabla \varphi\|_2^2$ on $C_0^\infty(\Omega) \subset L^2(\Omega)$. Define

$$e_D(\Omega) = \inf \sigma(-\Delta_\Omega^D) \tag{14.84}$$

so by the definition,

$$e_D(\Omega) = \inf\{\|\nabla\varphi\|_2^2 \mid \varphi \in C_0^\infty(\Omega),\ \|\varphi\|_2 = 1\} \tag{14.85}$$

If $\Omega$ is bounded, $e_D(\Omega)$ is an eigenvalue, called the *Dirichlet ground state energy.*

**Theorem 14.19** (Faber–Krahn Inequality) *Let $\Omega \subset \mathbb{R}^\nu$ be a bounded open set and let $\Omega^*$ be the open ball with $|\Omega| = |\Omega|^*$. Then*

$$e_D(\Omega^*) \le e_D(\Omega \tag{14.86}$$

*Remark* One can also obtain this from (14.74) by taking $V_\lambda(x) = \lambda \operatorname{dist}(x, \Omega)^2$ and taking $\lambda$ to infinity, and proving $e_D(\Omega^*) = \lim_{\lambda\to\infty} e(V_\lambda^*)$ and $e_D(\Omega) = \lim_{\lambda\to\infty} e(V)$.

*Proof* We have, by (14.85), that

$$e_D(\Omega) = \inf\{\|\nabla\varphi\|_2^2 \mid \varphi \in Q(H_0),\ \text{supp}(\varphi) \subset \Omega\}$$

Thus, by (14.81),

$$\begin{aligned} e_D(\Omega) &\ge \inf\{\|\nabla|\varphi|^*\|_2^2 \mid \varphi \subset Q(H_0),\ \text{supp}(\varphi) \in \Omega\} \\ &\ge \inf\{\|\nabla\psi\|_2^2 \mid \psi \in Q(H_0),\ \text{supp}(\psi) \subset \Omega^*\} \\ &= e_D(\Omega^*) \end{aligned} \tag{14.87}$$

since $\text{supp}(\varphi) \subset \Omega$ implies $\text{supp}(|\varphi|^*) \subset \Omega^*$.

The equality (14.87) says that

$$\inf(\|\nabla\psi\|_2^2 \mid \psi \in Q(H_0),\ \text{supp}(\psi) \in \Omega^*) = \inf(\|\nabla\psi\|_2^2 \mid \psi \in C_0^\infty(\Omega^*))$$

for $\Omega^*$ the open balls. That is, we can approximate $\psi \in Q(H_0)$ with $\text{supp}(\psi) \subset \Omega^*$ by functions in $C_0^\infty$. Since $\text{supp}(\psi)$ is compact, if $j_\delta$ is a standard approximate identity, then $\text{supp}(j_\delta * \psi) \in C_0^\infty(\Omega)$ for $\delta$ small. As $j_\delta * \psi \to \psi$ in $Q(H_0)$, so $\|\nabla(j_\delta * \psi)\|_2^2 \to \|\nabla\psi\|_2^2$. □

Along the same lines, we have an isoperimetric inequality for torsional rigidity – a quantity defined for two-dimensional regions in elasticity theory. One definition is the following: If $D$ is a bounded open region in $\mathbb{R}^2$, define $P(D)$ by

$$P(D)^{-1} = \inf\left(\frac{\|\nabla f\|_2^2}{4(\int_D f\, d^2x)^2}\right) \tag{14.88}$$

where the inf is taken over all $f \in Q(-\Delta)$ with $\text{supp}(f) \subset D$ and with $\int_D f \neq 0$.

**Theorem 14.20** *Let $D$ be an open bounded region in $\mathbb{R}^2$ and let $D^*$ be the disk of the same volume. Then*

$$P(D^*) \geq P(D) \tag{14.89}$$

*or equivalently,*

$$P(D) \leq \frac{|D|^2}{2\pi} \tag{14.90}$$

*Proof* We first claim that in (14.88), the inf need only be taken over positive $f$'s. For the first inequality in (14.83) and the argument leading to (14.81) implies that

$$\|\nabla |f|\,\|_2^2 \leq \|\nabla f\|_2^2 \tag{14.91}$$

Since $(\int |f|\, d^2x)^2 \geq (\int f\, d^2x)^2$, the ratio can only go down if $f$ is replaced by $|f|$.

For any positive $f$, $\|\nabla f\|_2^2 \geq \|\nabla f^*\|_2^2$ and $\int f\, d^2x = \int f^*\, d^2x$. Moreover, for $D^*$, the inf over symmetric decreasing $f$ is certainly no smaller than the inf over all $f$'s. Thus, $P(D)^{-1} \geq P(D^*)^{-1}$, which is (14.89).

To obtain (14.90), we note first that $P(D)$ scales like the square of the area (since $\|\nabla f\|_2^2$ is scaling invariant), so it suffices to prove $P(D_0) = 2\pi$ where $D_0$ is the unit disk (with $|D_0| = 2\pi$). A variational argument shows that the minimizer in (14.88) obeys $\Delta f = c$ with $f \restriction \partial D_0 = 0$. $c$ drops out of the ratio, so we can take $f(x,y) = 1 - x^2 - y^2$, and then direct calculation shows that $\|\nabla f\|_2^2/4(\int f)^2 = 1/2\pi$. □

Our final pair of isoperimetric inequalities concerns the Coulomb energy:

**Definition** Given $f \in L^1(\mathbb{R}^\nu)$ and $\nu \geq 3$ with $f \geq 0$, we define the *Coulomb energy* by

$$\mathcal{E}(f) = \frac{1}{2}\int \frac{f(x)f(y)}{|x-y|^{\nu-2}}\, d^\nu x\, d^\nu y \tag{14.92}$$

The Coulomb energy, $e_c(\Omega)$, of a region, $\Omega \subset \mathbb{R}^\nu$, is $\mathcal{E}(\chi_\Omega)$ with $\chi_\Omega$ the characteristic function of $\Omega$. We define the *capacity*, $\text{cap}(\Omega)$, of a bounded open set, $\Omega$, by

$$\frac{1}{2\,\text{cap}(\Omega)} = \inf\left\{\mathcal{E}(f) \mid f \geq 0,\ \int f(x)\, d^\nu x = 1\right\} \tag{14.93}$$

*Remarks* 1. In (14.92), the integral may be divergent, in which case we set $\mathcal{E}(f) = \infty$.

2. One can define $\mathcal{E}(f)$ when $f\, d^\nu x$ is replaced by a measure; indeed, if $\mu$ is a finite signed measure, one can define $\mathcal{E}(\mu)$ by going to Fourier transforms in such a way that if $\mathcal{E}(\mu) < \infty$, one can show $\mathcal{E}(\mu)$ is given by the integral $\int d\mu(x)\, d\mu(y) |x-y|^{-\nu-2}$.

3. The funny $(2\,\mathrm{cap}(\Omega))^{-1}$ in (14.93) comes from the well-known formula in physics that the energy stored in a capacitor storing charge $Q$ is $\frac{1}{2}Q^2/C$. There are other equivalent definitions (see the discussion in [231]) which look very different: for example, for a suitable $\nu$-dependent constant $d_\nu$, and $\Omega$ open,

$$\mathrm{cap}(\Omega) = \inf\left\{ d_\nu \int |\nabla\varphi|^2\, d^\nu x \;\middle|\; \varphi \in Q(-\Delta),\ \varphi \geq 1 \text{ on } \Omega \right\} \tag{14.94}$$

The usual definition of type (14.93) does not require $f \geq 0$, but one can show the inf is the same.

4. There is no minimum for (14.93) among $f \in L^1(\Omega)$. Rather, there will be a minimizing measure on $\bar{\Omega}$ supported on $\partial\Omega$.

5. In some ways, our decision to restrict capacity to open sets is "wrong." Capacity is most often associated to closed, or even general, sets that may have empty interior! We make this choice to have a very quick isoperimetric inequality. For a more complete result and discussion, see [231].

**Theorem 14.21** *For any region $\Omega \subset \mathbb{R}^\nu$ of finite volume,*

$$e_c(\Omega) \leq e_c(\Omega^*) \tag{14.95}$$

*where $\Omega^*$ is the ball of the same volume as $\Omega$.*

*Proof* This is an immediate consequence of the BLL inequality. □

**Theorem 14.22** *Let $\Omega$ be a bounded open set in $\mathbb{R}^\nu$ and let $\Omega^*$ be the ball of the same volume as $\Omega$. Then*

$$\mathrm{cap}(\Omega^*) \leq \mathrm{cap}(\Omega) \tag{14.96}$$

*Proof* $\mathcal{E}(f) \leq \mathcal{E}(f^*)$ by the BLL inequality. Thus, the inf over all $f \in L^1(\Omega)$ with $\int f\, d^\nu x = 1$ and $f > 0$ is less than the inf over all $\mathcal{E}(f^*)$. For a ball, the inf is over all $f^*$. □

The other traditional use of rearrangements concerns generalized Young inequalities. We saw (see Theorem 12.8) that Sobolev inequalities are a special case of generalized Young. The point is that, using BLL, one can go backwards from the special case:

**Theorem 14.23** *Suppose for some $\nu$, $\alpha < n$, $p$, $r$, and $C$,*

$$\int \frac{|f(x)|\,|g(y)|}{|x-y|^{\alpha}}\, d^{\nu}x\, d^{\nu}y \le C\|f\|_p\,\|g\|_r \tag{14.97}$$

*for all $f \in L^p(\mathbb{R}^\nu)$ and $g \in L^p(\mathbb{R}^\nu)$. Then for all $h \in L_w^{n/\alpha}$,*

$$\int |h(x-y)|\,|f(x)|\, d^{\nu}x\, d^{\nu}y \le C\tau_{\nu}^{-\alpha/\nu}\|f\|_p\,\|g\|_r\,\|h\|^*_{n/\alpha,w} \tag{14.98}$$

*where $\tau_\nu$ is the volume of the unit ball in $\nu$-dimensions.*

*Remark* By scaling, if (14.97) holds, one must have $\frac{1}{p}+\frac{1}{r}+\frac{\alpha}{\nu}=2$ and, in that case, it does hold as we have already seen. (14.97) relates optimal constants.

*Proof* By definition, (12.34) of $\|h\|^*_{p,w}$, if $q=\nu/\alpha$,

$$|\{x \mid h(x) > \lambda\}| \le (\|h\|^*_{p,w})^q \lambda^{-q}$$

The ball of radius $r$ has volume $\tau_\nu r^\nu$, so $h^*(x) > \lambda$ precisely in a ball of radius

$$r_\lambda \le (\|h\|^*_{p,w}\lambda^{-1}\tau_\nu^{-\alpha/\nu})^{1/\alpha}$$

which means

$$h^*(x) \le \tau_\nu^{-\alpha/\nu}\frac{\|h\|^*_{p,w}}{|x|^\alpha}$$

Thus, the BLL inequalities and (14.97) imply (14.98). □

# 15

# Rearrangement inequalities, II. Majorization

We turn to the second major theme of this pair of chapters, the study of inequalities like (14.9) and its implication for bounds like (14.11) and, in particular, to prove Theorem 14.2. Here is a motivating example. Let $A$ be an $n \times n$ matrix with eigenvalues $\{\lambda_i(A)\}_{i=1}^n$. Let $|A| = \sqrt{A^*A}$ and let $\{\mu_i(A)\}_{i=1}^n$ be its eigenvalues. Then

$$\sum_{i=1}^n \lambda_i(A)^2 = \operatorname{tr}(A^2) \tag{15.1}$$

$$\sum_{i=1}^n \mu_i(A)^2 = \operatorname{tr}(A^*A) \tag{15.2}$$

Since $\langle A, B\rangle = \operatorname{tr}(A^*B)$ is an inner product on matrices, by the Schwarz inequality,

$$|\operatorname{tr}(A^2)| \le \operatorname{tr}(A^*A)^{1/2} \operatorname{tr}(AA^*)^{1/2} = \operatorname{tr}(A^*A) \tag{15.3}$$

by the cyclicity of the trace. Thus, by (15.1) and (15.2),

$$\left|\sum_{i=1}^n \lambda_i(A)^2\right| \le \sum_{i=1}^n |\mu_i(A)|^2 \tag{15.4}$$

Two questions immediately come to mind: Can one take the absolute value inside the sum in (15.4)? Does a similar result hold when 2 is replaced by $p \in [1, \infty)$? In fact, we will prove a result for any $p \in (0, \infty)$ (see Theorem 15.20).

To state the general context for Theorem 14.2, we will need some preliminaries:

**Definition** An $n \times n$ matrix is called *doubly stochastic* (ds) if and only if (rows and columns sum to 1):

$$\text{(i)} \qquad \sum_{i=1}^n a_{ij} = 1, \qquad j = 1, \dots, n \tag{15.5}$$

$$\text{(ii)} \qquad \sum_{j=1}^{n} a_{ij} = 1, \qquad i = 1, \dots, n \tag{15.6}$$

$$\text{(iii)} \qquad a_{ij} \geq 0, \qquad i, j = 1, \dots, n \tag{15.7}$$

The set of all such matrices will be denoted $\mathbb{D}_n$.

**Example 15.1** Let $\{\varphi_i\}_{i=1}^n$ and $\{\psi_j\}_{j=1}^n$ be two orthonormal bases and let $u_{ij} = \langle \varphi_i, \psi_j \rangle$ be the unitary change of basis. Let $a_{ij} = |u_{ij}|^2$. Then by the Plancherel theorem, the matrix is doubly stochastic. In particular, if $B$ is a self-adjoint $n \times n$ matrix, $\{\lambda_i\}_{i=1}^n$ its eigenvalues, and $\{\psi_j\}_{i=1}^n$ the eigenvectors, then the diagonal matrix elements of $b$ obey

$$\begin{aligned} b_{ii} &= \langle \delta_i, B\delta_i \rangle \\ &= \sum a_{ij} \lambda_j \end{aligned}$$

where $a_{ij} = |\langle \delta_i, \psi_j \rangle|^2$. The inequalities below (see Theorem 15.5) imply, for example, that

$$\sum_{i=1}^{n} |b_{ii}|^p \leq \sum_{i=1}^{n} |\lambda_i|^p \tag{15.8}$$

an inequality of Schur. This example is discussed further in Theorem 15.40. □

We want to begin by finding the extreme points of $\mathbb{D}$. We will need the following preliminary of general interest:

**Proposition 15.2** *Let $\{\ell_\alpha\}_{\alpha=1}^m$ be a finite number of linear functionals on $\mathbb{R}^\nu$. Let $\beta_1, \dots, \beta_m \in \mathbb{R}$. Let $K$ be a finite intersection of closed half-planes*

$$K = \bigcap_{\alpha=1}^{m} \{x \mid \ell_\alpha(x) \geq \beta_\alpha\} \tag{15.9}$$

*Let $x \in \mathcal{E}(K)$ be an extreme point of $K$. Then $x$ obeys at least $\nu$ distinct equations*

$$\ell_\alpha(x) = \beta_\alpha \tag{15.10}$$

*Remarks* 1. $K$ may not be compact, so $\mathcal{E}(K)$ might be empty (e.g., if $K = \{x \mid \alpha \leq x_1 \leq \beta\}$ and $\nu \geq 2$).

2. What the proof shows is the $\ell_\alpha$'s for which equality holds must span $(\mathbb{R}^\nu)^*$.

*Proof* Given $x$, renumber the $\ell$'s so $\ell_1(x) = \beta_1, \dots, \ell_k(x) = \beta_k, \ell_{k+1}(x) > \beta_{k+1}, \dots, \ell_m(x) > \beta_m$ and suppose $k < \nu$. Then

$$X = \{y \mid \ell_\alpha(y) = \beta_\alpha;\ \alpha = 1, \dots, k\}$$

has dimension at least $\nu - k \geq 1$. Let

$$U = \{y \in X \mid \ell_\alpha(y) > \beta_\alpha;\ \alpha = k+1, \dots, m\}$$

so $x \in U \subset K$. But $U$ is open in $X$, so any $x \in U$ is a midpoint of a nontrival line segment in $U$, and so is $x \notin \mathcal{E}(K)$. $\square$

**Corollary 15.3** *Let $K$ be an intersection of finitely many closed half-spaces in $\mathbb{R}^\nu$ and suppose $K$ is compact. Then $\mathcal{E}(K)$ is finite; in fact, if $K$ is an intersection of $m$ hyperplanes,*

$$\#(\mathcal{E}(K)) \leq \frac{2}{\nu}\binom{m}{\nu-1} \tag{15.11}$$

*Conversely, if $K \subset \mathbb{R}^\nu$ is the convex hull of a finite set of points, P, then $K$ is the intersection of finitely many closed half-spaces; in fact, if $P$ has $p$ points and $K$ has dimension $d \leq \nu$, then the number of half-spaces, $m$, needed is bounded by*

$$m \leq 2(\nu - d) + \frac{2}{d}\binom{p}{d-1} \tag{15.12}$$

*Remarks* 1. The convex hull of a finite number of points in $\mathbb{R}^\nu$ is called a *convex polytope*.

2. (15.11) has equality for a convex polygon in $\mathbb{R}^2$ with $m$ vertices ($\#(\mathcal{E}(K)) = m$) and $m$ sides. It also has equality for the simplex, $S_\nu$, in $\mathbb{R}^\nu$ whose extreme points are $0, e_1, e_2, \dots, e_\nu$ with $m = \nu + 1$. $S_\nu = \{x_i \geq 0,\ i = 1, \dots, \nu;\ \sum_{i=1}^{\nu} x_i \leq 1\}$ and $\mathcal{E}(K) = \nu + 1$.

3. (15.12) has equality for a convex polygon in $\mathbb{R}^2$ ($\nu = d = 2$; $p = m$) and for the simplex $S_\nu$ of Remark 2 ($d = \nu$, $p = \nu + 1$, and $m = \nu + 1$).

4. (15.11) is well-defined since $m \geq \nu+1$. For if $m \leq \nu$, the $\{\ell_\alpha\}_{\alpha=1}^m$ either span $(\mathbb{R}^\nu)^*$ or a smaller subspace. In the latter case, $\{x \mid \ell_\alpha(x) = 0\}$ is a subspace of dimension at least 1, so $K$ contains a translate of the subspace and is not bounded. If the $\ell_\alpha$'s span $(\mathbb{R}^\nu)^*$, they define a coordinate system and $K$ is obtained from the orthant $\{x \mid \ell_\alpha(x) \geq 0\}$ by translation and reflection and is not bounded either.

5. The right side of (15.11) may not be an integer; for example, $m = 5$, $\nu = 3$ (as occurs for a triangular prism in $\mathbb{R}^3$).

6. By Proposition 15.2, each $x \in \mathcal{E}(K)$ is the unique point where some $\nu$ distinct $\ell$'s have a given value (unique because the $\ell$'s can be picked to be independent). Thus, $\#(\mathcal{E}(K)) \leq \binom{m}{\nu}$. (15.11) improves on this by a factor of $2/(m - \nu + 1)$ (always $\leq 1$ by Remark 4).

*Proof* Let $N_{\nu,m}$ be the sup of $\#(\mathcal{E}(K))$ over all $K$'s in $\mathbb{R}^\nu$ which are the intersection of $m$ half-spaces. We will prove (15.11) inductively – proving at the same time that $N_{\nu,m} < \infty$. If $\nu = 1$, $K$ is a closed interval and $\#(\mathcal{E}(K)) = 2$, giving equality in (15.11).

Let $\{H_\alpha\}_{\alpha=1}^m$ be the hyperplanes

$$H_\alpha = \{y \mid \ell_\alpha(y) = \beta_\alpha\}$$

$H_\alpha$ is a face of $K$, so $\mathcal{E}(K) \cap H_\alpha$ are the extreme points of $H_\alpha$ by Proposition 8.6. $H_\alpha$ lies in a $\nu - 1$-dimensional space, so by induction, $\mathcal{E}(H_\alpha)$ has at most $N_{\nu-1,\,m-1}$ extreme points. By the last proposition, each extreme point lies in at least $\nu$ of the $H_\alpha$'s and so

$$\nu N_{\nu,m} \le \sum_{\alpha=1}^{m} \#(\mathcal{E}(H_\alpha)) \le m N_{\nu-1,\,m-1}$$

proving (15.11).

The proof of the converse is a lovely use of the bipolar theorem. By translating the points, we can suppose 0 is an intrinsic interior point of $K$. Let $V$ be the space spanned by $\{y_i\}_{i=1}^p$ so $K \subset V$. Suppose we show $K$ viewed as a subset of $V$ is an intersection of closed half-spaces of $V$. Since $V$ is an intersection of finitely many half-spaces (if $V$ has codimension $\ell$, $2\ell$ half-spaces are needed), $K$ is a finite intersection of half-spaces. Thus, we are reduced to the case where $\text{cch}(P)$ has a nonempty interior.

The polar of $P$ is

$$P^\circ = \{\ell \mid \ell(y_\gamma) \ge -1,\ \gamma = 1, \dots, p\} \tag{15.13}$$

since $\ell(y)$ must take its extreme values at one of the $y_\gamma$'s.

Thus, $P^\circ$ is an intersection of closed half-spaces and it is bounded since $\text{cch}(P) = P^{\circ\circ}$ (since $0 \in K^{\text{iint}}$) is open. Thus, by the first half, $P^\circ$ has finitely many extreme points $\{\ell_\alpha\}_{\alpha=1}^m$ and

$$\text{cch}(P) = P^{\circ\circ} = \{y \mid \ell_\alpha(y) \ge -1\}$$

is an intersection of closed subspaces.

To obtain (15.12), we need $2(\nu - d)$ half-spaces to define the affine subspaces generated by $K$, so (15.12) holds if we prove it for the case $\nu = d$. In that case, by the above argument, $m \le \#(\mathcal{E}(P^\circ))$. $P^\circ$ is the intersection of $p$ half-spaces by (15.13) and is compact since $d = \nu$. Thus, (15.12) follows form (15.11). □

**Definition** Let $\pi \in \sum_n$, the permutations group for $\{1, \dots, n\}$. The *permutation matrix*, $M_\pi$, is the $m \times m$ matrix,

$$(M_\pi)_{ij} = \delta_{i\pi(j)} \tag{15.14}$$

This choice is made so

$$(M_{\pi_1} M_{\pi_2})_{ij} = \sum_k \delta_{i\pi_1(k)} \delta_{k\pi_2(j)}$$

$$= \delta_{i(\pi_1\pi_2)(j)} = (M_{\pi_1\pi_2})_{ij}$$

that is, $\pi \mapsto M_\pi$ is a group homomorphism. Notice

$$(M_\pi x)_i = x_{\pi^{-1}(i)} \tag{15.15}$$

**Theorem 15.4** (Birkhoff's Theorem) *Let $\mathbb{D}_n$ be the $n \times n$ doubly stochastic matrices. Then $\mathbb{D}_n$ is a compact convex set and*

$$\mathcal{E}(\mathbb{D}_n) = \{M_\pi \mid \pi \in \Sigma_n\} \tag{15.16}$$

*Proof* Clearly, (15.5)–(15.7) define the intersection of closed half-spaces (since $\{x \mid \ell(x) = \beta\} = \{x \mid \ell(x) \geq \beta\} \cap \{x \mid \ell(x) \leq \beta\}$) and so convex and closed. Since they imply $a_{ij} \in [0, 1]$, $\mathbb{D}_n$ is bounded, so compact.

Each $M_\pi$ is an extreme point since the functional

$$\ell_\pi(A) = \sum_{i=1}^{n} a_{i\pi^{-1}(i)}$$

has $0 \leq \ell_\pi \leq n$ on $\mathbb{D}_n$ and $M_\pi$ is the unique point with $\ell_\pi(A) = n$.

The conditions (15.5)/(15.6) are not independent since the $n$ conditions in (15.5) and the first $n - 1$ in (15.6) imply the last condition in (15.6). Thus, the set of matrices obeying (15.5)/(15.6) is a subspace of dimension $n^2 - (2n - 1) = n^2 - 2n + 1$. (The independence of the other $2n - 1$ is not important – the critical part is the dimension is at least $n^2 - 2n + 1$.)

Now let $A$ be an extreme point of $\mathbb{D}_\pi$. By Proposition 15.2, $A$ must have equality in (15.7) in at least $n^2 - 2n + 1 = n(n - 2) + 1$ of the $(i, j)$ pairs. If no row had at least $n - 1$ zeros, only $n(n - 2)$ of the conditions would hold. Thus, some row must have at least $n - 1$ zeros. Since the sum is 1, the row has $n - 1$ zeros and a single 1. The column with the 1 must have all other elements 0.

Now consider the $(n-1) \times (n-1)$ matrix with their row and column removed. Because of where the zeros are, the new matrix is in $\mathbb{D}_{n-1}$. By an induction argument, the new matrix has a single 1 in each row and each column, and so defines an $M_\pi$. □

We are now ready to turn to the analysis of (14.9) and its consequences. We will go through six variants:

(a) (14.9) holds with equality for $k = n$.
(b) An analog of this analysis for positive matrices.
(c) (14.9) holds without demanding equality if $k = n$.
(d) (14.9) holds for the absolute values of two sequences.
(e) Analysis for $n = \infty$ for discrete variables.
(f) Analogs for general measure spaces.

To be explicit, we introduce some notation:

**Definition** Given $a \in \mathbb{R}^\nu$ and $k = 1, \dots, \nu$, define

$$S_k(a) = \sup_{\substack{j_1, \dots, j_k \in \{1, \dots, \nu\} \\ \text{distinct}}} \sum_1^k a_{j_k} \tag{15.17}$$

If $a \in \mathbb{R}^\nu_+$, the points with nonnegative coordinates, then

$$S_k(a) = \sum_{j=1}^{k} a_j^*, \qquad (a \in \mathbb{R}^\nu_+) \tag{15.18}$$

**Definition** $\mathcal{C}_\Sigma(\mathbb{R}^\nu)$ is the set of convex functions, $\Phi$, with

$$\Phi(M_\pi x) = \Phi(x)$$

for all $x \in \mathbb{R}^\nu$ and $\pi \in \Sigma_\nu$, that is, $\Phi$ is permutation invariant.

**Theorem 15.5** (The Hardy–Littlewood–Pólya Theorem, HLP Theorem for short) *Let $a, b \in \mathbb{R}^\nu_+$. The following are equivalent:*

(i) $S_k(b) \leq S_k(a)$ *for* $k = 1, \dots, \nu - 1$ *and* $S_\nu(b) = S_\nu(a)$
(ii) $b \in \text{cch}(\{M_\pi a \mid \pi \in \Sigma_\nu\})$
(iii) $b = Da$ *for some* $D \in \mathbb{D}_\nu$
(iv) $\Phi(b) \leq \Phi(a)$ *for all* $\Phi \in C_\Sigma(\mathbb{R}^\nu)$
(v)

$$\sum_{j=1}^{\nu} \varphi(b_j) \leq \sum_{j=1}^{\nu} \varphi(a_j) \tag{15.19}$$

*for all convex* $\varphi\colon \mathbb{R} \to \mathbb{R}$

(vi) *For all* $s \in \mathbb{R}_+$,

$$\sum_{j=1}^{\nu} (b_j - s)_+ \leq \sum_{j=1}^{\nu} (a_j - s)_+ \tag{15.20}$$

*and*

$$\sum_{j=1}^{\nu} b_j = \sum_{j=1}^{\nu} a_j \tag{15.21}$$

*Remarks* 1. HLP only proved the equivalence of (i), (iii), and (v). See the Notes for a discussion of the history of this theorem.

2. This, of course, includes Theorem 14.2.

3. It is remarkable that the inequality for the special convex functions of (15.19) implies $\Phi(b) \leq \Phi(a)$ for all symmetric convex functions.

4. This theorem holds for $a, b \in \mathbb{R}^\nu$ without assuming positive components by the simple device of picking $\alpha \in \mathbb{R}$ so $a + \alpha(1, \dots, 1)$ and $b + \alpha(1, \dots, 1)$ lie in $\mathbb{R}^\nu_+$ and noting that each condition holds before the $\alpha$ addition if and only if it holds afterwards. We need to define $S_k(a)$ by (15.17), not (15.18), for this to be true.

*Proof* We will show (i) ⇒ (ii) ⇔ (iii) ⇒ (iv) ⇒ (v) ⇒ (vi) ⇒ (i).

<u>(i) ⇒ (ii)</u> Suppose (i) holds but $b$ is not in $K \equiv \text{cch}(\sum M_\pi a\})$. Then by Theorem 5.5, there exists a linear functional on $\mathbb{R}^\nu$ so

$$\ell(b) > \sup_K \ell(x) = \max_\pi \ell(M_\pi a) \tag{15.22}$$

Write $\ell(x) = \sum_{j=1}^\nu \ell_j x_j$. Since (15.21) holds, we can add a constant to each $\ell_j$ without changing the validity of (15.22) so we can suppose that $\ell \in \mathbb{R}_+^\nu$. By (14.2),

$$\ell^*(b^*) \geq \ell(b) > \max_\pi \ell(M_\pi a) = \ell^*(a^*) \tag{15.23}$$

by taking the permutation $\pi = \pi_1^{-1}\pi_2$ with $M_{\pi_2}(a) = a^*$ and $M_{\pi_1}(\ell) = \ell^*$. But by (14.3) and (15.18),

$$\begin{aligned}\ell^*(a^*) - \ell^*(b^*) &= \ell_n^*(S_n(a) - S_n(b)) + (\ell_{n-1}^* - \ell_n^*)(S_{n-1}(a) - S_{n-1}(b)) \\ &\qquad + \cdots + (\ell_1^* - \ell_2^*)(S_1(a^*) - S_1(b^*)) \\ &\geq 0\end{aligned}$$

by (i). This contradicts (15.23) and proves that $b \in \text{cch}(\{M_\pi a\})$.

<u>(ii) ⇔ (iii)</u> If $b \in \text{cch}(\{M_\pi a\})$, then by Theorem 8.11, $b = \sum_{i=1}^k (\theta_i M_{\pi_i})a$ with $0 \leq \theta_i \leq 1$ and $\sum_{i=1}^k \theta_i = 1$. Then $D = \sum_{i=1}^k \theta_i M_{\pi_i} \in \mathbb{D}_\nu$, so (iii) holds. Conversely, by Birkhoff's theorem (Theorem 15.4) and Theorem 8.11, if $b = Da$ with $D \in \mathbb{D}_\nu$, then $D = \sum_{i=1}^k \theta_i M_{\pi_i}$ and so $b = \sum_{i=1}^k \theta_i M_{\pi_i} a \in \text{cch}(\{M_\pi a\})$.

<u>(ii) ⇒ (iv)</u> By convexity,

$$\sup_{b \in \text{cch}(\{M_\pi a\})} \Phi(b) \leq \max_{\pi \in \Sigma_\nu} \Phi(M_\pi a) = \Phi(a)$$

since $\Phi$ is symmetric.

<u>(iv) ⇒ (v)</u> Trivial since $\Phi(x_1, \dots, x_\nu) = \sum_{i=1}^\nu \varphi(x_i)$ lies in $C_\Sigma$.

<u>(v) ⇒ (vi)</u> (15.20) is trivial since $\varphi_s(x) = (x - s)_+$ is a convex function of $x$. (15.21) holds because $\varphi_\pm(x) = \pm x$ are both convex functions.

<u>(vi) ⇒ (i)</u> Suppose (vi) holds. Let $k \leq \nu - 1$ and $s = a_k^*$. Then

$$\begin{aligned}S_k(b) - sk &= \sum_{j=1}^k (b_j^* - s) && \text{(by (15.18))} \\ &\leq \sum_{j=1}^k (b_j^* - s)_+ && \text{(by } x \leq x_+) \\ &\leq \sum_j (b_j - s)_+ && \text{(by } (b_j - s)_+ \geq 0) \\ &\leq \sum_j (a_j - s)_+ && \text{(by (15.20))}\end{aligned}$$

$$= \sum_{j=1}^{k} (a_j^* - s) \qquad \text{(by choice of } s\text{)}$$

$$= S_k(a) - sk \qquad \text{(by (15.18))}$$

so $S_k(b) \leq S_k(a)$ for $k = 1, \dots, \nu - 1$. By (15.21), we have equality for $k = \nu$. □

For an interesting application of the HLP theorem, see Muirhead's theorem (Proposition 17.1) in the Notes. It generalizes the arithmetic-geometric mean inequality.

It turns out that $\Phi(b) \leq \Phi(a)$ holds for even more functions than $\Phi \in C_\Sigma(\mathbb{R}^\nu)$. One of the most important examples is $\Phi(x_1, \dots, x_\nu) = -x_1 x_2 \dots x_\nu$ with $x \in \mathbb{R}_+^\nu$, which is very far from convex (indeed, $f(t) = \Phi(tx_1, \dots, tx_\nu) = -at^\nu$ is concave!). It pays to consider the full class of functions, so we make several definitions.

**Definition** Let $a, b \in \mathbb{R}^\nu$. Define $S_k(\cdot)$ by (15.17). If $S_\nu(a) = S_\nu(b)$ and $S_k(b) \leq S_k(a)$ for $k = 1, \dots, \nu - 1$, we say *a majorizes* $b$ and write $b \prec_{\text{HLP}} a$.

**Definition** We call $K \subset \mathbb{R}^\nu$ a *permutation invariant set* if and only if $x \in K \Rightarrow M_\pi x \in K$ for all $\pi \in \Sigma_\nu$.

If $K$ is a permutation invariant convex set, $x \in K$ and $y \prec_{\text{HLP}} x$, then $y \in K$ by (ii) ⇒ (i) in Theorem 15.5, and the fact that for any finite set, $F$, $\text{ch}(F) = \text{cch}(F)$.

**Definition** Let $K$ be a permutation invariant convex set in $\mathbb{R}^\nu$. Let $\Phi\colon K \to \mathbb{R}$. $\Phi$ is called *Schur convex* (resp. *Schur concave*) if and only if $x, y \in K$ and $y \prec_{\text{HLP}} x \Rightarrow \Phi(y) \leq \Phi(x)$ (resp. $\Phi(y) \geq \Phi(x)$).

**Proposition 15.6** (i) *Any Schur convex or Schur concave function is permutation invariant.*

(ii) *Let $I \subset \mathbb{R}$ be an interval and let $K = I^\nu$. If $\varphi$ is a nonnegative log convex (resp. log concave) function on $\mathbb{R}$, then*

$$\Phi(x) = \varphi(x_1)\varphi(x_2)\dots\varphi(x_\nu) \tag{15.24}$$

*is Schur convex (resp. Schur concave) on $K$.*

(iii) *Let $K \subset \mathbb{R}^2$ be a permutation invariant set. Let $\Phi\colon K \to \mathbb{R}$. Define $f$ by $\Phi(x_1, x_2) = f(x_1 + x_2, x_1 - x_2)$ that is,* $\text{dom}(f) = \{(a, b) \mid \frac{1}{2}(a + b, a - b) \in K\}$ *so permutation invariance is equivalent to* $\text{dom}(f)$ *invariant under $(a, b) \to (a, -b)$ and*

$$f(a, b) = \Phi\left(\frac{a+b}{2}, \frac{a-b}{2}\right) \tag{15.25}$$

*Then $\Phi$ is Schur convex (resp. Schur concave) if and only if $f(a, b)$ is a function only of $|b|$ and is monotone increasing (decreasing) in $|b|$.*

*Proof* (i) If $y = M_\pi x$, then $y \prec_{\rm HLP} x$ and $x \prec_{HLP} y$, so in either case, $\Phi(x) = \Phi(y)$.

(ii) $\log \Phi(x) = \sum_{j=1}^{\nu} \log \varphi(x_j)$, so by (i) $\Rightarrow$ (v) in the HLP theorem, if $\log \varphi$ is convex (resp. concave), then $y \prec_{\rm HLP} x \Rightarrow \log \Phi(y) \le \log \Phi(x)$ (resp. $\log \Phi(y) \ge \log \Phi(x)$). Since exp is order preserving, $\Phi$ is Schur convex (resp. Schur concave).

(iii) As noted in the statement, $\Phi$ is permutation invariant if and only if $f$ is even in $b$ for $a$ fixed. Moreover, if $0 \le b_1 \le b_2$ and $a$ is fixed,

$$y \equiv \left( \frac{a+b_1}{2}, \frac{a-b_1}{2} \right) \prec_{\rm HLP} x \equiv \left( \frac{a+b_2}{2}, \frac{a-b_2}{2} \right) \tag{15.26}$$

(since $S_2(y) = a = S_2(x)$ and $S_1(y) = (a+b_1)/2 \le (a+b_2)/2 = S_1(x)$.) Conversely, if $y \prec_{\rm HLP} x$, then $y^*$ and $x^*$ have the form (15.26) with $a = S_2(x) = S_2(y)$ and $b_2 - b_1 = 2[S_1(x) - S_1(y)] \ge 0$. Thus, Schur convexity is equivalent to monotonicity of $f$ in $|b|$. □

**Example 15.7** Let

$$\Phi(x_1, \dots, x_\nu) = x_1 \dots x_\nu \tag{15.27}$$

on $\mathbb{R}_+^\nu$. Since $x, y \in \mathbb{R}_+^\nu$ and $y \prec_{\rm HLP} x$ and some $y_j = 0$ implies some $x_j = 0$ (for $y_\nu^* = S_\nu(y) - S_{\nu-1}(y)$), we need only consider $(0,\infty)^\nu$. On that set, $x \mapsto \log x$ is concave, so $\Phi$ is Schur concave. It is worth considering the case $\nu = 2$. First, $f$ given by (15.25) has

$$f(a,b) = \tfrac{1}{4}\,(a^2 - b^2)$$

This is a function only of $|b|$ and is monotone decreasing in $|b|$, making the Schur concavity clear. Also,

$$x_1 x_2 = \tfrac{1}{2}\,[(x_1 + x_2)^2 - x_1^2 - x_2^2]$$

Since $(x_1 + x_2)^2$ is constant on $\{y \mid y \prec_{\rm HLP} x\}$ and, by (i) $\Rightarrow$ (v) of the HLP theorem, $x_1^2 + x_2^2$ is Schur convex, we again see that $x_1 x_2$ is Schur concave. □

*Remark* We will see later (Theorem 15.40) that Schur concavity of (15.27) implies Hadamard's determinantal inequality that if $A$ is a positive matrix, then $\det(A) \le \prod_{j=1}^{n} a_{jj}$.

**Example 15.8** Let

$$\Phi(x_1, x_2) = |x_1 - x_2|^{1/2}$$

Then, by (iii) of Proposition 15.6, $\Phi$ is Schur convex on $\mathbb{R}^2$, although it is far from being convex; indeed, on $\{x \mid x_1 \ge x_2\}$, it is concave!

Let $\varphi$ be defined on $\mathbb{R}_+$ and suppose

$$\Phi(x_1, \dots, x_\nu) = \sum_{j=1}^{\nu} \varphi(x_j) \tag{15.28}$$

is Schur concave on $\mathbb{R}_+$. Then for $(x_3, \dots, x_\nu)$ fixed, $\Phi$ is Schur concave on $(x_1, x_2)$ (for $(y_1, y_2, x_3, \dots, x_\nu) \prec_{\text{HLP}} (x_1, x_2, \dots, x_\nu)$ if and only if $(y_1, y_2) \prec_{\text{HLP}} (x_1, x_2)$ by, e.g., (i) $\Leftrightarrow$ (v) in the HLP theorem). Thus,

$$\varphi\left(\frac{a+b}{2}\right) + \varphi\left(\frac{a-b}{2}\right) \le \varphi\left(\frac{a+\tilde{b}}{2}\right) + \varphi\left(\frac{a-\tilde{b}}{2}\right)$$

if $0 \le b \le \tilde{b} \le a$, which implies that $\varphi$ is convex. (For taking $b = 0$, $\varphi$ is midpoint-convex. That means for each $a$, $g(b) = \varphi((a+b)/2) + \varphi((a-b)/2)$ is midpoint-convex and monotone on $[0, a]$.) Thus, on $\mathbb{R}^\nu_+$, the only Schur convex functions of the form (15.28) are those covered by (v) of the HLP theorem.

On the other hand, let $K = \{(x, y) \in \mathbb{R}^2_+ \mid 0 < x + y < 1\}$. Let $\varphi(t) = \log(1-t) - \log t$ on $(0, 1)$. So

$$\varphi'(t) = -\frac{1}{t(1-t)}$$

and

$$\varphi''(t) = -\frac{1}{(1-t)^2} + \frac{1}{t^2}$$

In the region $x \ge y$, $f$ monotone in $|b|$ is equivalent to $\varphi'(x) - \varphi'(y) \ge 0$. This holds since

$$\begin{aligned} |y - \tfrac{1}{2}| - |x - \tfrac{1}{2}| &= x - y, && \text{if } y \le x \le \tfrac{1}{2} \\ &= 1 - (x+y), && \text{if } y \le \tfrac{1}{2} \le x \end{aligned}$$

This is always positive so $x$ is closer to $\frac{1}{2}$ than $y$ is, and the above shows $\varphi(t)$ is a function of $|t - \frac{1}{2}|$ monotone increasing on $(0, \frac{1}{2})$. Thus, $\varphi$ is Schur convex. But while $\varphi$ is concave on $(0, \frac{1}{2})$, it is strictly concave on $(\frac{1}{2}, 1)$, so if $K$ is not a product set, $\varphi$ need not be convex for (15.28) to be Schur convex. □

**Lemma 15.9** *Let $x, y \in \mathbb{R}^\nu$ with $y \prec_{\text{HLP}} x$. Then there exists $z^{(1)}, \dots, z^{(k-1)}$ with*

$$y \equiv z^{(0)} \prec_{\text{HLP}} z^{(1)} \prec_{\text{HLP}} z^{(2)} \prec_{\text{HLP}} \cdots \prec_{\text{HLP}} z^{(k-1)} \prec_{\text{HLP}} z^{(k)} \equiv x$$

*so for $j = 0, 1, \dots, k-1$, $z^{(j)}$ and $z^{(j+1)}$ differ in only two coordinate spots.*

*Remark* If $y \prec_{\text{HLP}} x$ and $y_j = x_j$ for $j = 3, 4, \dots, \nu$, then $y_1, y_2 \le \max(x_1, x_2)$ and $y_1 + y_2 = x_1 + x_2$, so for some $\theta$, $y_1 = \theta x_1 + (1-\theta) x_2$, $y_2 = (1-\theta) x_1 + \theta x_2$, and $y = (\theta \mathbf{1} + (1-\theta) M_{(12)}) x$, where $(12)$ is the transposition of 1 and 2. Thus, this lemma implies, without using Birkhoff's theorem, that if $y \prec_{\text{HLP}} x$, then $y = Dx$ with $D$ doubly stochastic (i.e., (i) $\Rightarrow$ (iii) in the HLP theorem). Indeed, this is how HLP proved (i) $\Rightarrow$ (iii).

*Proof* By induction on $\nu$. $\nu = 2$ is immediate, so suppose the theorem is true for $\mathbb{R}^\mu$, $\mu = 2, \dots, \nu - 1$. Since any permutation is a product of elementary transpositions which only change two coordinates, we can go from $x$ to $x^*$ and $y^*$ to $y$ ($x^*$ defined as $x_1^* \geq \dots \geq x_\nu^*$, a permutation of $x$, not $|x|$) by two-coordinate changes. So we can suppose without loss that $x_1 > \dots > x_\nu$, $y_1 > \dots > y_\nu$.

If $y \prec_{\text{HLP}} x$ and $S_k(y) = S_k(x)$ for some $k \in \{1, 2, \dots, \nu - 1\}$, then $(y_1, \dots, y_k) \prec_{\text{HLP}} (x_1, \dots x_k)$ and $(y_{k+1}, \dots, y_\nu) \prec (x_{k+1}, \dots, x_\nu)$. So by induction, we can find the required chain of $z$'s. If $S_k(y) < S_k(x)$ for all $k \in \{1, \dots, \nu - 1\}$, in particular, $y_1 = S_1(y) < S_1(x) = x_1$ and $y_\nu = S_\nu(y) - S_{\nu-1}(y) > S_\nu(x) - S_{\nu-1}(x) = x_\nu$, so define

$$x_\alpha = (x_1 - \alpha, x_2, \dots, x_{\nu-1}, x_\nu + \alpha)$$

For small $\alpha$,

$$y \prec_{\text{HLP}} x_\alpha \tag{15.29}$$

Pick $\alpha_0$, the largest $\alpha$ for which (15.29) holds. Then $x \succ_{\text{HLP}} x_{\alpha_0} \succ_{\text{HLP}} y$, $x$ and $x_{\alpha_0}$ differ in only two slots and for some $k \leq \nu + 1$, $S_k(x_{\alpha_0}) = S_k(y)$. So by induction, we can get from $x_{\alpha_0}$ to $y$ by a succession of two slot changes. □

**Theorem 15.10** *Let $K$ be a permutation invariant convex subset of $\mathbb{R}^\nu$ and $\Phi\colon K \to \mathbb{R}$. Then $\Phi$ is a Schur convex function if and only if*

(i) *$\Phi$ is permutation invariant.*

(ii) *For each fixed $x_3, \dots, x_\nu$ and $a$ with $(a, a, x_3, \dots, x_\nu) \in K$,*

$$g(b) = \Phi(a + b, a - b, x_3, \dots, x_\nu) \tag{15.30}$$

*is monotone increasing in $b$ for $b > 0$ and in the set where $(a + b, a - b, x_3, \dots, x_\nu) \in K$.*

*In particular, if $K$ is open and $\Phi$ is $C^1$,* (ii) *holds if and only if*

$$(x_2 - x_1)\left(\frac{\partial\Phi}{\partial x_2} - \frac{\partial\Phi}{\partial x_1}\right)(x) \geq 0 \tag{15.31}$$

*on all of $K$.*

*Proof* Since

$$bg'(b) = b\left(\frac{\partial\Phi}{\partial x_1} - \frac{\partial\Phi}{\partial x_2}\right) = \frac{1}{2}(x_1 - x_2)\left(\frac{\partial\Phi}{\partial x_1} - \frac{\partial\Phi}{\partial x_2}\right)$$

if $\Phi$ is $C^1$, (15.31) is equivalent to monotonicity of $g$. Thus, we need only show that $\Phi$ is Schur convex $\Leftrightarrow$ (i), (ii).

If $\Phi$ is Schur convex, (i) holds by Proposition 15.6(i). Moreover, by Proposition 15.6(iii) and the fact that $\Phi$ Schur convex implies $(x_1, x_2) \mapsto \Phi(x_1, x_2, x_3, \dots, x_\nu)$ is Schur convex for $(x_3, \dots, x_\nu)$ fixed, (ii) holds.

Conversely, if (i) and (ii) hold, Proposition 15.6(iii) implies $\Phi(y) \leq \Phi(x)$ if $y \prec_{\text{HLP}} x$, $y$ and $x$ differ in only two slots. Then by Lemma 15.9, $\Phi$ is Schur convex. □

**Example 15.11** The elementary symmetric functions $\sigma_1, \dots, \sigma_\nu$ on $\mathbb{R}^\nu$ are defined by

$$\sigma_\ell(x) = \sum_{i_1 < i_2 < \dots < i_\ell} x_{i_1} \dots x_{i_\ell} \tag{15.32}$$

so, for example, on $\mathbb{R}^3$,

$$\sigma_1(x) = x_1 + x_2 + x_3$$
$$\sigma_2(x) = x_1 x_2 + x_1 x_3 + x_2 x_3$$
$$\sigma_3(x) = x_1 x_2 x_3$$

We claim all the $\sigma_\ell$ are Schur concave functions on $\mathbb{R}_+^\nu$ (generalizing Example 15.7 and providing a new proof for (15.27)).

Obviously, $\sigma_\ell$ is symmetric in its arguments. Moreover, if $x_3, \dots, x_\nu$ are fixed in $\mathbb{R}_+$, then

$$\sigma_\ell(x_1, \dots, x_\nu) \equiv \alpha x_1 x_2 + \beta(x_1 + x_2) + \gamma$$

with $\alpha, \beta, \gamma \geq 0$. Thus,

$$(x_2 - x_1)\left(-\frac{\partial \sigma_\ell}{\partial x_2} + \frac{\partial \sigma_\ell}{\partial x_1}\right) = \alpha(x_2 - x_1)^2 \geq 0$$

and so Schur convex by Theorem 15.10. Thus, $\sigma_\ell$ is Schur concave.

We will apply this in Theorem 15.40. □

This completes our initial look at the classical case. We turn next to finite sequences being replaced by finite matrices.

Given a positive $n \times n$ matrix, $A$, we let $\mu_1^+(A) \geq \mu_2^+(A) \geq \dots \geq \mu_n^+(A)$ be its eigenvalues listed in decreasing order and $S_k^+ = \sum_{j=1}^k \mu_j^+(A)$ so $S_n^+(A) = \sum_{j=1}^n \mu_j^+(A) = \operatorname{tr}(A)$.

Let $P$ be a rank $k$ orthogonal projection. Then, by a variational principle argument,

$$\mu_\ell^+(A) \geq \mu_\ell^+(PAP), \qquad \ell = 1, \dots, k \tag{15.33}$$

so

$$S_k^+(A) \geq S_k^+(PAP) = \operatorname{tr}(PAP) = \operatorname{tr}(AP) \tag{15.34}$$

Since we have equality in (15.33) if $P$ is the projection onto the eigenvectors for $\{\mu_\ell^+(A)\}_{\ell=1}^k$, we obtain

$$S_k^+(A) = \sup\{\operatorname{tr}(AP) \mid P \text{ is a rank } k \text{ orthogonal projection}\} \tag{15.35}$$

We will let $\mathbb{M}_n$ denote the set of all $n \times n$ complex Hermitian matrices and $\mathbb{M}_n^+ = \{A \in \mathbb{M}_n \mid A \geq 0\}$. A *doubly stochastic* map on $\mathbb{M}_n$ is a linear map, $\Psi$, from $\mathbb{M}_n$ to itself so that

(i) $\Psi[\mathbb{M}_n^+] \subset \mathbb{M}_n^+$
(ii) $\Psi(\mathbf{1}) = \mathbf{1}$
(iii) $\operatorname{tr}(\Psi(A)) = \operatorname{tr}(A)$

We denote the set of all such maps by $\mathbb{D}(\mathbb{M}_n)$. To understand why this is analogous to $\mathbb{D}_\nu$, note $D\colon \mathbb{R}^\nu \to \mathbb{R}^\nu$ is doubly stochastic if and only if $D[\mathbb{R}_+^\nu] \subset \mathbb{R}_\nu^+$, $D(1,\dots,1) = (1,\dots,1)$, and $\sum_{j=1}^\nu (Da)_j = \sum_{j=1}^\nu a_j$. Note that if $\mathbb{M}_n$ is made into a Hilbert space with the inner product $\langle A, B\rangle = \operatorname{tr}(A^*B)$ and $\Psi \in \mathbb{D}(\mathbb{M}_n)$, then $\Psi^* \in \mathbb{D}(\mathbb{M}_n)$.

By $C_U(\mathbb{M}_n)$, we mean all convex maps $\Phi\colon \mathbb{M}_n \to \mathbb{R}$ with $\Phi(UAU^{-1}) = \Phi(U)$ for all unitaries $U \in \mathbb{U}_n$, the $n \times n$ unitaries.

We are heading towards

**Theorem 15.12** *Let $A, B \in \mathbb{M}_n^+$. The following are equivalent:*

(i) $S_k^+(B) \leq S_k^+(A)$ *for* $k = 1,\dots,n-1$; $\operatorname{tr}(B) = \operatorname{tr}(A)$
(ii) $B \in \operatorname{cch}(\{UAU^{-1} \mid U \in \mathbb{U}_n\})$
(iii) $B = \Psi(A)$ *for some* $\Psi \in \mathbb{D}(\mathbb{M}_n)$
(iv) $\Phi(B) \leq \Phi(A)$ *for all* $\Phi \in C_U(\mathbb{M}_n)$
(v) $\operatorname{tr}(\varphi(B)) \leq \operatorname{tr}(\varphi(A))$ *for all convex* $\varphi\colon \mathbb{R} \to \mathbb{R}$
(vi)

$$\operatorname{tr}((B-\alpha)P_{[\alpha,\infty)}(B)) \leq \operatorname{tr}((A-\alpha)P_{[\alpha,\infty)}(A)) \tag{15.36}$$

*for all $\alpha \in \mathbb{R}_+$ where $P_{[\alpha,\infty)}(\cdot)$ is the spectral projection and*

$$\operatorname{tr}(B) = \operatorname{tr}(A) \tag{15.37}$$

*Remark* As with the HLP theorem, by replacing $A$ and $B$ by $A + \alpha\mathbf{1}$, $B + \alpha\mathbf{1}$, one sees the theorem holds for $A, B \in \mathbb{M}_n$.

As a preliminary to the proof of Theorem 15.12, we need to generalize (14.2):

**Proposition 15.13** *For $A \in \mathbb{M}_n$, we have*

$$S_j(A) = \sup\{\operatorname{tr}(AB) \mid 0 \leq B \leq 1,\ B = B^*,\ \operatorname{tr}(B) = j\} \tag{15.38}$$

*Proof* We claim any $B$ with $B = B^*$, $0 \leq B \leq 1$, and $\operatorname{tr}(B) = j$ can be written $\sum_{k=1}^\ell \theta_k P_k$ with $P_k$ an orthogonal projection of rank $j$ and $\theta_k \geq 0$, $\sum_{k=1}^\ell \theta_k = 1$. Given the claim, we clearly have

$$\text{RHS of (15.38)} = \text{RHS of (15.35)}$$

and so (15.35) implies (15.38).

The claim is a nice application of the HLP theorem! For pass to a basis in which $B$ is diagonal, $0 \le B \le 1$, and $\mathrm{tr}(B) = j$ is equivalent to $b \prec_{\mathrm{HLP}} a$ where $b_k = B_{kk}$ is the diagonal of $B$ and $a$ is the vector

$$\begin{aligned} a_k &= 1, \qquad k = 1, 2, \dots, j \\ &= 0, \qquad k = j+1, \dots, n \end{aligned}$$

By the HLP theorem,

$$b = \sum_{k=1}^{\ell} \theta_k M_\pi a \tag{15.39}$$

Let $P_k$ be the diagonal matrix with 1's in the positions $\pi[\{1, \dots, j\}]$ and 0's in the positions $\pi[\{j+1, \dots, n\}]$. Thus, $P_k$ is a projection of rank $j$ and (15.39) is

$$A = \sum_{k=1}^{\ell} \theta_k P_k \qquad \square$$

*Proof of Theorem 15.12* We will show (i) ⇒ (ii) ⇒ (iii) ⇒ (i) and (ii) ⇒ (iv) ⇒ (v) ⇒ (vi) ⇒ (i).

(i) ⇒ (ii) Pick $V, W \in \mathbb{U}_n$ so $VAV^{-1}$ is the diagonal matrix with $a = (a_1, \dots, a_n)$ along the diagonal and $WBW^{-1}$ a diagonal matrix with $b = (b_1, \dots, b_n)$ along the diagonal. By (i), $b \prec_{\mathrm{HLP}} a$, so by the HLP theorem, $b = \sum_{k=1}^{\ell} \theta_k (M_{\pi_k} a)$. Let $M_\pi \in \mathbb{U}_n$ as a unitary matrix. Then $M_\pi VAV^{-1} M_\pi^{-1}$ is the diagonal matrix with $M_\pi a$ along the diagonal. Thus,

$$B = \sum_{k=1}^{\ell} \theta_k U_k A U_k^{-1}$$

where $U_k = W^{-1} M_\pi V$.

(ii) ⇒ (iii) $\Psi(C) = \sum_{k=1}^{\ell} \theta_k U_k C U_k^{-1}$ is a map in $\mathbb{D}(M_n)$.

(iii) ⇒ (i) If $\Gamma \in \mathbb{D}(M_n)$ and $0 \le C \le \mathbf{1}$ with $\mathrm{tr}(C) = j$, then $0 \le \Gamma(C) \le \mathbf{1}$ (since $\Gamma$ is positivity preserving and $\Gamma(\mathbf{1}) = \mathbf{1}$) and $\mathrm{tr}(\Gamma(C)) = j$. Picking $\Gamma = \Psi^* \in \mathbb{D}(M_n)$, we have by Proposition 15.13,

$$\begin{aligned} S_k^+(\Psi(A)) &= \sup\{\mathrm{tr}(\Gamma(C)A) \mid 0 \le C \le 1,\ \mathrm{tr}(C) = k\} \\ &\le \sup\{\mathrm{tr}(CA) \mid 0 \le C \le 1,\ \mathrm{tr}(C) = k\} \\ &= S_k^+(A) \end{aligned}$$

and $\mathrm{tr}(\Psi(A)) = \mathrm{tr}(A)$, so (iii) ⇒ (i).

(ii) ⇒ (iv) Immediate.

(iv) ⇒ (v) Follows since $\Phi(A) = \mathrm{tr}(\varphi(A))$ is convex and clearly symmetric. To see it is convex, suppose $Af_j = \alpha_j f_j$ for an orthonormal basis $\{f_j\}_{j=1}^n$. Suppose

$e$ is a unit vector and note

$$\begin{aligned}\varphi(\langle e, Ae\rangle) &= \varphi\Big(\sum_j \alpha_j |\langle e, f_j\rangle|^2\Big) \\ &\le \sum_j \varphi(\alpha_j)|\langle e, f_j\rangle|^2\end{aligned}$$

by Jensen's inequality. Thus, if $\{e_i\}_{i=1}^n$ is any orthonormal basis, then

$$\sum_i \varphi(\langle e_i, Ae_i\rangle) \le \Phi(A) \tag{15.40}$$

Given $A, B \in \mathbb{M}_n$ and $\theta \in [0,1]$, let $\{e_i\}_{i=1}^n$ be an orthonormal basis of eigenvectors for $\theta A + (1-\theta)B$. Then

$$\begin{aligned}\Phi(\theta A + (1-\theta)B) &= \sum_i \varphi(\langle e_i, (\theta A + (1-\theta)B)e_j\rangle \\ &\le \theta \sum_i \varphi(\langle e_i, Ae_j\rangle) + (1-\theta)\sum_i \varphi(\langle e_i, Be_j\rangle) \\ &\le \theta\Phi(A) + (1-\theta)\Phi(B)\end{aligned}$$

by (15.40).

(v) $\Rightarrow$ (vi) $\varphi(x) = (x-\alpha)_+$ is a convex function.

(vi) $\Rightarrow$ (i) Writing out $\operatorname{tr}((B-\alpha)P_{[\alpha,\infty)}(A))$ in terms of eigenvalues $b_j = \mu_j^+(B)$, (15.36)/(15.37) is equivalent to $b \prec_{\text{HLP}} a$ so the HLP theorem implies that (i) holds. □

**Corollary 15.14** *Let $A$ be an arbitrary matrix in $\mathbb{M}_n$. Then as sequences in $\mathbb{R}^n$,*

$$a_{jj} \prec_{\text{HLP}} \mu_j^+(A) \tag{15.41}$$

*Proof* Let $B$ be the diagonal matrix with $b_{jj} = a_{jj}$. Then as matrices, (15.41) is equivalent to $A$ majorizing $B$ in the sense of Theorem 15.12. Pick $\alpha_1, \dots, \alpha_n \in \mathbb{Z}$ distinct so that

$$\frac{1}{2\pi}\int_0^{2\pi} \exp(i[\alpha_j - \alpha_k]\theta)\, d\theta = \delta_{jk} \tag{15.42}$$

Let $U(\theta)$ be the diagonal matrix $U(\theta)_{jj} = \exp(i\alpha_j\theta)$. By (15.42), $B = 1/2\pi \int d\theta\, U(\theta)AU(\theta)^{-1}$ so (15.41) follows from Theorem 15.12. □

*Remark* We will provide another proof of this in Theorem 15.40. With that proof, this is a result of Schur.

This completes the discussion of the finite matrix case. We turn to the case where $S_\nu(b) = S_\nu(a)$ is weakened to $S_\nu(b) \le S_\nu(a)$.

**Definition** An $n \times n$ matrix is called *doubly substochastic* (dss) if and only if

$$\text{(i)} \qquad \sum_{i=1}^{n} a_{ij} \leq 1, \qquad j = 1, \dots, n \tag{15.43}$$

$$\text{(ii)} \qquad \sum_{j=1}^{n} a_{ij} \leq 1, \qquad i = 1, \dots, n \tag{15.44}$$

$$\text{(iii)} \qquad a_{ij} \geq 0, \qquad i, j = 1, \dots, n \tag{15.45}$$

The set of all such matrices will be denoted $\mathbb{S}_n$.

*Remark* Some authors use doubly substochastic for what we later call complex substochastic.

**Definition** A *subpermutation* of $\{1, \dots, n\}$ is a one-one map, $\pi$, defined on a subset $D(\pi)$ of $\{1, \dots, n\}$. We denote $\{1, \dots, n\} \backslash D(\pi)$ by $K(\pi)$, the range of $\pi$ by $R(\pi)$, and $C(\pi) = \{1, \dots, n\} \backslash R(\pi)$.

**Definition** Given a subpermutation, $\pi$, the subpermutation matrix $S_\pi$ is defined by

$$(S_\pi)_{ij} = \begin{cases} \delta_{i\pi(j)}, & \text{if } j \in D(\pi) \\ 0, & \text{if } j \in K(\pi) \end{cases} \tag{15.46}$$

Thus, the rows in $C(\pi)$ and columns in $K(\pi)$ are all zeros, but after they are removed, what is left is a permutation matrix.

**Theorem 15.15** *$\mathbb{S}_n$ is a compact convex subset and $\mathcal{E}(\mathbb{S}_n) = \{S_\pi \mid \pi$ is a subpermutation$\}$.*

*Proof* $\mathbb{S}_n$ is clearly closed and convex and, since $0 \leq a_{ij} \leq 1$, bounded, it is compact. Given a subpermutation $\pi$, define the linear functional

$$\ell_\pi(A) = \sum_{j \in D(\pi)} \Big[ A_{\pi(j)j} - \sum_{i \neq \pi(j)} A_{ij} \Big] - \sum_{j \in K} \sum_{i=1}^{n} A_{ij} \tag{15.47}$$

$S_\pi$ is the unique $a \in \mathbb{S}_n$ with $\ell_\pi(A) = \#D(\pi) = \sup_{A \in \mathbb{S}_n} \ell_\pi(A)$, so $S_\pi$ is an extreme point.

Conversely, suppose $A \in \mathcal{E}(\mathbb{S}_n)$. $\mathbb{D}_n$ is a face of $\mathbb{S}_n$ so if $A \in \mathbb{D}_n$, we have $A \in \mathcal{E}(\mathbb{D}_n)$, and thus, $A$ is a permutation matrix by Birkhoff's theorem.

So let $A \notin \mathbb{D}_n$. Each row has $n+1$ possible equalities among (15.44) and (15.45). Suppose no row has $n$ equalities associated with it. Thus, there are at most $n(n-1)$ equalities among the relations (15.44) and (15.45). Since Proposition 15.2 says $A$ must obey $n^2$ equalities, all $n$ relations (15.43) must hold. But then $\sum_{i,j} a_{ij} = n$ and so equality is forced in (15.44) and $A \in \mathbb{D}_n$. This construction shows that if $A \notin \mathbb{D}_n$, some row must obey $n$ relations, meaning it is either a row of zeros or a row with a single 1 and the rest zeros. If it has single 1, cross out that row and the column with the 1; what is left is $(n-1) \times (n-1)$ substochastic. So by an induction argument, $A$ is a subpermutation matrix.

If there is a row of zeros, repeat the argument on columns. If there is a column with a single 1, as above, by induction $A$ is an $S_\pi$. If not, there is a row of zeros and a column of zeros. Cross both out and use induction to see again that $A$ is an $S_\pi$. □

**Theorem 15.16** *Let $a, b \in \mathbb{R}^\nu_+$. The following are equivalent:*

(i) $S_k(b) \le S_k(a)$ *for* $k = 1, \dots, \nu$
(ii) $b \in \text{cch}(\{S_\pi a \mid \pi \text{ a subpermutation}\})$
(iii) $b = Sa$ *for some* $S \in \mathbb{S}_\nu$
(iv) $\Phi(b) \le \Phi(a)$ *for all* $\Phi \in C_\Sigma(\mathbb{R}^\nu)$ *which are monotone increasing in each variable.*
(v) (15.19) *holds for all convex* $\varphi\colon \mathbb{R} \to \mathbb{R}$ *which are monotone increasing.*
(vi) *For all* $s \in \mathbb{R}_+$, (15.20) *holds.*

*Proof* We will prove (i) ⇒ (ii) ⇔ (iii) ⇒ (iv) ⇒ (v) ⇒ (vi) ⇒ (i). We will only indicate the places where there are differences from the proof of Theorem 15.5.

<u>(i) ⇒ (ii)</u> As in the earlier proof, if $b \notin \text{cch}(\{S_\pi a\})$, there is $\ell \in (\mathbb{R}^\nu)^*$ so

$$\ell(b) > \max_\pi \ell(S_\pi a) \tag{15.48}$$

Write $\ell(x) = \sum_i \ell_i x_i$. Let $\tilde{\ell}(x) = \sum_i \max(\ell_i, 0) x_i$, where we set negative $\ell_i$'s to zero. Since $b \in \mathbb{R}^\nu_+$, $\tilde{\ell}(b) \ge \ell(b)$. Moreover, we claim that

$$\max_\pi \ell(S_\pi a) = \max_\pi \tilde{\ell}(S_\pi a) \tag{15.49}$$

Since $S_\pi a \in \mathbb{R}^\nu_+$, $\tilde{\ell}(S_\pi a) \ge \ell(S_\pi a)$. On the other hand, for any $S_\pi$ and subset $I \subset \{1, \dots, n\}$, there is an $S_{\tilde{\pi}}$ so

$$(S_{\tilde{\pi}} a)_i = \begin{cases} (S_\pi a)_i, & \text{if } i \notin I \\ 0, & \text{if } i \in I \end{cases}$$

by just zeroing out the rows with $i \in I$. Picking $I = \{i \mid \ell_i < 0\}$, we see $\tilde{\ell}(S_\pi a) = \ell(S_{\tilde{\pi}} a)$, proving (15.49).

Thus, without loss, we can suppose the $\ell$ in (15.48) has $\ell_i \ge 0$. For this $\ell$, the argument in the proof of Theorem 15.5 works.

<u>(ii) ⇔ (iii)</u> Identical to the argument in Theorem 15.5 using Theorem 15.15 to replace Theorem 15.4.

<u>(iii) ⇒ (iv)</u> By the monotonicity of $\Phi$, if $S_\pi$ is a proper subpermutation, that is, not a permutation, $\Phi(S_\pi a) \le \Phi(M_{\tilde{\pi}} a) = \Phi(a)$ for a $\tilde{\pi}$ obtained from $\pi$ by replacing

the zero elements in $S_\pi a$ by missing components of $a$. Thus, $\max_\pi \Phi(S_\pi a) = \Phi(a)$ and $\Phi(b) \le \Phi(a)$ holds on $\operatorname{cch}(\{S_\pi a\})$.

(iv) $\Rightarrow$ (v) Trivial as in Theorem 15.5.

(v) $\Rightarrow$ (vi) $(x - s)_+$ is a monotone convex function, so this is immediate.

(vi) $\Rightarrow$ (i) Identical to Theorem 15.5. □

That completes the analysis when $S_\nu(b) = S_\nu(a)$ is replaced by $S_\nu(b) \le S_\nu(a)$. We turn next to the case when $S_j(|b|) \le S_j(|a|)$. We turn to rearrangements of $|b|^*$ where $b \in \mathbb{R}^\nu$ or $b \in \mathbb{C}^\nu$. We define:

**Definition** The *complex substochastic matrices*, $\mathbb{S}_{n;\mathbb{C}}$, are $n \times n$ matrices with $a_{ij} \in \mathbb{C}$,

$$\text{(i)} \qquad \sum_{i=1}^n |a_{ij}| \le 1, \qquad j = 1, \dots, n \tag{15.50}$$

$$\text{(ii)} \qquad \sum_{j=1}^n |a_{ij}| \le 1, \qquad i = 1, \dots, n \tag{15.51}$$

The real substochastic matrices, $\mathbb{S}_{n;\mathbb{R}}$, are $n \times n$ matrices with $a_{ij} \in \mathbb{R}$ and with (15.50) and (15.51).

We skip identifying the extreme points of $\mathbb{S}_{n;\mathbb{R}}$ and $\mathbb{S}_{n;\mathbb{C}}$ because our method above does not work directly for $\mathbb{S}_{n;\mathbb{C}}$. (The extreme points for $\mathbb{S}_{n;\mathbb{R}}$ are the $2^n n!$ matrices with matrix elements $\sigma_i \delta_{i\pi(j)}$ with each $\sigma_i = \pm 1$ and $\pi$ a permutation. For $\mathbb{S}_{n;\mathbb{C}}$, it has the same form, but now each $\sigma_i$ is a complex number of magnitude 1.) This means we will need another argument for (iii) $\Rightarrow$ (ii) (in fact, we will give a direct proof of (iii) $\Rightarrow$ (iv)).

Let $W(\mathbb{R}^\nu)$ be the group of $2^\nu \nu!$ elements of linear maps on $\mathbb{R}^n$ generated by permutations of coordinates and arbitrary sign flips $x_i \to \sigma_i x_i$ with each $\sigma_i = +1$ or $\sigma_i = -1$. Let $W(\mathbb{C}^\nu)$ be the group of linear maps on $\mathbb{C}^n$ of coordinate permutations and phase changes $z_i \to \omega_i z_i$ with $\omega_i \in \partial D$. Let $\mathcal{C}_W(\mathbb{R}^\nu)$ be the convex functions on $\mathbb{R}^\nu$ invariant under $W(\mathbb{R}^\nu)$ and $\mathcal{C}_W(\mathbb{C}^\nu)$ the convex functions on $\mathbb{C}^\nu$ invariant under $W(\mathbb{C}^\nu)$.

**Theorem 15.17** *Fix $\mathbb{K} = \mathbb{R}$ or $\mathbb{K} = \mathbb{C}$. Let $a, b \in \mathbb{K}^\nu$. The following are equivalent:*

(i) $S_k(|b|) \le S_k(|a|)$ *for* $k = 1, 2, \dots, \nu$
(ii) $b \in \operatorname{cch}(\{Aa \mid A \in W(\mathbb{K}^\nu)\})$
(iii) $b = Da$ *for* $S \in \mathbb{S}_{\nu;\mathbb{K}}$
(iv) $\Phi(b) \le \Phi(a)$ *for all* $\Phi \in \mathcal{C}_W(\mathbb{K}^\nu)$

(v)

$$\sum_{j=1}^{\nu} \varphi(|b_j|) \le \sum_{j=1}^{\nu} \varphi(|a_j|) \tag{15.52}$$

*for all convex even functions of* $\mathbb{R}$ *to* $\mathbb{R}$.

(vi) *For all* $s \in \mathbb{R}_+$,

$$\sum_{j=1}^{\nu} (|b_j| - s)_+ \le \sum_{j=1}^{\nu} (|a_j| - s)_+ \tag{15.53}$$

*Proof* We will give the proof in case $\mathbb{K} = \mathbb{C}$; the real case is essentially identical. We will prove (i) $\Rightarrow$ (ii) $\Rightarrow$ (iii) $\Rightarrow$ (i) and (ii) $\Rightarrow$ (iv) $\Rightarrow$ (v) $\Rightarrow$ (vi) $\Rightarrow$ (i).

(i) $\Rightarrow$ (ii) Suppose $b, a$ obey (i) but $b \notin K \equiv \text{cch}(\{Aa\})$. Then there exists $\ell \in (\mathbb{C}^n)^*$ so

$$\operatorname{Re} \ell(b) > \sup_{A \in W(\mathbb{C}^\nu)} \operatorname{Re} \ell(Aa) \tag{15.54}$$

Suppose $b_j = e^{i\theta_j} |b_j|$ and define $\tilde{\ell}$ by $\tilde{\ell}(x) = \sum_{j=1}^{\nu} |\ell_j| e^{-i\theta_j} x_j$. Then $\tilde{\ell}(b)$ is real, $\tilde{\ell}(b) \ge \ell(b)$, and since $\tilde{\ell} = \ell \circ A_0$ for $A_0 \in W(\mathbb{C}^\nu)$, $\sup \operatorname{Re} \ell(Aa) = \sup \operatorname{Re} \tilde{\ell}(Aa)$. Moreover, $\tilde{\ell}(b) = \sum_{j=1}^{\nu} |\ell_j| \, |b_j|$. The argument in the proof of Theorem 15.5 then proves that (15.54) cannot hold.

(ii) $\Rightarrow$ (iii) By the Minkowski–Carathéodory theorem (Theorem 8.11), $b = (\sum_{j=1}^{2\nu} \theta_j A_j) a$ for some $A_j \in W(\mathbb{C}^\nu)$ and $\theta_j \ge 0$, with $\sum \theta_j = 1$. $D = \sum_{j=1}^{2\nu} \theta_j A_j \in \mathbb{S}_{\nu;\mathbb{C}}$.

(iii) $\Rightarrow$ (i) Suppose (iii) holds. Since phase factors turn $b$ into $|b|$ and $a$ into $|a|$ and permutations map $|a|$ to $|a|^*$ and $|b|$ to $|b|^*$, we have

$$|b|^* = \tilde{D} |a|^*$$

for a $\tilde{D}$ in $\mathbb{S}_{\nu;\mathbb{R}}$. Thus,

$$\begin{aligned} \sum_{i=1}^{k} |b|_i^* &\le \sum_{i=1}^{k} \sum_{j=1}^{\nu} |\tilde{D}_{ij}| \, |a|_j^* \\ &= \sum_{j=1}^{\nu} \gamma_j |a|_j^* \end{aligned}$$

where

$$\gamma_j = \sum_{i=1}^{k} |\tilde{D}_{ij}|$$

From the fact that $\tilde{D} \in \mathbb{S}_{\nu;\mathbb{C}}$, we have

$$0 \le \gamma_j \le 1, \qquad \sum_{j=1}^{\nu} \gamma_j \le k \tag{15.55}$$

By the lemma below,

$$\sum_{\gamma=1}^{k} |b|_i^* \le \sum_{i=1}^{k} |a|_i^*$$

that is, $S_k(|b|) \le S_k(|a|)$ so (i) holds.

<u>(ii) ⇒ (iv)</u> Follows from convexity since $\Phi(Aa) = \Phi(a)$ for all $A \in W(\mathbb{C}^\nu)$.

<u>(iv) ⇒ (v)</u> $\Phi(x_1, \dots, x_\nu) = \sum_{i=1}^{\nu} \varphi(|x_i|)$ is in $\mathcal{C}_W(\mathbb{C}^\nu)$ by Proposition 1.7.

<u>(v) ⇒ (vi)</u> $\varphi(x) = (|x| - s)_+$ is an even convex function.

<u>(vi) ⇒ (i)</u> This is the same as in the proof of Theorem 15.5. □

**Lemma 15.18** *Let $a_1 \ge a_2 \ge \cdots \ge a_\nu \ge 0$ in $\mathbb{R}$. Let $\gamma_1, \gamma_2, \dots, \gamma_\nu \in [0,1]$ with $\sum_{j=1}^{\nu} \gamma_j \le k \in \{1, 2, \dots, \nu\}$. Then*

$$\sum_{j=1}^{\nu} \gamma_j a_j \le \sum_{j=1}^{k} a_j \tag{15.56}$$

*Proof* Let $\eta \in \mathbb{R}^\nu$ be $\eta = (1, \dots, 1, 0, \dots, 0)$ with $k$ ones and $\nu - k$ zeros. Then clearly, $S_j(\gamma) \le S_j(\eta)$ for $j = 1, 2, \dots, \nu$. By (i) ⇒ (ii) in Theorem 15.16, $\gamma = \sum_{\ell=1}^{L} \theta_\ell S_{\pi_\ell}(\eta)$ with $\theta_\ell \ge 0$, $\sum_{\ell=1}^{L} \theta_\ell = 1$, and $\prod_\ell$ subpermutations. Thus,

$$\sum_{j=1}^{\nu} \gamma_j a_j = \sum_{\ell=1}^{L} \theta_\ell \sum_{j \in B_\ell} a_j \tag{15.57}$$

where $B_\ell = \{j \mid [S_{\pi_\ell}(\eta)]_j = 1\}$ is a set with at most $k$ elements.

Since $a_1 \ge a_2 \ge \cdots \ge a_\nu \ge 0$, we have for each $\ell$ that

$$\sum_{j \in B_\ell} a_j \le \sum_{j=1}^{k} a_j$$

so (15.56) follows from (15.57) and $\sum_{\ell=1}^{L} \theta_\ell = 1$. □

With this machinery under our belt, we can turn to improving the motivating inequality (15.4). Recall $\{\lambda_j(A)\}_{j=1}^n$ are the eigenvalues of an $n \times n$ matrix $A$ (not necessarily self-adjoint) and $\{\mu_j(A)\}_{j=1}^n$ are the eigenvalues of $|A| = \sqrt{A^*A}$. Order the $\lambda$'s so $|\lambda_1| \ge |\lambda_2| \ge \cdots$ and $\mu$'s so $|\mu_1| \ge |\mu_2| \ge \cdots$

**Lemma 15.19** *For any $k = 1, 2, \dots, n$,*

$$|\lambda_1(A) \dots \lambda_k(A)| \le \mu_1(A) \dots \mu_k(A) \tag{15.58}$$

*Proof* For $k = 1$, this is just that $|\lambda_1(A)|$ as an eigenvalue of $A$ is bounded by $\mu_1(A) = \|\sqrt{A^*A}\| = \|A^*A\|^{1/2} = \|A\|$. For arbitrary $k$, consider the wedge product (see [352, Appendix A]) $\wedge^k(A)$. Then $\lambda_1(\wedge^k(A)) = \lambda_1(A) \dots \lambda_k(A)$ since the eigenvalues of $\wedge^k(A)$ are precisely $\{\lambda_{j_1}(A)\dots\lambda_{j_k}(A) \mid j_1,\dots,j_k \text{ distinct}\}$ while, for the same reason, $\mu_1(\wedge^k(A)) = \mu_1(A)\dots\mu_k(A)$. Thus, (15.58) is just $|\lambda_1(A^k(A))| \le \mu_1(\wedge^k(A)) = \| \wedge^k (A)\|$. □

**Theorem 15.20** (Weyl's Inequality) *Let $\phi$ be a function on $(0,\infty)$ which is monotone increasing and so that*

$$t \mapsto \phi(e^t)$$

*is convex in t on $(-\infty,\infty)$. Then*

$$\sum_{j=1}^{n} \phi(\lambda_j(A)) \le \sum_{j=1}^{n} \phi(\mu_j(A)) \tag{15.59}$$

*In particular, for any $p \in (0,\infty)$,*

$$\sum_{j=1}^{n} |\lambda_j(A)|^p \le \sum_{j=1}^{n} |\mu_j(A)|^p \tag{15.60}$$

*Proof* Let $b_j = \log(\lambda_j(A))$ and $a_j = \log(\mu_j(A))$ so $a_1 \ge \dots \ge a_n$ and $b_1 \ge \dots \ge b_n$. (15.58) implies

$$S_k(b) \le S_k(a)$$

By Theorem 15.16 (i) ⇒ (v), $t \mapsto \phi(e^t)$ convex implies

$$\sum_{j=1}^{n} \phi(e^{b_j}) \le \sum_{j=1}^{n} \phi(e^{a_j})$$

which is (15.59). (15.60) follows if we note that $t \to e^{tp}$ is convex for all $p > 0$. □

*Remark* We discuss (15.60) for $p \in [1,\infty]$ further at the end of the chapter (see Proposition 15.38).

Of course, $\|AB\| \le \|A\|\,\|B\|$, which applied to $\wedge^k(A)$, yields

$$\prod_{j=1}^{k} \mu_j(AB) \le \prod_{j=1}^{k} \mu_j(A)\mu_j(B)$$

which as above implies

**Theorem 15.21** (Horn's Inequality) *For any $n \times n$ matrices,*

$$\sum_{j=1}^{n} \mu_j(AB)^p \le \sum_{j=1}^{n} \mu_j(A)^p \mu_j(B)^p \tag{15.61}$$

*Remark* If we follow (15.61), by Hölder's inequality, we get Hölder's inequality for matrices

$$\|AB\|_r \le \|A\|_p\,\|B\|_q \tag{15.62}$$

where $r^{-1} = p^{-1} + q^{-1}$ and

$$\|A\|_r = \left( \sum_{j=1}^{n} \mu_j(A)^r \right)^{1/r} \tag{15.63}$$

Next, we turn to infinite-dimensional analogs of the HLP theorem. We look at extending the complex case of Theorem 15.17. There are extensions of the other results as well. We begin with the natural sequence space and extension of doubly substochastic matrices:

**Definition** $c_0$ is the family of sequences $\{a_n\}_{n=1}^{\infty}$ of complex numbers with

$$\lim_{n\to\infty} |a_n| = 0 \tag{15.64}$$

We put the norm $\|a\|_\infty = \sup_n |a_n|$ on $c_0$. $\ell^1$ is the subset of $c_0$ with

$$\|a\|_1 \equiv \sum_{n=1}^{\infty} |a_n| < \infty \tag{15.65}$$

It is easy to see that $c_0$ and $\ell^1$ are Banach spaces and $c_0^* = \ell^1$. $c_0$ is the natural space on which $|a^*|$ is defined by

$$\sum_{j=1}^{n} |a|_j^* = \max_{k_1,\dots,k_n} \sum_{j=1}^{n} |a_{k_j}|$$

so $|a|_1^* = \max|a_j|$, $|a|_2^*$ is the second largest counting multiplicity. Since (15.64) holds, it is easy to see $|a|_j^*$ is well defined and there is a bijection $\pi\colon \{1,\dots,n,\dots\} \to \{j \mid a_j \neq 0\}$ so

$$|a|_j^* = |a_{\pi(j)}| \tag{15.66}$$

**Definition** A *complex doubly substochastic infinite matrix* is a collection of complex numbers $\{\alpha_{ij}\}_{1\le i,j\le\infty}$ with

$$\sum_{i=1}^{\infty} |\alpha_{ij}| \le 1, \qquad j = 1,2,\dots \tag{15.67}$$

$$\sum_{j=1}^{\infty} |\alpha_{ij}| \le 1, \qquad i = 1,2,\dots \tag{15.68}$$

The set of such matrices is denoted by $\mathbb{S}_\infty$.

**Theorem 15.22** (i) *$\mathbb{S}_\infty$ is compact in the topology of weak convergence, that is, $\alpha^{(n)} \to \alpha^{(\infty)}$ if $\alpha_{ij}^{(n)} \to \alpha_{ij}^{(\infty)}$ for each $i, j$.*

(ii) *Given* $\alpha \in \mathbb{S}_\infty$, *define* $\mathcal{O}(\alpha)\colon c_0 \to c_0$ *by*

$$\mathcal{O}(\alpha)(a)_i = \sum_{j=1}^{\infty} \alpha_{ij} a_j \tag{15.69}$$

*where the sum in* (15.69) *converges absolutely.*

$$\|\mathcal{O}(\alpha)a\|_\infty \le \|a\|_\infty \tag{15.70}$$

*Moreover,* $\mathcal{O}(\alpha)$ *maps* $\ell^1$ *to* $\ell^1$ *and*

$$\|\mathcal{O}(\alpha)a\|_1 \le \|a\|_1 \tag{15.71}$$

(iii) *If* $\mathcal{O}$ *is a linear map of* $c_0$ *to* $c_0$ *that obeys* (15.70) *and* $\mathcal{O}$ *maps* $\ell^1$ *to* $\ell^1$ *and obeys* (15.71)*, then*

$$\alpha_{ij} = (\mathcal{O}\delta_j)_i \tag{15.72}$$

*is a matrix in* $\mathbb{S}_\infty$ *and* $\mathcal{O} = \mathcal{O}(a)$.

*Proof* (i) $\mathbb{S}_\infty$ is a closed subset of a product of $D$'s, and so, compact by Tychonoff's theorem.

(ii) The convergence of (15.69) and (15.70) follows from (15.68) and

$$|\mathcal{O}(\alpha)(a)_i| \le \left[\sum_{j=1}^{\infty} |\alpha_{ij}|\right] \sup |a_j|$$

(15.71) follows from (15.67) and

$$\begin{aligned}\|\mathcal{O}(\alpha)a\|_1 &\le \sum_{i,j} |\alpha_{ij}|\, |a_j| \\ &\le \left[\sup_j \sum_{i=1}^{\infty} |\alpha_{ij}|\right] \|a\|_1\end{aligned}$$

(iii) By (15.71), (15.72), and $\|\delta_j\|_1 = 1$, (15.67) holds. Fix $i$ and taking

$$(a^N)_j = \begin{cases} \bar{\alpha}_{ij}/|\alpha_{ij}|, & j \le N \text{ and } a_{ij} \ne 0 \\ 0, & j > N \text{ or } \alpha_{ij} = 0 \end{cases}$$

shows, by (15.70), that

$$\sum_{j=1}^{N} |\alpha_{ij}| \le 1$$

Taking $N \to \infty$ proves $\alpha \in S_\infty$. A simple argument proves $\mathcal{O} = \mathcal{O}(\alpha)$. □

We will let $C_W(c_0)$ denote the set of all convex functions $\Phi$ on $c_0$ with values in $\mathbb{R} \cup \{\infty\}$:

(i) $\Phi$ is lsc
(ii) $\Phi(x) = \Phi(a)$ if $|x|^* = |a|^*$
(iii) $\Phi(0) = 0$

Condition (iii) and $\Phi(-x) = \Phi(x)$ implies by convexity that $\Phi(x) \geq 0$. Condition (ii) is strong because if $x$ has infinitely many nonzero elements, all the zeros get lost in $|x|^*$, so (ii) means if all $x_j \neq 0$, then

$$\Phi(x_1, x_2, \dots) = \Phi(x_1, 0, x_2, 0, x_3, 0, \dots)$$

**Lemma 15.23** *Let $\ell \in \ell^1$ and $a \in c_0$. Then*

$$\left| \sum_{j=1}^{\infty} \ell_j a_j \right| \leq \sum_{j=1}^{\infty} |\ell|_j^* \, |a|_j^* \tag{15.73}$$

*Proof* Let $\pi$ be defined by (15.66). Since the sum is absolutely convergent and all $a_j = 0$ if $j \notin \operatorname{Ran}(\pi)$,

$$\left| \sum_{j=1}^{\infty} \ell_j a_j \right| \leq \sum_{j=1}^{\infty} |\ell_{\pi(j)}| \, |a|_j^*$$

But by the absolute convergence again,

$$\sum_{j=1}^{\infty} |\ell_j| \, |a|_j^* = \sum_{j=1}^{\infty} (|a|_j^* - |a|_{j+1}^*) \sum_{k=1}^{j} |\ell_k|$$

(15.73) then follows from

$$\sum_{k=1}^{j} |\ell_k| \leq \sum_{k=1}^{j} |\ell|_k^*$$

□

**Theorem 15.24** *Let $a, b \in c_0$. Then the following are equivalent:*

(i) $S_k(|b|) \leq S_k(|c|)$ *for* $k = 1, 2, \dots$
(ii) $b \in \operatorname{cch}(\{x \in c_0 \mid |x|^* = |a|^*\})$
(iii) $b = Da$ *for some* $D \in \mathbb{S}_\infty$
(iv) $\Phi(b) \leq \Phi(a)$ *for all* $\Phi \in C_W(c_0)$
(v)

$$\sum_{j=1}^{\infty} \varphi(b_j) \leq \sum_{j=1}^{\infty} \varphi(a_j) \tag{15.74}$$

*for all functions $\varphi$ on $\mathbb{C}$ with $\varphi(z) = f(|z|)$, where $f$ is an even convex function in $\mathbb{R}$ with $f(0) = 0$.*

(vi) *For all $s > 0$,*

$$\sum_{j=1}^{\infty}(|b_j| - s)_+ \le \sum_{j=1}^{\infty}(|a_j| - s)_+ \tag{15.75}$$

*Proof* We will prove (i) ⇒ (ii) ⇒ (iv) ⇒ (v) ⇒ (i) and (ii) ⇒ (iii) ⇒ (i).

<u>(i) ⇒ (ii)</u> Suppose (i) holds. If $b$ is not in the $\mathrm{cch}(\{x \mid |x|^* = |a|^*\})$, then there is an $\ell \in c_0^* = \ell^1$ so

$$\mathrm{Re}[\ell(b)] > \sup_x \{\mathrm{Re}\,\ell(x) \mid |x|^* = |a|^*\} \tag{15.76}$$

Pick $\pi$ so (15.66) holds for $\ell$. Define

$$x_j = \begin{cases} a^*_{\pi^{-1}(j)}\ell_j/|\ell_j|, & \text{if } \ell_j \ne 0 \\ 0, & \text{if } \ell_j = 0 \end{cases}$$

Then $|x|^* = |a|^*$ and $\ell(x) = \sum |\ell|_j^* |a|_j^*$. Thus, by (15.73),

$$\begin{aligned} \sup_x \{\mathrm{Re}\,\ell(x) \mid |x|^* = |a|^*\} &= \sum_j |\ell|_j^*\,|a|_j^* \\ &< \mathrm{Re}[\ell(b)] \le |\ell(b)| \le \sum_j |\ell|_j^*\,|b|_j^* \end{aligned} \tag{15.77}$$

by (15.73) again.

On the other hand, by (14.3),

$$\begin{aligned} &\sum_{j=1}^{n} |\ell|_j^*\,|b|_j^* \\ &\quad = |\ell|_n^* S_n(|b|) + (|\ell|_{n-1}^* - |\ell|_n^*)S_{n-1}(|b|) + \cdots + (|\ell|_1^* - |\ell|_2^*)S_1(|b|) \\ &\quad \le |\ell|_n^* S_n(|a|) + \cdots + (|\ell|_1^* - |\ell|_2^*)S_1(a) \\ &\quad = \sum_{j=1}^{n} |\ell|_j^*\,|a|_j^* \end{aligned}$$

if (i) holds. Thus, (i) is inconsistent with (15.77) and so with (15.76) and $b \in \mathrm{cch}(\{x \mid |x|^* = |a|^*\})$.

<u>(ii) ⇒ (iii)</u> If $|x|^* = |a|^*$, then

$$x = \mathcal{O}(\alpha)a \tag{15.78}$$

where $\alpha_{ij}$ is defined inductively in $i$ as follows: If $x_i = 0$, $\alpha_{ij} = 0$ for all $j$. If $x_i \ne 0$, find $\pi(i)$ so $|x_i| = |a_{\pi(i)}|$ and $\pi(i) \ne \pi(1), \dots, \pi(i-1)$. This can be done since $|x|^* = |a|^*$. Then set

$$\alpha_{ij} = \begin{cases} x_i/a_{\pi(i)}, & j = \pi(i) \\ 0, & j \ne \pi(i) \end{cases}$$

Thus, $\alpha \in \mathbb{S}_\infty$ and (15.78) holds. Now let $b \in \text{cch}(\{x \in c_0 \mid |x|^* = |a|^*\})$ and let $\{x \in c_0 \mid |x|^* = |a|^*\} \equiv K$ so there are $\theta^{(\ell)} \in \mathbb{R}_+^{\nu_\ell}$ with $\sum_{j=1}^{\nu_\ell} \theta_j^{(\ell)} = 1$ and $\{x_j^{(\ell)}\}_{j=1}^{\nu_\ell} \subset K$ so

$$b = \lim_{\ell\to\infty} \sum_{j=1}^{\nu_\ell} \theta_j^{(\ell)} x_j^{(\ell)} = \lim_{\ell\to\infty} D^{(\ell)} a$$

where $D^\ell = \sum_{j=1}^{\nu} \theta_j^{(\ell)} \mathbb{O}(\alpha \text{ for } a \mapsto x_j^{(\ell)}) \in \mathbb{S}_\infty$. Since $\mathbb{S}_\infty$ is compact after passing to a subsequence, there is a $D \in \mathbb{S}_\infty$ so $D_{ij}^{(\ell)} \to D_{ij}$ for each $i, j$. Since $|a_j| \to 0$ as $j \to \infty$, this implies $(D^{(\ell)}a)_i \to (Da)_i$ for each $i$, and thus, $b = Da$.

(iii) $\Rightarrow$ (i) By taking limits from the case $\nu < \infty$, Lemma 15.18 extends to the case where $\nu = \infty$, that is, $a \in c_0$ and $\gamma$ is an infinite sequence in $[0, 1]$ with $\sum_{j=1}^{\infty} \gamma_j \leq k$. With that extension, the proof is the same as (iii) $\Rightarrow$ (i) in Theorem 15.17.

(ii) $\Rightarrow$ (iv) Immediate since $\Phi(x) = \Phi(a)$ for all $x$ with $|x|^* = |a|^*$, so by convexity, $\Phi(c) \leq \Phi(a)$ for all $c$ in $\text{ch}(\{x \mid |x|^* = |a|^*\})$, and thus, by lsc, $\Phi(b) \leq \Phi(a)$ for all $b$ in $\text{cch}(\{x \mid |x|^* = |a|^*\})$.

(iv) $\Rightarrow$ (v) $\Phi(x) = \sum_{j=1}^{\infty} \varphi(x_i) = \sup_n \sum_{j=1}^{N} \varphi(x_j)$ is an lsc convex function. It is symmetric, and thus, (iv) $\Rightarrow$ (v).

(v) $\Rightarrow$ (vi) $\varphi(y) = (|y| - s)_+$ is of the requisite form.

(v) $\Rightarrow$ (i) The proof in Theorem 15.5 extends with no change. □

We turn next to general measure spaces. To avoid having to discuss atoms in detail, we will take $\mu$ to a Baire measure on a locally compact metric space where atoms are pure points. We will focus on the inequality aspect first and then turn to the issues of substochastic maps and convex hulls. At least initially, we will look at functions $f$ which obey the sole condition

$$\mu(\{x \mid |f(x)| > \lambda\}) \equiv m_f(\lambda) < \infty \tag{15.79}$$

the distribution function of $f$. Note that if $F$ is a monotone, piecewise $C^1$ function with $F(0) = 0$, then

$$\begin{aligned} \int F(f(x))\, d\mu(x) &= \int \left( \int_0^{f(x)} F'(t)\, dt \right) d\mu(x) \\ &= (\mu \otimes F'(t)\, dt)(\{x, t\} \mid t \leq f(x)) \\ &= \int F'(t) m_f(t)\, dt \end{aligned} \tag{15.80}$$

It will also be of interest to define a function $f^*$ on $[0, \infty)$ by

$$f^*(s) = \inf\{\lambda \mid m_f(\lambda) \leq s\} \tag{15.81}$$

Notice that

$$\begin{aligned} f^*(s) > \lambda &\Leftrightarrow m_f(\lambda) > s \\ &\Leftrightarrow \mu(\{x \mid |f(x)| > \lambda) > s \\ &\Leftrightarrow 0 \le s < \mu(\{x \mid |f(x)| > \lambda\}) \end{aligned} \tag{15.82}$$

so

$$|\{s \mid f^*(s) > \lambda\}| = \mu(\{x \mid |f(x)| > \lambda\}) \tag{15.83}$$

so $f^*$ is the unique function on $[0, \infty)$ equimeasurable with $|f|$ with $f^*$ decreasing and lsc. It is thus a kind of rearrangement of $f$ on a potentially different space.

To define an analog of $S_k(a)$, we would like to take a definition like

$$S_\lambda(f) = \sup\biggl(\int_A |f(x)|\, d\mu \biggm| \mu(A) = \lambda\biggr) \qquad \text{(provisional!)} \tag{15.84}$$

There are two issues with this definition. If $\int (|f(x)| - 1)_+ \, d\mu(x) = \infty$, it is easy to see that $S_\lambda(f)$ is everywhere infinite, so we will add a condition that

$$\int_1^\infty m(\lambda)\, d\lambda = \int (|f(x)| - 1)_+ \, d\mu(x) < \infty \tag{15.85}$$

The issue, of course, is convergence at $\lambda = \infty$; the finiteness holds with 1 replaced by any $s > 0$ if and only if (15.85) holds.

The other problem with (15.84) is that it is not a good definition if $\mu$ has pure points. The extreme case is $X = \{1, 2, \dots\}$ with counting measure, that is, the sequences we have considered earlier. In that case, (15.84) gives if $n \in \{0, 1, 2, \dots\}$,

$$S_{\lambda=n}(a) = |a|_1^* + |a|_2^* + \cdots + |a|_n^*$$

but if $\lambda$ is not an integer, $S_\lambda(f)$ is either not defined or $-\infty$, depending on how one defines the sup of an empty set! One could try to replace $\mu(A) = \lambda$ by $\mu(A) \le \lambda$, in which case $S_\lambda(a)$ jumps from $S_{n-1}(a)$ to $S_n(a)$ as $\lambda$ passes through $n$, that is, $S_\lambda(a) = S_{[\lambda]}(a)$ where $[\lambda]$ is the greatest integer less than $\lambda$. This loses something, as can be seen by

**Proposition 15.25** *For sequences $a$, let $S_k$ be given by*

$$S_k(a) = \sum_{j=1}^k |a|_j^*$$

*for $k = 1, 2, 3, \dots$ $S_k$ obeys*

$$S_k(a) \ge \tfrac{1}{2}(S_{k-1}(a) + S_{k+1}(a)) \tag{15.86}$$

*Proof*

$$\begin{aligned} S_k(a) - \tfrac{1}{2}\, S_{k-1}(a) - \tfrac{1}{2}\, S_{k+1}(a) &= |a|_k^* - \tfrac{1}{2}\,(|a_k|^* + |a|_{k+1}^*) \\ &= \tfrac{1}{2}\,(|a_k|^* - |a|_{k+1}^*) \geq 0 \end{aligned}$$ □

(15.86) is a concavity of $S_k$ defined at the integers and implies concavity in the ordinary sense if $S$ is interpolated linearly between the integers. So we will take the following definition which allows us to split up pure points:

**Definition** Suppose (15.85) holds. Define for $0 < \lambda < \mu(X)$,

$$S_\lambda(f) = \sup\left\{\theta|f(x_0)| + \int_A |f(y)|\, d\mu(y) \;\middle|\; \theta \leq \mu(\{x_0\});\ x_0 \notin A,\ \theta + \mu(A) = \lambda\right\} \tag{15.87}$$

where included in the sup is the case where $\theta = 0$, there is no $x_0$ and $\mu(A) = \lambda$ (for just take any $x_0$, even one with $\mu(\{x_0\}) = 0$). If $\mu(X) < \infty$ and $\lambda \geq \mu(X)$, define

$$S_\lambda(f) = \int |f(y)|\, d\mu(y), \qquad (\lambda \geq \mu(X)) \tag{15.88}$$

which is finite if $\mu(X) < \infty$ and (15.85) holds. If $\lambda = 0$, we set $S_\lambda(f) = 0$ and if $\lambda < 0$, we set $S_\lambda(f) = -\infty$.

*Remarks* 1. With this definition in the case of sequences with counting measure, $S_\lambda(a)$ is the linear interpolation of $S_k(a)$ and is concave (as we will see is always the case).

2. One might wonder why we do not consider splitting multiple points. It is not hard to see that nothing is gained by doing so – the sup in (15.87) is unchanged if we do allow multiple points to be split.

3. It is unfortunate we decided not to discuss atoms since we could talk about splitting atoms otherwise.

The final object we will consider is

$$Q_f(s) = \int (|f(x)| - s)_+\, d\mu(x) \tag{15.89}$$

where we use this definition, even if $s < 0$, in which case if $\mu(X) = \infty$, $Q_f(s) = \infty$.

Here are the relations among the four objects $m_f(\lambda)$, $f^*(s)$, $S_\lambda(f)$, and $Q_f(\lambda)$, any of which determines the other three (!).

**Proposition 15.26** (i)

$$Q_f(s) = \int_s^\infty m_f(\lambda)\, d\lambda \tag{15.90}$$

(ii) *For* $\lambda \geq 0$,

$$S_\lambda(f) = \int_0^\lambda f^*(s)\, ds \tag{15.91}$$

(iii) $Q_f$ *is a convex, monotone decreasing function with*

$$(D_s^+ Q_f)(s) = -m_f(s) \tag{15.92}$$

(iv) $S_\lambda(f)$ *is a concave, monotone increasing function with*

$$(D_\lambda^+ S_\lambda(f)) = f^*(\lambda) \tag{15.93}$$

(v) *If* $\lambda_0$ *is a point of continuity of* $m_f$ *and* $s_0 = m_f(\lambda_0)$ *is a point of continuity of* $f^*$, *then*

$$f^*(m_f(\lambda_0)) = \lambda_0, \qquad m_f(f^*(s_0)) = s_0 \tag{15.94}$$

*More generally,*

$$f^*(m_f(\lambda_0)) = \inf\{\lambda \mid m_f(\lambda) = m_f(\lambda_0)\} \tag{15.95}$$

$$m_f(f^*(s_0)) = \inf\{s \mid f^*(s) = f^*(s_0)\} \tag{15.96}$$

(v)

$$S_\lambda(f) = \inf_s\, (Q_f(s) + \lambda s) \tag{15.97}$$

(vi)

$$Q_f(s) = \sup_\lambda\, (S_\lambda(f) - \lambda s) \tag{15.98}$$

*Remarks* 1. (15.97) says that in terms of Legendre transforms (see Chapter 5 and Theorem 5.23, in particular),

$$S_\lambda(f) = -Q_f^*(-\lambda)$$

Thus, (15.98) follows from (15.97) by Fenchel's theorem.

2. The picture is much like the one for conjugate convex functions, except $m_f$ is decreasing, not increasing.

*Proof* (i) By (15.80) and $F'(s) = \chi_{[s,\infty]}(\lambda)\, d\lambda$ as distributions, we obtain (15.90).

(ii) It is clear that to maximize $\int_A |f(x)|\, d\mu(x)$ subject to $\mu(A) = \lambda$, we find $s$, so $m_f(s) \leq \lambda \leq m_f(s-0)$ and take $A = \{x \mid f(x) > s\} \cup \tilde{A}$, where $\tilde{A}$ is a set of measures $\lambda - m_f(s)$ in $\{x \mid f(x) = s\}$. If $\mu$ has pure points, we may not be able to find $\tilde{A}$ and need to split a point. By (15.83), $|f|$ and $f^*$ are equimeasurable, and since we can split points, for any set $S \subset [0, \mu(X)]$, we can find $A \subset X$, $x_0 \notin A$, and $0 \leq \theta \leq \mu(x_0)$ so $|S| = \mu(A) + \theta$ and

$$\int_S f^*(t)\, dt = \int_A |f(x)|\, d\mu(x) + \theta f(x_0)$$

Thus, repeating the construction for $f^*$, the way to maximize $\int_S f^*(t)\,dt$ with $|S| = \lambda$ is to take $S = [0, \lambda]$, that is, (15.91) holds.

(iii) $Q_f$ is convex since $m_f$ is monotone decreasing. (15.92) is immediate from (15.90) and the fact that $m_f$ is continuous from above.

(iv) $\lambda \mapsto S_\lambda(f)$ is concave since $f^*$ is decreasing (the difference in convex vs. concave comes from $D^+Q_f = -m$ but $D^+S(f) = f^*$). (15.93) follows from the fact that $f^*$ is continuous from the right.

(v) By (15.82), if $f^*$ and $m_f$ are both continuous and strictly monotone, then they are inverses. The problem areas involve, first, intervals of constancy where $m_f(\lambda) = s_0$ for $\lambda \in [\lambda_1, \lambda_2]$ or $\lambda \in [\lambda_1, \lambda_2)$, in which case $f^*(s)$ is discontinuous and $f^*(s_0) = \lambda_1 = \inf\{\lambda \mid m_f(\lambda) = s_0\}$, so (15.95) holds, consistent with the definition (15.81).

The second problem area is where $m_f(\lambda_0 - 0) > m_f(\lambda_0)$ (the difference is, of course, $\mu(\{x \mid f(x) = \lambda_0\})$). In that case $f^*(s) = \lambda_0$ in the interval $[m_f(\lambda_0), m_f(\lambda_0 - 0)]$ or $(m_f(\lambda_0), m_f(\lambda_0 - 0)]$, so in either case, use $m_f(\lambda_0) = \inf(\{s \mid f^*(s) = \lambda_0\})$ and (15.96) holds.

(vi) We begin with considering a set $A$, point $x_0 \notin A$, and $0 \le \theta \le \mu(\{x_0\})$. Let $\lambda = \mu(A) + \theta$. Then

$$\begin{aligned}
\theta|f(x_0)| &+ \int_A |f(x)|\,d\mu(x) \\
&= \theta(|f(x_0)| - s) + \int_A (|f(x)| - s)\,d\mu(x) + s\lambda \\
&\le \theta(|f(x)| - s)_+ + \int_A (|f(x)| - s)_+\,d\mu(x) + s\lambda \\
&\le \int (|f(x)| - s)_+\,d\mu(x) + s\lambda \\
&= Q_f(s) + s\lambda
\end{aligned}$$

Taking the sup over all such $A, x_0, \theta$, we see that for all $\lambda > 0$ and all $s$,

$$S_\lambda(f) \le Q_f(s) + s\lambda \tag{15.99}$$

so

$$S_\lambda(f) \le \inf_s\,(Q_f(s) + s\lambda) \tag{15.100}$$

To see equality, we consider the trivial regions of $\lambda$ first. If $\lambda \ge \mu(X)$, take $s = 0$ and see $S_\lambda(f) = \|f\|_1 = Q_f(0) + 0\lambda \ge \inf_s(Q_f(s) + s\lambda)$. If $\lambda = 0$, the inf is zero since $\lim_{s\to\infty} Q_f(s) = 0$. If $\lambda < 0$, taking $s \to \infty$, we see the inf is $-\infty$. That leaves the case $\lambda \in [0, \mu(X))$. Find $s$ so $m_f(s) \le \lambda \le m_f(s - 0)$. Let $B = \{s \mid |f(x)| = s\}$ and find $A_0 \subset B$, $x_0 \in B \backslash A_0$, and $0 \le \theta \le \mu(\{x\})$ so

$\mu(A_0) + \theta = \lambda - m_f(s)$. Let $A = A_0 \cup \{x \mid |f(x)| > s\}$. Then since $S_\lambda$ is a sup,

$$\begin{aligned} S_\lambda(f) &\geq \int_A |f(x)|\, d\mu(x) + \theta s \\ &= \int_A (|f(x)| - s)\, d\mu(x) + \lambda s \\ &= Q_f(s) + \lambda s \end{aligned}$$

proving equality in (15.100) and so (15.97).

(vii) As noted, this follows from (15.97) and Fenchel's theorem. Here is a direct proof: By (15.99),

$$Q_f(s) \geq \sup_\lambda (S_\lambda(f) - s\lambda) \tag{15.101}$$

To get equality, let $A = \{x \mid |f(x)| > s\}$ and suppose $\mu(X) < \infty$. Let $\lambda = \mu(A)$. Then

$$S_\lambda(f) \geq \int_A |f(x)|\, d\mu(x) = Q_f(s) + s\lambda$$

proving equality in (15.101). If $\mu(A) = \infty$ (only possible if $\mu(X) = \infty$ and $s \leq 0$), take $\lambda \to \infty$. Then $\lim_{\lambda\to\infty} S_\lambda(f) = \|f\|_1$ so

$$\lim_{\lambda\to\infty} [S_\lambda(f) - s\lambda] = \|f\|_1 + |s|\infty$$

which is infinite if $s < 0$ or if $s = 0$ and $Q_f(0) = \infty$. Either way, we get equality in (15.101). □

With this under our belt, the following is easy:

**Theorem 15.27** *Let $\mu$ be a Baire measure on a locally compact metric space, $X$. Let $f, g$ be Baire functions on $X$ so $\int(|f(x)|-1)_+\, d\mu(x) + \int(|g(x)|-1)_+\, d\mu(x) < \infty$. Then the following are equivalent:*

(i) $Q_g(s) \leq Q_f(s)$ *for all* $s$

(ii) $S_\lambda(g) \leq S_\lambda(f)$ *for all* $\lambda$

(iii) *For all monotone increasing, convex $\varphi$ on $[0, \infty)$ with $\varphi(0) = 0$,*

$$\int \varphi(|g(x)|)\, d\mu(x) \leq \int \varphi(|f(x)|)\, d\mu(x)$$

*Proof* The equivalence of (i) and (iii) is (15.80) (essentially (iii) ⇒ (i) taking $\varphi(x) = (x - s)_+$ and (i) ⇒ (iii) because any $\varphi$ is an integral of such $(x - s)_+$). The equivalence of (i) and (ii) follows immediately from (15.97) and (15.98). □

Next, we turn to the study of substochastic maps and convex hulls, where we will need to suppose that $\mu$ has no pure points. We will discuss this issue in the context of general measure spaces.

**Definition** Let $(M, \Sigma, \mu)$ be a $\sigma$-finite measure space. A measurable set $A \subset M$ is called an *atom* if

(i) $\mu(A) > 0$

(ii) $B \subset A$, $B$ measurable $\Rightarrow \mu(B) = 0$ or $\mu(A \backslash B) = 0$

Points of positive mass are atoms but there are examples of nonpoint atoms. Recall from Chapter 2 that a measure space is called nonatomic if for any measure $B \subset M$ and $0 < \alpha < \mu(B)$, there is a $C \subset B$ with $\mu(C) = \alpha$. As the name suggests

**Theorem 15.28** *A $\sigma$-finite measure space $(M, \Sigma, \mu)$ is nonatomic if and only if it has no atoms.*

We also proved this earlier as Corollary 8.24.

*Proof* Obviously, if $B$ is an atom, there are no $C \subset B$ with $0 < \mu(C) < \mu(B)$, so $M$ is not nonatomic. The subtle half of this theorem is the converse: that if $M$ has no atoms, then for any $B \subset M$ with $\mu(B) < 0$ and any $\alpha$ with $0 < \alpha < \mu(B)$, there is $C \subset B$ with $\mu(C) = \alpha$. First, since $A$ is $\sigma$-finite, $B$ is a countable union of disjoint sets of finite measure, so we can find $B_1 \subset B$ with $\alpha < \mu(B_1) < \infty$. Thus, it suffices to suppose $\mu(B) < \infty$.

We first claim that any such $B$ has subsets of arbitrarily small measure. For since $B$ is not an atom, we can find $E \subset B$ with $\mu(E), \mu(B \backslash E) > 0$. It follows that one of the two sets $E$ or $B \backslash E$ has measure in $(0, \frac{1}{2}\mu(B)]$, so we can find $F_1 \subset B$ with $0 < \mu(E_1) \leq \frac{1}{2}\mu(B)$. By induction, we find $E_1 \supset E_2 \supset \dots$ with $0 < \mu(E_n) \leq 2^{-n}\mu(B)$.

Given $\alpha$, we define numbers $\alpha \geq \lambda_1 \geq \lambda_2 \geq \dots$ and sets $F_1 \subset F_2 \subset \cdots \subset B$ as follows:

$$\lambda_1 = \sup\{\mu(F) \mid F \subset B,\ \mu(F) \leq \alpha\}$$

$F_1$ is a set with

$$F_1 \subset B, \qquad \mu(F_1) \in [\lambda_1 - 1, \lambda_1]$$

Assuming we have $\lambda_1, \dots, \lambda_{n-1}$ and $F_1 \subset \cdots \subset F_{n-1}$, define

$$\lambda_n = \sup\{\mu(F) \mid F_{n-1} \subset F \subset B,\ \mu(F) \leq \alpha\}$$

and pick $F_n$ a set with

$$F_{n-1} \subset F_n \subset B, \qquad \mu(F_n) \in [\lambda_n - 2^{-n}, \lambda_n]$$

Now, let $F_\infty = \cup_{n=1}^\infty F_n$ and $\lambda_\infty = \lim_{n\to\infty} \lambda_n$ which exists since $\lambda_1 \geq \lambda_2 \geq \cdots \geq \mu(F_1)$. $\mu(F_n)$ is increasing and $\mu(F) = \lim_n \mu(F_n)$. Since $|\mu(F_n) - \lambda_n| \leq 2^{-n}$, we see $\mu(F) = \lambda_\infty$.

We claim $\lambda_\infty = \alpha$, proving that $\mu$ is nonatomic. For suppose $\lambda_\infty < \alpha$. By the fact that $B\backslash F$ has sets of arbitrarily small measure, we can find $H \subset B\backslash F$ so that

$$0 < \mu(H) < \alpha - \lambda_\infty \tag{15.102}$$

Let $G_n = F_n \cup H$. Then $F_{n-1} \subset G_n \subset B$ and

$$\mu(G_n) = \mu(F_n) + \mu(H) \le \lambda_\infty + \alpha - \lambda_\infty = \alpha$$

so, by definition of $\lambda_n$,

$$\mu(F_n) + \mu(H) \le \lambda_n$$

Taking $n$ to infinity, $\mu(H) = 0$, which is a contradiction with (15.102). □

It is worth noting what the absence of atoms means for Baire measure on separable locally compact metric spaces.

**Proposition 15.29** *Let $\mu$ be a Baire measure on a locally compact separable metric space, $M$. Then any point $x$ with $\mu(\{x\}) > 0$ is an atom and, conversely, if $A \subset M$ is an atom, there is $x \in A$ with $\mu(A\backslash\{x\}) = 0$. In particular, $\mu$ is nonatomic if and only if $\mu(\{x\}) = 0$ for all $x \in M$.*

*Proof* Obviously, if $\mu(\{x\}) > 0$, $A = \{x\}$ is an atom. Conversely, suppose $A$ is an atom. $M$ can be covered by a countable family $\{M_n\}_{n=1}^\infty$ of compact sets since $M$ is separable and locally compact (pick a countable dense set $x_n$ and let $M_n$ be a compact neighborhood of $x_n$). Thus, $\mu(A) = \sum_{n=1}^\infty \mu(A \cap M_n)$, so some $\mu(A \cap M_n) > 0$. It then follows $\mu(A\backslash M_n) = 0$ so without loss, we can suppose $A \subset M_n$ a compact set.

For each $k$, $M_n$ is covered by finitely many $2^{-k}$ balls so we can inductively find compact balls $\overline{B_{2^{-k}}^{x_k}}$ of radius $2^{-k}$ so $\mu(A\backslash \overline{B_{2^{-k}}^{x_k}}) = 0$ and $x_{k+1} \subset \overline{B_{2^{-k}}^{x_k}}$. It follows that $x_k$ has a limit point $x$ and $\mu(A\backslash\{x\}) = 0$. □

If you look back at the various proofs that $S_n(b) \le S_n(a)$ implies $b \in$ cch (some transforms of $a$), they all depended on the ability to take one sequence (usually defined by a linear functional) and move it to a transformed sequence where $|a|$ is large. So we are heading to being able to move all functions around freely. The key will be maps from $M$ to $[0, \mu(X)]$ that move $f$ to $f^*$, the symmetric rearrangement. We will need the following notion:

**Definition** Let $(M, \Sigma, \mu)$ and $(N, \Gamma, \nu)$ be two measure spaces. A measurable map $T\colon M \to N$ is called *measure preserving* if and only if $\nu(A) = \mu(T^{-1}[A])$ for all $A \in \Gamma$.

This definition is clearly close to the idea of invariant measure used in Example 8.17. $T$ mapping a measure space to itself is measure preserving if and only if $\mu$ is an invariant measure for $T$.

Measure-preserving maps are almost surjective in the sense that $\nu(N \setminus \operatorname{Ran} T) = \mu(\phi) = 0$. But they can be very, very far from injective, as the following example shows. Let $\mu$ be Lebesgue measure on $\mathbb{R}^\nu$. Let $\nu$ be the measure $\tau_\nu x^{\nu-1}\, dx$ on $[0, \infty)$ where $\tau_\nu$ is the surface area of the unit sphere in $\mathbb{R}^\nu$. Then $\varphi(x) = |x|$ is measure preserving, but only injective on $\{0\}$. Nevertheless, as we will see, measure-preserving maps can be very useful.

**Proposition 15.30** *Let $(M, \mu, \Sigma)$ and $(N, \nu, \Gamma)$ be two $\sigma$-finite measure spaces. Let $T$ be a measure-preserving map from $M_0 \subset M$ to $N$. Let $f$, a function on $N$, obeying*

$$\nu(\{x \mid f(x) > t\}) < \infty$$

*for all $t > 0$. Define $\pi_T f$ on $M$ by*

$$(\pi_T f)(m) = \begin{cases} f(T(m)), & m \in M_0 \\ 0, & \textit{otherwise} \end{cases} \tag{15.103}$$

*Then*

(i) *$\pi_T f$ and $f$ are equimeasurable.*
(ii) *$\|\pi_T f\|_p = \|f\|_p$ for all $p \in [1, \infty]$*
(iii)

$$\int_M (\pi_T f)(m)(\pi_T g)(m)\, d\mu(m) = \int_N f(n) g(n)\, d\nu(n) \tag{15.104}$$

*for all nonnegative functions $f, g$ on $N$. If $f, g$ are arbitrary complex functions with $fg \in L^1(N, d\nu)$, then $(\pi_T f)(\pi_T g) \in L^1(M, d\mu)$ and (15.104) holds.*

*Remark* In the case of the map $T\colon (\mathbb{R}^\nu, d^\nu x) \to ((0, \infty), \tau_\nu x^{\nu-1}\, dx)$, $\pi_T$ maps onto spherically symmetric functions.

*Proof* (i) This follows from the fact that $T$ is measure preserving and

$$\{m \mid |\pi_T f(m)| > t\} = T^{-1}[\{n \mid |f(n)| > t\}]$$

(ii) Immediate from (i).

(iii) By the wedding cake representation, we need only prove the result when $f$ and $g$ are characteristic functions of sets $A, B$ of finite measure. In that case, (15.104) says

$$\mu(T^{-1}[A] \cap T^{-1}[B]) = \nu(A \cap B)$$

which follows from the measure-preserving property since $T^{-1}[A] \cap T^{-1}[B] = T^{-1}[A \cap B]$. The passage from positive to $L^1$ functions is standard writing $f$ and $g$ as Real and Imaginary parts and then further breaking $\operatorname{Re} f = \operatorname{Re} f_+ - \operatorname{Re} f_-$. $\square$

Here is how the nonatomic condition enters:

**Proposition 15.31** *Let $(M, \Sigma, \mu)$ be a nonatomic $\sigma$-finite measure space. Then there exists a measure-preserving map $\varphi\colon M \to [0, \mu(M)) \subset \mathbb{R}$, where the Borel field and Lebesgue measure are put on $[0, \mu(M))$.*

*Proof* If $\mu(M) = \infty$, we can, by the fact that $M$ is $\sigma$-finite, write $M = \cup_{n=1}^{\infty} M_n$ with $\mu(M_n) < \infty$ and the $M_n$'s disjoint. By putting together maps $\varphi_n$ on each $M_n$ by

$$\varphi(x) = \sum_{j=1}^{n-k} \mu(M_j) + \varphi_n(x), \qquad \text{if } x \in M_n$$

we see that we need only consider the case $\mu(M) < \infty$. By scaling, we can take $\mu(M) = 1$.

Since $M$ is nonatomic, find $M_{1/2} \subset M$ with $\mu(M_{1/2}) = 1/2$. Then find $M_{1/4} \subset M_{1/2}$ and $M_{3/4}$ with $M_{1/2} \subset M_{3/4} \subset M$ and $\mu(M_{j/4}) = j/4$. By induction, we find for each dyadic rational $\alpha$, $M_\alpha$ with $\mu(M_\alpha) = \alpha$ and $M_\alpha \subset M_\beta$ if $\alpha < \beta$.

Define $\varphi_0 : M \to [0, 1]$ by

$$\varphi_0(x) = \inf\{\alpha \mid x \in M_\alpha\}$$

Thus, for any dyadic rational,

$$\varphi_0^{-1}([0, \alpha_0]) = M_{\alpha_0}$$

so

$$\mu(\varphi_0^{-1}([0, \alpha_0)) = \alpha_0$$

Since $\{[0, \alpha]\}$ generates the Borel $\sigma$-algebra, $\varphi_0$ is measure preserving. The range of $\varphi_0$ is $[0, 1]$, so define $\varphi(x) = \varphi_0(x)$ if $\varphi_0(x) < 1$ and $= 0$ if $\varphi_0(x) = 1$ and get a measure-preserving map to $[0, 1)$. □

For any measurable function, $f$, obeying (15.79), we define $f^*$ by (15.81). Here is the key to extending the HLP theorem to general nonatomic measures:

**Theorem 15.32** (Lorentz–Ryff Lemma) *Let $(M, \Sigma, \mu)$ be a $\sigma$-finite nonatomic measure space and let $f$ be a measurable function from $M$ to $\mathbb{C}$ so that*

$$\mu(\{x \mid |f(x)| > t\}) < \infty, \qquad \textit{all } t > 0 \tag{15.105}$$

*Let $M_f = \{x \mid |f(x)| > 0\}$. Then there exists a measure-preserving map $\psi\colon M_f \to [0, \mu(M_f))$ so that for a.e. $m \in M$ with $f(m) \neq 0$,*

$$|f(m)| = f^*(\psi(m)) \tag{15.106}$$

*Remark* If $\mu(\{x \mid |f(x)| > 0\}) < \infty$ and, in particular, if $\mu(M) < \infty$, $\psi$ can be extended so that (15.106) will hold even when $f(m) = 0$, but if $\mu(\{x \mid |f(x)| > 0\}) = \infty$, $f^*(t)$ is never zero, so (15.106) cannot hold for $m$'s with $f(m) = 0$, which could be on a set of positive measure!

*Proof* $f^*$ is defined by (15.81) for all $t$, so let $R = \{f^*(t) \mid t \in [0, \mu(M_f))\}$. Since $f$ and $f^*$ are equimeasurable, $\mu(\{x \in M_f \mid f(x) \notin R\}) = 0$, and thus, we can redefine $f$ on a set of measure zero so $|f(x)| \in R$ for all $x$.

Since $|f(x)|$ always takes values in $R$, (15.95) says that if $|f(x)| = \lambda_0$, then $\lambda_0 = \inf\{\lambda \mid m_f(\lambda) = m_f(\lambda_0)\}$ (since $f^*(t)$ has jumps at $t = m_f(\lambda_0)$ and the only value it takes in $\{\lambda \mid m_f(\lambda) = m_f(\lambda_0)\}$ is the inf). Thus, by (15.95),

$$f^*(m_f(|f(x)|)) = |f(x)| \tag{15.107}$$

for all $x$.

Call $\lambda > 0$ exceptional if $\gamma(\lambda) \equiv \mu(\{x \mid |f(x)| = \lambda\}) > 0$. There are at most countably many exceptional values. If $\lambda$ is an exceptional value, $f^*(t)$ has the value $\lambda$ on an interval $[\beta(\lambda), \beta(\lambda) + \gamma(\lambda))$ (and perhaps also at $\beta(\lambda) + \gamma(\lambda)$). For each exceptional value of $\lambda$, use Proposition 15.31 to define

$$\eta_\lambda = \{x \mid |f(x)| = \lambda\} \mapsto [\beta(\lambda), \beta(\lambda) + \gamma(\lambda))$$

which is measure preserving. Define $\psi$ as follows:

$$\psi(x) = \begin{cases} \eta_\lambda(x), & \text{if } |f(x)| = \lambda \text{ is exceptional} \\ m_f(|f(x)|), & \text{if } |f(x)| = \lambda \text{ is not exceptional} \end{cases}$$

If $\lambda_0$ is not exceptional, then by (15.107),

$$f^*(\psi(x)) = f^*(m_f(|f(x)|)) = |f(x)|$$

If $\lambda$ is exceptional, then $\eta_\lambda(x) \in [\beta(\lambda), \beta(\lambda) + \gamma(\lambda))$, the set on which $f^*$ is $\lambda$, so again (15.106) holds. Thus, $\psi$ obeys (15.106).

Since $\{[0, s) \mid s > 0\}$ generates the Borel $\sigma$-algebra, we need only show

$$\mu(\psi^{-1}([0, s))) = s \tag{15.108}$$

to conclude that $\psi$ is measure preserving. If $s$ is such that $\lambda \equiv f^*(s)$ is nonexceptional, then since $m_f$ is monotone,

$$\psi^{-1}([0, s)) = \{x \mid |f(x)| > \lambda\} \tag{15.109}$$

$\lambda$ nonexceptional means $\{s \mid f^*(s) = \lambda\}$ has Lebesgue measure zero, hence it is a single point since $f^*$ is monotone. Therefore, by (15.96),

$$m_f(f^*(s)) = s$$

Thus, by (15.109),

$$\mu(\psi^{-1}[(0,s)) = m_f(f^*(s)) = s$$

proving (15.108) for $s \in \{t \mid f^*(t) \text{ nonexceptional}\}$.

Let $s \in [\beta(\lambda), \beta(\lambda)+\gamma(\lambda))$ for some exceptional $\lambda$. Then

$$\psi^{-1}([0,s)) = \{x \mid |f(x)| > \lambda\} \cup \{x \mid \eta_\lambda(x) < s\} \tag{15.110}$$

Since $\{t \mid f^*(t) = \lambda\} = \beta(\lambda)$ implies via (15.96) that $m_f(\lambda) = \beta(\lambda)$, we have

$$\mu(\{x \mid |f(x)| > \lambda\} = \beta(\lambda) \tag{15.111}$$

Since $\eta_\lambda$ is measure preserving,

$$\{x \mid \eta_\lambda(x) \in [\beta(\lambda), s)\} = s - \beta(\lambda) \tag{15.112}$$

(15.110)–(15.112) imply that (15.108) holds for exceptional $\lambda$'s also. □

The usefulness of the Lorentz–Ryff lemma can be seen from the following, which should be viewed as the ultimate version of (14.2) which started the last chapter:

**Theorem 15.33** *Let $f, g$ be positive functions on a $\sigma$-finite measure space $(M, \Sigma, \mu)$. Then*

$$\int_M f(m)g(m)\,d\mu(m) \le \int_{[0,\infty)} f^*(t)g^*(t)\,dt \tag{15.113}$$

*Moreover, if $M$ is nonatomic and $f, g_1, \dots, g_\ell$ are functions obeying* (15.79), *then there exist functions $\tilde{g}_1, \dots, \tilde{g}_\ell$ on $M$ so*

(a) *$g_j$ is equimeasurable with $\tilde{g}_j$ for $j = 1, \dots, \ell$.*

(b) *For $j = 1, \dots, \ell$,*

$$\int_M f(m)\tilde{g}_j(m)\,d\mu(m) = \int_{[0,\infty)} f^*(t)g_j^*(t)\,dt \tag{15.114}$$

*Proof* If $A \subset M$ is a set of finite measure, then

$$\chi_A^*(t) = \begin{cases} 1, & 0 \le t < \mu(A) \\ 0, & t \ge \mu(A) \end{cases}$$

Moreover, since $f^*$ is lsc and decreasing, $\{x \mid f^*(t) > \lambda\} = [0, m_f(\lambda))$. Thus, by the wedding cake representation,

$$f^*(t) = \int \chi^*_{\{x \mid f(x) > \lambda\}}(t)\,d\lambda$$

That means, by the wedding cake representation again, we need only prove (15.113) when $f = \chi_A$ and $g = \chi_B$. But then (15.113) says

$$\mu(A \cap B) \le \min(\mu(A), \mu(B)) \tag{15.115}$$

which is obvious.

For the second assertion, let $\psi$ be the measure-preserving map guaranteed by the Lorentz–Ryff lemma, so $f^* \circ \psi = f$. Define

$$\tilde{g}_j = (g_j^* \circ \psi)\,[\bar{f}/|f|]$$

where we interpret $\bar{f}/|f|$ as 1 if $f(m) = 0$.

By Proposition 15.30(i), $\tilde{g}_j$ and $g_j^*$ are equimeasurable so (a) holds since $g_j^*$ is equimeasurable with $g_j$. Moreover, by Proposition 15.30(iii),

$$\begin{aligned} \int g_j(m) f(m)\, d\mu(m) &= \int (g_j^* \circ \psi)(m)(f^* \circ \psi)(m)\, d\mu(m) \\ &= \int g_j^*(t) f^*(t)\, dt \end{aligned}$$

so (b) holds. □

As a final preliminary, we need to define the analog of substochastic matrices. Theorem 15.22 is the motivation for the following definition:

**Definition** Let $(M, \Sigma, \mu)$ and $(N, \Gamma, \nu)$ be two $\sigma$-finite measure spaces. A *substochastic map* is a linear mapping $\zeta\colon L^1(M, d\mu) \to L^1(N, d\nu)$ that obeys

$$\|\zeta(f)\|_1 \le \|f\|_1 \tag{15.116}$$

for all $f \in L^1(M, d\mu)$ and

$$\|\zeta(f)\|_\infty \le \|f\|_\infty \tag{15.117}$$

for all $f \in L^1 \cap L^\infty(M, d\mu)$. $\mathbb{S}(M, N)$ is the set of substochastic maps of $M$ to $N$ (of course, depending on $\mu, \nu$).

**Proposition 15.34** (i) *If $\zeta \in \mathbb{S}(M, N)$ and $\gamma \in \mathbb{S}(N, Q)$, then $\gamma \circ \zeta \in \mathbb{S}(M, Q)$.*

(ii) *If $T$ is a measure-preserving map of $M_0 \subset M$ onto $N$, then $\pi_T \in \mathbb{S}(N, M)$ (given by* (15.103)*).*

(iii) *If $\zeta \in \mathbb{S}(M, N)$, there is a map $\zeta^\dagger \in \mathbb{S}(N, M)$ so for all $f \in L^1 \cap L^\infty(M)$ and $g \in L^1 \cap L^\infty(N)$:*

$$\int_N g(n)(\zeta f)(n)\, d\nu(n) = \int_M (\zeta^\dagger g)(m) f(m)\, d\mu(m) \tag{15.118}$$

(iv)

$$\pi_T^\dagger \pi_T = \mathbf{1} \tag{15.119}$$

*In particular, if $\psi$ is given by the Lorentz–Ryff lemma, so*

$$\pi_\psi f^* = |f| \tag{15.120}$$

*we also have*

$$\pi_\psi^\dagger |f| = f^* \tag{15.121}$$

(v) *Put the weak topology on $\mathbb{S}(M,N)$ defined by the linear functions $L_{f,g}(\zeta) = \int g(n)(\zeta f)(n)\, d\nu(n)$ for all $g \in L^1 \cap L^\infty(N, d\nu)$ and $f \in L^1 \cap L^\infty(M, d\mu)$. Then $\mathbb{S}$ is compact.*

*Remarks* 1. We use $^\dagger$ for adjoint maps to avoid confusion with the $^*$ in $f^*$.

2. While (15.119) holds, it may not be true that $\pi_T \pi_T^\dagger = \mathbf{1}$. In fact, if $T$ is the example we gave above of $(\mathbb{R}^\nu, d^\nu x)$ to $([0,\infty), \tau_\nu x^{\nu-1}\, dx)$ with $T(x) = |x|$, then $\pi_T \pi_T^\dagger$ is $L^2$ projection onto the spherically symmetric functions.

*Proof* (i) Trivial.

(ii) Follows immediately from Proposition 15.30(ii).

(iii) Since $\zeta\colon L^1(M,d\mu) \to L^1(N,d\nu)$, by duality there is a map $\zeta^\dagger\colon L^\infty(N,d\nu) \to L^\infty(M,d\mu)$ and (15.118) holds for all $f \in L^1$ and $g \in L^\infty$ and so, certainly for $f \in L^1 \cap L^\infty(M,d\mu)$ and $g \in L^1 \cap L^\infty(N,d\nu)$. Since $\|\zeta f\|_1 \le \|f\|_1$, we have $\|\zeta^\dagger g\|_\infty \le \|g\|_\infty$. Moreover, if $g \in L^1 \cap L^\infty(N)$ and $f \in L^1 \cap L^\infty(M)$, by (15.118),

$$\left| \int_M (\zeta^\dagger g)(x) f(x)\, dx \right| \le \|g\|_1 \, \|\zeta f\|_\infty \le \|g\|_1 \, \|f\|_\infty$$

In particular, if $A \subset M$ has finite measure and we take

$$f(x) = \chi_A(x)\, \frac{\overline{(\zeta^\dagger g)(x)}}{|\zeta^\dagger g(x)|}$$

(where $\bar{w}/|w|$ is interpreted as 0 if $w = 0$), then

$$\int_A |(\zeta^\dagger g)(x)|\, dx \le \|g\|_1$$

We can then take $A \uparrow M$ and obtain $\|\zeta^\dagger g\|_1 \le \|g\|_1$ so $\zeta^\dagger \in \mathbb{S}(N,M)$.

(iv) By (15.118) if $f, g \in L^1 \cap L^\infty(N, d\nu)$,

$$\begin{aligned} \int_N g(x)[\pi_T^\dagger \pi_T f](x)\, d\nu(x) &= \int_M (\pi_T g)(m)(\pi_T f)(m)\, d\mu(m) \\ &= \int_N g(x) f(x)\, d\nu(x) \end{aligned}$$

by (15.104). It follows that $\pi_T^\dagger \pi_T f = f$.

(v) This is a simple variant of the argument behind the Banach–Alaoglu theorem. For each $f \in L^1 \cap L^\infty(M, d\mu)$ and $g \in L^1 \cap L^\infty(N, d\nu)$, let $P_{f,g}$ be the closed disk in $\mathbb{C}$ of radius $\delta_{f,g} \equiv \min(\|f\|_1 \|g\|_\infty, \|f\|_\infty \|g\|_1)$. By Tychonoff's theorem, $\Omega = \times_{f,g} D_{f,g}$ is compact in the coordinate topology. Any $\eta \in \mathbb{S}(M, N)$ defines a point in $\Omega$ with

$$\eta_{f,g} = \int g(n)(\eta f)(n)\, d\nu(n)$$

and the map $\mathbb{S}(M, N) \to \Omega$ is clearly one-one. Its range is closed because linearity (i.e., $\eta_{f,g+h} = \eta_{f,g} + \eta_{f,h}$, etc.) is preserved under limits and $|\eta_{f,g}| \leq \delta_{f,g}$ implies the associated map is a contraction on $L^1$ and $L^\infty$. The map is clearly a homeomorphism if $\mathbb{S}$ is given the weak topology. It follows that $\mathbb{S}$ is compact in this topology. □

We need one final lemma that can be viewed as the ultimate version of Lemma 15.18.

**Lemma 15.35** *Let $(M, \Sigma, \mu)$ be a $\sigma$-finite measure space. Suppose $g$ has $\mu(\{x \mid |g(x)| > \lambda\})$. Let $h \in L^1 \cap L^\infty(M, d\mu)$. Then*

$$\int |h(m)g(m)|\, d\mu(m) \leq \|h\|_\infty S_{\|h\|_1/\|h\|_\infty}(g) \tag{15.122}$$

*Proof* By replacing $h$ by $h/\|h\|_\infty$, we can suppose $\|h\|_\infty = 1$. Since $|h^*(t)| \leq 1$, for any $\alpha$, $\int_0^\alpha h^*(t)\, dt \leq \alpha$. Also, clearly, $\int_0^\alpha h^*(t)\, dt \leq \int_0^\infty h^*(t)\, dt = \|h\|_1$. Thus,

$$\begin{aligned}\int \chi_{[0,a)}(t) h^*(t)\, dt &\leq \min(\alpha, \|h|_1) \\ &\leq \int \chi_{[0,\alpha)}(t) \chi_{0,\|h\|_1)}(t)\, dt\end{aligned}$$

Since this is true for any $\alpha$, the wedding cake representation shows it holds if $\chi_{[0,\alpha)}$ is replaced by any monotone decreasing function. In particular,

$$\begin{aligned}\int g^*(t) h^*(t)\, dt &\leq \int_0^{\|h\|_1} g^*(t)\, dt \\ &= S_{\|h\|_1}(g)\end{aligned}$$

(15.122) now follows from (15.113). □

With these lengthy preliminaries out of the way, we can state and prove:

**Theorem 15.36** *Let $p \in [1, \infty)$ and let $(M, \Sigma, \mu)$ be a $\sigma$-finite nonatomic measure space. Let $f, g \in L^p(M, d\mu)$. Then the following are equivalent:*

(i) $S_\lambda(g) \leq S_\lambda(f)$ *for all* $\lambda > 0$

(ii) $g \in \operatorname{cch}(\{h \in L^p \mid h$ *is equimeasurable with* $f\})$

(iii) *$g = \zeta f$ for some $\zeta \in \mathbb{S}(M, M)$*

(iv) *$\Phi(g) \le \Phi(f)$ for any lsc convex function $\Phi$ on $L^p$ (with values in $[0, \infty]$) with the property that $\Phi(k) = \Phi(h)$ if $h$ and $k$ are equimeasurable functions in $L^p(M, d\mu)$.*

(v) *$\int \varphi(|g(m)|)\, d\mu(m) \le \int \varphi(|f(m)|)\, d\mu(m)$ for all monotone increasing convex $\varphi\colon [0, \infty) \to [0, \infty)$.*

(vi) *For all $s \in \mathbb{R}_+$,*

$$\int (|g(m)| - s)_+ \, d\mu(m) \le \int (|f(m)| - s)_+ \, d\mu(m) \tag{15.123}$$

*Proof* We will show (i) ⇒ (ii) ⇒ (iii) ⇒ (i) and (ii) ⇒ (iv) ⇒ (v) ⇒ (vi) ⇒ (i).

(i) ⇒ (ii) If $g$ is not in the $\text{cch}\{h \in L^p \mid h \text{ equimeasurable to} f\}$, then by the separating hyperplane theorem, there exists a function $\ell \in L^q$ so that

$$\operatorname{Re} \ell(g) > \sup\{\ell(h) \mid h \text{ equimeasurable to } f\} \tag{15.124}$$

By Theorem 15.33, the sup in (15.124) is $\int \ell^*(t) f^*(t)\, dt$. Moreover, by the same theorem,

$$\operatorname{Re} \ell(g) \le \int \ell^*(t) g^*(t)\, dt$$

Thus, if (15.124) holds,

$$\int \ell^*(t) g^*(t)\, dt > \int \ell^*(t) f^*(t)\, dt \tag{15.125}$$

But, by (15.91), if $S_\lambda(g) \le S_\lambda(t)$, we have

$$\int_0^\lambda g^*(t)\, dt \le \int_0^\lambda f^*(t)\, dt \tag{15.126}$$

Since $\ell^*(t)$ is decreasing, the set $\{t \mid \ell^*(t) > \alpha\} = [0, \lambda(\alpha))$ so the wedding cake representation for $\ell^*$ and (15.126) implies

$$\int \ell^*(t) g^*(t)\, dt \le \int \ell^*(t) f^*(t)\, dt$$

This contradicts (15.125) and so (15.124) which implies that $g$ is in the convex hull.

(ii) ⇒ (iii) Given $f$, let $\pi_f$ denote the elements of $\mathbb{S}([0, \infty), M)$ induced by the measure-preserving $\psi$ with $f^* \circ \psi = |f|$ (see Theorem 15.32 and Proposition 15.34). As proven in Proposition 15.34, $\pi_f f^* = |f|$ and $\pi_f^\dagger |f| = f^*$. Let $\theta_f$ be a function with $|\theta_f| \le 1$ so $\theta_f f = |f|$. Then if $h$ and $f$ are equimeasurable, $f^* = h^*$ so

$$h = (\bar{\theta}_h \pi_h \pi_f^\dagger \theta_f) f$$

But $\bar{\theta}_h \pi_h \pi_f^\dagger \theta_f$ is a product of elements in $\mathbb{S}$ and so it lies in $\mathbb{S}(M,M)$. We have thus shown that

$$\mathcal{D}_f = \{\zeta f \mid \zeta \in \mathbb{S}(M,M)\} \supset \{h \mid h \text{ is equimeasurable with } f\}$$

But $\mathbb{S}(M,M)$ is a compact convex set, so $\mathcal{D}_f$ is also. It follows $\mathcal{D}_f$ contains the weak cch of the $h$'s, and so the norm cch of the $h$'s.

<u>(iii) ⇒ (i)</u> Let $g = \eta f$ with $\eta \in \mathbb{S}(M,M)$. Let $\mu(A) = \lambda$. Then

$$\int_A |g(m)|\, d\mu(m) = \int g(m)h(m)\, d\mu(m) \tag{15.127}$$

where $h = \theta\chi_A$ with $|\theta| \equiv 1$. Thus, $\|h\|_\infty = 1$ and $\|h\|_1 = \lambda$. But

$$\int g(m)h(m)\, d\mu(m) = \int f(m)(\eta^\dagger h)(m)\, d\mu(m) \tag{15.128}$$

Since $\eta^\dagger \in \mathbb{S}$, $\|\eta^\dagger h\|_\infty \le 1$ and $\|\eta^\dagger h\|_1 \le \lambda$. By (15.122),

$$\int |f(m)(\eta^\dagger h)(m)|\, d\mu(m) \le S_\lambda(f) \tag{15.129}$$

(15.127)–(15.129) imply that if $\mu(A) = \lambda$, then

$$\int_A |g(m)|\, d\mu(m) \le S_\lambda(f)$$

Taking the sup over $A$, we see $S_\lambda(g) \le S_\lambda(f)$.

<u>(ii) ⇒ (iv)</u> Immediate since $\Phi(g) \le \Phi(f)$ for $g \in \text{ch}\{\dots\}$ by convexity and invariance of $\Phi$ and then for $g \in \text{cch}\{\dots\}$ since $\Phi$ is lsc.

<u>(iv) ⇒ (v)</u> Each such $\Phi(f) = \int \varphi(|f(m)|)\, d\mu(m)$ is a special case of the $\Phi$'s of (iv).

<u>(v) ⇒ (vi)</u> Immediate since $\varphi(x) = (|x| - s)_+$ is monotone in $|x|$ and convex.

<u>(vi) ⇒ (i)</u> Follows from (15.97). □

HLP theorems assert that, under certain circumstances (namely, $a \prec_{\text{HLP}} b$), we have $\sum_{j=1}^n \varphi(a_j) \le \sum_{j=1}^n \varphi(b)$ for all convex $\varphi$. But this is precisely the notion of Choquet order discussed in Chapter 10. Thus, the HLP theorem provides necessary and sufficient conditions for $\frac{1}{n}\sum_{j=1}^n \delta_{a_j} \prec \frac{1}{n}\sum_{j=1}^n \delta_{b_j}$ (Choquet order) as measures on $[\min(b_j), \max(b_j)]$. We can generalize this to general one-dimensional measures.

Given a probability measure $\mu$ on $[0,1]$, define

$$Q_\mu(s) = \int (x-s)_+\, d\mu(x) \tag{15.130}$$

If

$$m_\mu(\lambda) = \mu((\lambda, 1]) \tag{15.131}$$

then writing

$$(x - s)_+ = \int_s^1 \chi_{(\lambda,1]}(x)\, d\lambda \tag{15.132}$$

we see

$$Q_\mu(s) = \int_s^1 m_\mu(\lambda)\, d\lambda$$

Define for $0 \le s \le 1$,

$$\mu^*(s) = \inf\{\lambda \mid \mu((x, 1]) \le s\} \tag{15.133}$$

and

$$S_\lambda(\mu) = \int_0^\lambda \mu^*(s)\, ds \tag{15.134}$$

Then, as with Proposition 15.26,

$$S_\lambda(\mu) = s\,[\lambda - \mu(s, 1])] + \int_{s+0}^1 x\, d\mu(x) \tag{15.135}$$

where $s$ is determined by $s = \inf\{t \mid \mu((t, 1]) \le \lambda\}$ and $\int_{s+0}^1 x\, d\mu(x)$ means over the set $(s, 1]$ and

$$S_\lambda(\mu) = \sup_s\, [s\lambda + Q_\mu(s)] \tag{15.136}$$

$$Q_\mu(s) = \sup_\lambda [S_\lambda(\mu) - s\lambda] \tag{15.137}$$

**Theorem 15.37** *Let $\mu, \nu$ be two probability measures on $[0, 1]$. Then the following are equivalent:*

(i) *$\mu \prec \nu$ in Choquet order*
(ii) *$Q_\mu(s) \le Q_\nu(s)$, $0 \le s \le 1$; $Q_\mu(0) = Q_\nu(0)$*
(iii) *$S_\lambda(\mu) \le S_\lambda(\nu)$, $0 \le s \le 1$; $S_1(\mu) = S_1(\nu)$*

*Proof* (i) ⇔ (ii) As usual, any monotone increasing convex function has the form $\varphi(0) + \int (x - t)_+\, d\gamma(t)$, so given that $\mu([0, 1]) = \nu([0, 1])$, $Q_\mu(s) \le Q_\nu(s)$ for $0 \le s \le 1$ is equivalent to $\int \varphi(x)\, d\mu(x) \le \int \varphi(x)\, d\nu(x)$ for all monotone convex functions. If also $Q_\mu(0) = Q_\nu(0)$, then the inequality holds for any convex function since any convex function has the form $\alpha x + \eta$ where $\eta$ is monotone and convex. Conversely, if $\mu \prec \nu$, they have the same barycenter, so $Q_\mu(0) = Q_\nu(0)$.

(ii) ⇔ (iii) By (15.136) and (15.137), $Q_\mu(s) \le Q_\nu(s)$ is equivalent to $S_\lambda(\mu) \le S_\lambda(\nu)$. Moreover, since $Q_\mu(0) = S_\lambda(\mu) = \int x\, d\mu$, the equality conditions are equivalent. □

One can ask about the analog of a doubly stochastic relation. In fact, such an analog exists not only in the one-dimensional case but in general – the analog is known as Cartier's theorem and is stated and discussed but not proven in the Notes; see Theorem 17.8.

Finally, we turn to some applications of the machinery. We saw in the definition of $\mathbb{S}(M, N)$ that doubly substochastic matrices are connected to contractions on $L^p$ for $p = 1, \infty$. This will allow a new proof of (15.60) for $1 \le p \le \infty$ without using Weyl's lemma (15.58) or the HLP theorem!

**Proposition 15.38** *Let $A$ be an $n \times n$ complex substochastic matrix. Let $y = Ax$. Then for any $p \in [1, \infty)$,*

$$\sum_{j=1}^{n} |y_j|^p \le \sum_{j=1}^{n} |x_j|^p \tag{15.138}$$

*Proof*

$$\begin{aligned} |y_j| &\le \sum_{k=1}^{n} |a_{jk}|\,|x_k| \\ &\le \sup_k |x_k| \biggl(\sum_{j=1}^{n} |a_{jk}|\biggr) \end{aligned}$$

so

$$\sup_j |y_j| \le \sup_k |x_k|$$

which is (15.138) for $p \to \infty$. Moreover,

$$\begin{aligned} \sum_j |y_k| &\le \sum_{j,k} |a_{jk}|\,|x_k| \\ &\le \sum_{k=1}^{n} |x_k| \biggl(\sum_{j=1}^{n} |a_{jk}|\biggr) \\ &\le \sum_{k=1}^{k} |x_k| \end{aligned}$$

which is (15.138) for $p = 1$. Now use complex interpolation (see Chapter 12). □

**Theorem 15.39** *Let $B$ be an $n \times n$ matrix with $\{\lambda_j(B)\}_{j=1}^n$ the eigenvalues of $B$ counting geometric multiplicity (i.e., roots of* $\det(B - \lambda\mathbf{1}) = 0$) *and $\{\mu_j(B)\}_{j=1}^n$ the eigenvalues of $|B|$. Then, there is a complex substochastic matrix $A$ so*

$$\lambda_j(A) = \sum_{k=1}^{n} a_{jk}\mu_k(A) \tag{15.139}$$

*In particular, for $p \in [1, \infty)$,* (15.60) *holds.*

*Proof* We first claim that there is an orthonormal basis, $\{\varphi_j\}_{j=1}^n$, called a *Schur basis*, with

$$A\varphi_j = \lambda_j(A)\varphi_j + \sum_{k=1}^{j-1} \beta_{jk}\varphi_k \tag{15.140}$$

For the Jordan normal form says there is a basis of generalized eigenvalues $\eta_j$ with

$$A\eta_j = \lambda_j(A)\eta_j + x_j(A)\eta_{j-1}$$

with $x_j(A) = 0, 1$. If $\{\varphi_j\}$ is obtained from $\{\eta_j\}$ by a Gram–Schmidt process, (15.140) follows. In particular,

$$\langle \varphi_j, A\varphi_j\rangle = \lambda_j(A) \tag{15.141}$$

Let $A = U|A|$ with $U$ a partial isometry and

$$|A|\psi_j = \mu_j(A)\psi_j \tag{15.142}$$

with $\psi_j$ the eigenvectors of the self-adjoint operator $|A|$. Then

$$\begin{aligned}\lambda_j(A) &= \langle U^*\varphi_j, |A|\varphi_j\rangle \\ &= \sum_{k=1}^{n} a_{jk}\mu_k(A)\end{aligned}$$

where

$$a_{jk} = \langle U^*\varphi_j, \psi_k\rangle\langle\psi_k, \varphi_j\rangle \tag{15.143}$$

By the Schwarz inequality, we claim $a_{jk}$ is complex doubly substochastic for

$$\begin{aligned}\sum_{k=1}^{n} |a_{jk}| &\le \left(\sum_{k=1}^{n} |\langle U^*\varphi_j, \psi_k\rangle|^2\right)^{1/2} \left(\sum_{k=1}^{n} \langle\psi_k, \varphi_j\rangle^2\right)^{1/2} \\ &= \|U^*\varphi_j\|\,\|\varphi_j\| \le 1\end{aligned}$$

by the fact that $\{\psi_k\}_{k=1}^N$ is an $ON$ basis and

$$\begin{aligned}\sum_{j=1}^{n} |a_{jk}| &\le \left(\sum_{j=1}^{n} |\langle \varphi_j, U\psi_k\rangle|^2\right)^{1/2} \left(\sum_{j=1}^{n} \langle\psi_k, \varphi_j\rangle^2\right)^{1/2} \\ &= \|U\psi_k\|\,\|\psi_k\| \le 1\end{aligned}$$

Thus, (15.60) for $p \in [1, \infty)$ follows Proposition 15.38. □

**Theorem 15.40** (Hadamard's Determinantal Inequality) *Let $A$ be a positive $n \times n$ matrix. Then*

$$\det(A) \le \prod_{j=1}^{n} a_{jj} \tag{15.144}$$

*More generally, if* $\sigma_\ell$ *is the elementary symmetric function given by* (15.32)*, then for* $1 \le \ell \le n$*,*

$$\operatorname{tr}(\wedge^n(A)) \le \sigma_\ell(a_{11}, \dots, a_{nn}) \tag{15.145}$$

*Proof* We will show that

$$a_{jj} = \sum_{k=1}^{n} \alpha_{jk} \lambda_k(A) \tag{15.146}$$

with $\alpha_{jk}$ doubly stochastic. Thus, by the HLP theorem, $a_{jj} \prec_{\text{HLP}} \lambda_j(A)$ so (15.144) follows from the fact that on $\mathbb{R}^n_+$, $f(x_1, \dots, x_n) = x_1 \dots x_n$ is Schur concave (Example 15.7), and (15.145) follows from the fact that $\sigma_\ell$ is Schur concave (Example 15.11).

Let $D$ be the diagonal matrix with diagonal entries $D_{kk} = \lambda_k(A)$. Then $A = UDU^*$ for some unitary matrix $U$. Thus, (15.146) holds where

$$\alpha_{jk} = U_{jk}(U^*)_{kj} = |U_{jk}|^2$$

The unitarity of $U$ implies that $\alpha$ is doubly stochastic. □

There is an equivalent inequality to (15.144) also often called Hadamard's inequality:

**Theorem 15.41** *Let* $A$ *be an arbitrary* $n \times n$ *matrix. Then*

$$|\det(A)|^2 \le \prod_{j=1}^{n} \left( \sum_{k=1}^{n} |a_{kj}|^2 \right) \tag{15.147}$$

*Proof* Let $B = A^* A$. Then

$$b_{jj} = \sum_{k=1}^{n} |a_{kj}|^2$$

and (15.144) implies that

$$\det(B) \le \text{RHS of (15.147)}$$

since $B \ge 0$. Since $\det(B) = |\det(A)|^2$, (15.147) follows. □

*Remarks* 1. Conversely, applying (15.147) to $\sqrt{A}$, we see (15.147) implies (15.144).

2. There is an elementary direct proof of (15.147); see the Notes.

The following is related to the Brunn–Minkowski inequality:

**Corollary 15.42** (Minkowski's Determinantal Inequality) *If* $A$ *and* $B$ *are two positive* $n \times n$ *matrices, then*

$$\det(A + B)^{1/n} \ge \det(A)^{1/n} + \det(B)^{1/n} \tag{15.148}$$

*Proof* We start by recalling for positive numbers $\{a_j\}_{j=1}^n$, $\{b_j\}_{j=1}^n$, one has

$$\prod_{j=1}^{n}(a_j+b_j)^{1/n} \geq \prod_{j=1}^{n} a_j^{1/n} + \prod_{j=1}^{n} b_j^{1/n} \tag{15.149}$$

This is a special case of the Brunn–Minkowski theorem – indeed, we began our proof of the general theorem by noting that this follows from the arithmetic-geometric inequality (see (13.46)).

Pick a basis in which $A+B$ is diagonal. Let $\{a_j\}_{j=1}^n$ and $\{b_j\}_{j=1}^n$ be the diagonal elements of $A$ and $B$. Then since $A+B$ is diagonal, $\det(A+B)^{1/n} = \prod_{j=1}^n (a_j + b_j)^{1/n}$ so (15.149) and $\det(A) \leq \prod_{j=1}^n a_j$, $\det(B) \leq \prod_{j=1}^n b_j$ (which is (15.144)) imply (15.148). $\square$

# 16
# The relative entropy

This brief chapter discusses a final convexity inequality, emphasizing the close connection between convexity and entropy. Let $X$ be a compact metric space and let $\mu, \nu \in \mathcal{M}_{+,1}(X)$ be two probability measures. Their *relative entropy* (in $\mathbb{R} \cup \{-\infty\}$) is defined by

$$S(\mu \mid \nu) = \begin{cases} -\infty, & \text{if } \mu \text{ is not } \nu\text{-a.c.} \\ -\int \log(d\mu/d\nu)\, d\mu, & \text{if } \mu \text{ is } \nu\text{-a.c.} \end{cases} \tag{16.1}$$

*Remarks* 1. The entropy is normally defined for a specific $\nu$ like counting measures on a finite set, normalized Lebesgue measures on a bounded open set in $\mathbb{R}^\ell$, or the natural normalized measure on some compact Riemannian manifold. It can be useful to also consider variations of $\nu$.

2. As we will show below, the positive part of the integral in (16.1) is always convergent, so the integral (including the minus sign) can only diverge to $-\infty$, in which case we set $S = -\infty$.

3. If $\mu$ is $\nu$-a.c. and $d\mu = \eta\, d\nu$, (16.1) becomes

$$S(\eta\, d\nu \mid \nu) = -\int \eta \log \eta\, d\nu \tag{16.2}$$

which is the more usual formula for $S$ in terms of the concave function $H(x) = -x \log x$ on $[0, \infty)$.

4. In some of the information theory literature, the relative entropy is defined with the opposite sign.

Our main goal in this chapter is to prove the following theorem:

**Theorem 16.1** *$S(\mu \mid \nu) \leq 0$ and $(\mu, \nu) \mapsto S(\mu \mid \nu)$ is a jointly concave and weakly usc function on $\mathcal{M}_{+,1} \times M_{+,1}$, that is, if $\mu_n \to \mu$ and $\nu_n \to \nu$ weakly, then*

$$S(\mu \mid \nu) \leq \limsup_n S(\mu_n \mid \nu_n) \tag{16.3}$$

This result fits in here because convex functions and especially the theory of Legendre transforms play a critical role. While the main interest in this theorem is in statistical mechanics and information theory, it is also useful in the spectral theory of Jacobi matrices (see Simon [353]).

Rather than prove the result only for $S$ built from $\log(\cdot)$, we will define a class of $S$'s built from suitable convex functions. These generalizations do not have applications that I know of, but the generality makes clear certain aspects of the argument obscured by two special features of log: namely, $\log(x^{-1}) = -\log x$ and the fact that if $F(x) = -\log x$, then $F^*(-x) = -1 - \log x$.

**Definition** A function $F$ on $(0, \infty)$ is called *entropic* if and only if

(i) $F$ is convex and monotone decreasing.

(ii)

$$\lim_{y \downarrow 0} F(y) = \infty \tag{16.4}$$

(iii)

$$\lim_{y \to \infty} \frac{F(y)}{y} = 0 \tag{16.5}$$

(iv)

$$F(1) = 0 \tag{16.6}$$

Given an entropic function, $F$, we will show below its Legendre transform is finite on $(-\infty, 0)$ (or perhaps $(-\infty, 0]$). We define $G$, its *entropic conjugate* on $(0, \infty)$, by

$$G(x) = F^*(-x) \tag{16.7}$$

**Example 16.2** Let

$$F(x) = -\log x \tag{16.8}$$

which is easily seen to be entropic. As noted in Example 5.19(iv),

$$G(x) = -\log x - 1 \tag{16.9}$$

Given an entropic function, $F$, we define the relative $F$-entropy of the measure by

$$S_F(\mu \mid \nu) = \begin{cases} -\infty, & \text{if } \mu \text{ is not } \nu\text{-a.c.} \\ -\int F((d\mu/d\nu)^{-1})\, d\mu, & \text{if } \mu \text{ is } \nu\text{-a.c.} \end{cases} \tag{16.10}$$

□

*Remark* If $F$ is given by (16.8), since $F(x^{-1}) = \log x$, $S_F$ agrees with the standard relative entropy.

We will prove the following theorem of which Theorem 16.1 is a special case.

**Theorem 16.3** *$S_F(\mu \mid \nu) \le 0$ and $(\mu, \nu) \mapsto S_F(\mu \mid \nu)$ is a jointly concave and weakly usc function of $(\mu, \nu)$.*

We will obtain this from the following variational principle.

**Theorem 16.4** *Let $\mathcal{E}(X)$ be the set of real-valued continuous functions on $X$ with $\inf_{x \in X} f(x) > 0$. Then*

$$S_F(\mu \mid \nu) = \inf_{f \in \mathcal{E}(X)} \left[ \int f(x)\, d\nu(x) + \int G(f(x))\, d\mu(x) \right] \tag{16.11}$$

*Remark* By (16.9), the usual relative entropy obeys

$$S(\mu \mid \nu) = \inf_{f \in \mathcal{E}(X)} \left[ \int f(x)\, d\nu(x) - \int (1 + \log(f(x)))\, d\mu(x) \right] \tag{16.12}$$

The log in (16.12) is not a direct transcription of the log in (16.1), but rather taken from the Legendre transform of log.

**Example 16.5** Let $a > 0$ and define

$$F(x) = x^{-a} - 1$$

which is entropic. By Example 5.18(iii),

$$G(x) = 1 - (a+1)x^p$$

where $p = a/(a+1) \in (0, 1)$.

$$\begin{aligned} S_F(\mu \mid \nu) &= 1 - \int \left( \frac{d\mu}{d\nu} \right)^{a+1} d\nu \\ &= \inf_{f \in \mathcal{E}(X)} \left[ \int f(x)\, d\nu(x) + 1 - \int (a+1) f(x)^p\, d\mu \right] \end{aligned}$$

*Proof of Theorem 16.3 given Theorem 16.4* Let us first show $S_F(\mu \mid \nu) \le 0$. If $\mu$ is not $\nu$-a.c. or the integral is $-\infty$, there is nothing to prove. So suppose $d\mu/d\nu = \eta$. Let $d\tilde{\nu} = \eta^{-1}\, d\mu$ so

$$d\tilde{\nu} = \chi_{\{x \mid d\eta(x) \neq 0\}}\, d\nu \tag{16.13}$$

Then, by Jensen's inequality,

$$\begin{aligned} S_F(\mu \mid \nu) &\le -F\left( \int \eta^{-1}\, d\mu \right) \\ &= -F(\nu\{x \mid \eta(x) \neq 0\}) \\ &\le -F(0) && \text{(since } F \text{ is decreasing)} \\ &= 0 && \text{(by (16.6))} \end{aligned}$$

proving $S_F(\mu \mid \nu) \le 0$.

Define for $f \in \mathcal{E}(X)$,

$$\mathcal{S}_F(f;\mu,\nu) = \int f(x)\,d\nu(x) + \int G(f(x))\,d\mu(x) \tag{16.14}$$

Since $G \circ f$ is also continuous,

$$(\mu \mid \nu) \mapsto \mathcal{S}_F(f;\mu,\nu)$$

is a weakly continuous affine function jointly in $(\mu,\nu)$. By (5.19), $S_F$ as an inf of such functions is concave and usc. □

We begin the proof of Theorem 16.4 with a preliminary:

**Proposition 16.6** *Let $F$ be an entropic function. Let $G$ be its entropic conjugate. Then*

(i) *Its Legendre transform is finite on either $(-\infty,0)$ or $(-\infty,0]$ depending on whether $\lim_{x\to\infty} F(x) = -\infty$ or $\lim_{x\to\infty} F(x) > -\infty$.*

(ii) *For all $x, y > 0$,*

$$xy^{-1} \geq -F(y^{-1}) - G(x) \tag{16.15}$$

(iii)

$$\lim_{x\to\infty} G(x) = -\infty \tag{16.16}$$

(iv) *$G$ is convex and monotone decreasing.*

(v)

$$-F(y) = \inf_{x>0} (xy + G(x)) \tag{16.17}$$

*for any $y \in (0,\infty)$.*

*Proof* (i) If $x > 0$, $\lim_{y\to\infty} xy - F(y) = \infty$ since $F$ is monotone decreasing, and thus, $F^*(x) = \infty$. If $x < 0$, $\lim_{y\to\infty} xy - F(y) = -\infty$ since $F(y)/y \to 0$ as $y \to \infty$ by (16.5) and $\lim_{y\downarrow 0} xy - F(y) = -\infty$ by (16.4), so $\sup(xy - F(y)) < \infty$, that is, $x < 0 \Rightarrow F^*(x) < \infty$.

Finally, by monotonicity of $F$,

$$\begin{aligned} F^*(0) &= \sup_y\,[-F(y)] \\ &= -\lim_{y\to\infty} F(y) \end{aligned}$$

proving (i)

(ii) Young's inequality says

$$wz \leq F(z) + F^*(w)$$

Letting $z = y^{-1}$, $w = -x$, and multiplying by $(-1)$ yields (16.15).

(iii) Given $x \in (0, \infty)$, let $q(x)$ be that value of $y$ where $-x$ is tangent to $F(y)$ so, by definition of $F^*$,

$$G(x) = -xq(x) - F(q(x))$$

Since $F$ is convex and (16.4) and (16.5) hold, as $x$ runs from 0 to $\infty$, $q(x)$ runs from $\infty$ to 0. Thus, since $xq(x) > 0$,

$$\begin{aligned}\limsup_{x\to\infty} G(x) &\le \limsup_{x\to\infty} -F(q(x)) \\ &= \limsup_{y\downarrow 0} -F(y) \\ &= -\infty\end{aligned}$$

(iv) $G(-x)$ is convex so $G$ is convex. Since $-xy$ is monotone decreasing in $x$ for $y \ge 0$, $G(x) = \sup_y -xy - F(y)$ is monotone.

(v) Writing (with $w = -x$)

$$-\inf_{x>0} (xy + G(x)) = \sup_{w<0} (wy - G(-w))$$

(and (16.17)) follows from Fenchel's theorem on the double Legendre transform (Theorem 5.23). $\square$

*Proof of Theorem 16.4* Suppose first $\mu$ is $\nu$-a.c. Let $f \in \mathcal{E}(X)$. Let $\eta = d\mu/d\nu$ and let $A = \{x \mid \eta(x) > 0\}$. On $A$, we have by (16.15) that

$$f(x)\eta(x)^{-1} \ge -F(\eta(x)^{-1}) - G(f(x)) \tag{16.18}$$

Thus,

$$\begin{aligned}\int f(x)\,d\nu &\ge \int_A f(x)\,d\nu \\ &= \int_A f(x)\eta(x)^{-1}\,d\mu \\ &\ge -\int F(\eta(x)^{-1})\,d\mu(x) - \int G(f(x))\,d\mu(x)\end{aligned} \tag{16.19}$$

by (16.18) where we used $\mu(X\backslash A) = 0$. With $\mathcal{S}$ given by (16.14), we thus have

$$\mathcal{S}_F(f;\mu,\nu) \ge S_F(\mu \mid \nu) \tag{16.20}$$

so

$$\inf_{f\in\mathcal{E}(X)} \mathcal{S}_F(f;\mu,\nu) \ge S_F(\mu \mid \nu) \tag{16.21}$$

The idea behind the other side is the following: Given $y \in (0, \infty)$, define $p(y)$ so $-p(y)$ is a tangent to $F$ at $y$. $p(y)$ is monotone decreasing from $\infty$ to 0 as $y$ goes from 0 to $\infty$ and is continuous except perhaps at a finite number of points and

$$yp(y) = -F(y) - G(p(y)) \tag{16.22}$$

Define for $x \in A$,

$$f_\infty(x) = p(\eta(x)^{-1})$$

and $f_0(x) = 0$ for $x \notin A$. Thus, by (16.22), integrating $d\mu(x)$,

$$\int f_\infty \, d\nu = S_F(\mu \mid \nu) - \int G(f_\infty(x)) \, d\mu(x) \tag{16.23}$$

so it appears $\mathcal{S}_F(f_\infty; \mu, \nu) = S_F(\mu \mid \nu)$ and we have equality in the variational principle. Alas, things are not that simple since $f_\infty$ may not lie in $\mathcal{E}(X)$. First, it is zero on $X \backslash A$ and it may not be continuous. Worse, it can happen that $S_F(\mu \mid \nu)$ is finite but $\int f_\infty(x) \, d\nu = \infty$ and $\int G(f_\infty(x)) \, d\mu(x) = -\infty$ with a cancellation of infinities leading to a finite value for $S_F(\mu \mid \nu)$.

*Note* For the case $F(x) = -\log x$, $p(y) = y^{-1}$ and $f_\infty(x) = \eta(x)$, so this cancellation of infinities cannot happen. But if, for example, $F(x) = x^{-1} - 1$, $p(y) = y^{-2}$ and $f_\infty(x) = \eta(x)^2$ may not be in $L^1(X, d\nu)$ even if $\log(\eta(x))$ and $\eta$ are both in $L^1(X, d\nu)$.

So, while the intuition is correct, we will need to exercise some care. Begin by defining $\mathcal{S}_f(f; \mu, \nu)$ by (16.14) for general $f \in L^1(X, d(\mu + \nu))$ where for some $\varepsilon > 0$,

$$\varepsilon \le f(x) \le \varepsilon^{-1} \tag{16.24}$$

(i.e., $f$ is bounded above and away from zero). Call the set of such $f$, $\tilde{\mathcal{E}}(X; \mu + \nu)$. The same argument that led to (16.20) applies, so (16.21) holds if $\mathcal{E}$ is replaced by $\tilde{\mathcal{E}}(X; \mu + \nu)$.

Given any $f \in \tilde{\mathcal{E}}(X; \mu + \nu)$, we can find $f_n \in C(X)$ so $f_n \to f$ pointwise a.e., and then by replacing $f_n$ by $\min(\varepsilon^{-1}, \max(f_n, \varepsilon))$, we can suppose for the same $\varepsilon$ for which (16.24) holds, $f_n$ obeys the same estimates. Since $G$ is bounded on $[\varepsilon, \varepsilon^{-1}]$, we have, by the dominated convergence theorem, that

$$\mathcal{S}(f_n; \mu, \nu) \to \mathcal{S}(f; \mu, \nu)$$

and so

$$\inf_{f \in \tilde{\mathcal{E}}(X; \mu, \nu)} \mathcal{S}_F(f; \mu, \nu) = \inf_{f \in \mathcal{E}(X)} \mathcal{S}_F(f; \mu, \nu) \tag{16.25}$$

and therefore, it suffices to find $f_n$ in $\tilde{\mathcal{E}}(X; \mu + \nu)$ so

$$\mathcal{S}(f_n; \mu, \nu) \to S_F(\mu \mid \nu) \tag{16.26}$$

Since

$$d\nu = \chi_{X \backslash A} \, d\nu + \eta^{-1} \, d\mu$$

we have

$$\mathcal{S}(f_n;\mu,\nu) = \int_{X\setminus A} f_n(x)\,d\nu(x) + \int (\eta^{-1}(x)f_n(x) + G(f_n(x))\,d\mu(x) \quad (16.27)$$

Define $f_n$ by

$$f_n(x) = \begin{cases} n^{-1}, & x \in X\setminus A \text{ or } f_\infty(x) \le n^{-1} \\ f_\infty(x), & \text{if } n^{-1} \le f_\infty(x) \le n \\ n, & \text{if } f_\infty(x) \ge n \end{cases} \quad (16.28)$$

Clearly, by the monotone convergence theorem,

$$\lim_{n\to\infty} \int_{X\setminus A} f_n(x)\,d\nu(x) = 0 \quad (16.29)$$

Moreover, in

$$-F(y) = \inf_{w>0} (yw + G(w))$$

with the inf taken at $w = p(y)$, $yw + G(w)$ is convex in $w$, so $w \mapsto yw + G(w)$ is monotone decreasing in $w$ for $w \in [1, p(y)]$ if $p(y) > 1$ and monotone increasing in $w$ for $w \in [p(y), 1]$ if $p(y) < 1$. By (16.28), $f_n(x)$ increases monotonically to $f_\infty(x)$ if $f_\infty(x) > 1$ and decreases monotonically to $f_\infty(x)$ if $f_\infty(x) < 1$, so either way,

$$\begin{aligned} \eta^{-1}(x)f_n(x) + G(f_n(x)) &\searrow \eta^{-1}(x)f_\infty(x) + G(f_\infty(x)) \\ &= -F(\eta^{-1}(x)) \end{aligned}$$

Since the integral is absolutely convergent for $n = 1$, the monotone convergence theorem implies the second integral in (16.27) converges to $S$ and (16.26) holds. We have thus proven (16.14) in case $\mu$ is $\nu$-a.c.

If $\mu$ is not $\nu$-a.c., find $A \subset X$ with $\mu(A) > 0$ and $\nu(A) = 0$ and then, by regularity of measures, $K \subset A$ compact with $\mu(K) > 0$ and $U_\varepsilon \supset A$ open with

$$\nu(U_\varepsilon) < \varepsilon \quad (16.30)$$

By Urysohn's lemma, find $f_\varepsilon \in C(X)$ so that

$$f_\varepsilon(x) = \begin{cases} \varepsilon^{-1}, & x \in K \\ 1, & x \in X\setminus U_\varepsilon \\ \in [1, \varepsilon^{-1}], & \text{all } x \end{cases}$$

Then

$$\begin{aligned} \mathcal{S}_F(f_\varepsilon;\mu,\nu) &= \int f\,d\nu + \int G(f(\cdot))\,d\mu \\ &\le \nu(X\setminus U_\varepsilon) + \varepsilon^{-1}\nu(U_\varepsilon) + G(\varepsilon^{-1})\mu(K) + G(1)\mu(X\setminus K) \end{aligned}$$

by using the definition of $f_\varepsilon$, writing the $\nu$ integral as contributions over $X\backslash U_\varepsilon$ and $U_\varepsilon$, and the $\mu$ integral as contributions over $K$ and $X\backslash K$. Since (16.30) holds and $\nu(X) = \mu(X) = 1$, we have

$$\begin{aligned} \mathcal{S}_f(f_\varepsilon; \mu, \nu) &\le 1 + 1 + G(1) + G(\varepsilon^{-1})\mu(K) \\ &\to -\infty \end{aligned}$$

by (16.16). Thus, (16.11) holds in this case also. □

*Remarks* 1. What the proof shows is why the variational principle holds. Essentially, it follows from

$$\inf_f \int [\eta^{-1}(x) f(x) + G(f(x))]\, dx = \int \inf_y [\eta^{-1}(x) y + G(y)]\, dx$$

and Fenchel's theorem.

2. Closely related to Theorem 16.1 is a theorem of Denisov–Kiselev [97] that for $\mu$ fixed, $\nu \to \exp(S(\mu \mid \nu))$ is concave. For discussion and proofs, see [353, Sect. 10.6].

We close with an illuminating comment about the classical case of (16.1). If $\mu_0, \mu_1, \nu$ are three probability measures, then concavity says

$$S((1-\theta)\mu_0 + \theta\mu_1 \mid \nu) \ge (1-\theta)(S(\mu_0 \mid \nu) + \theta S(\mu_1 \mid \nu) \tag{16.31}$$

In the other direction, we have almost convexity:

**Theorem 16.7** (Robinson–Ruelle [318]) *Let $\mu_0, \mu_1, \nu \in \mathcal{M}_{+,1}(X)$. One has*

$$S((1-\theta)\mu_0 + \theta\mu_1 \mid \nu) \le (1-\theta)S(\mu_0 \mid \nu) + \theta S(\mu_1 \mid \nu) - g(\theta) \tag{16.32}$$

*where*

$$g(\theta) = -(1-\theta)\log(1-\theta) - \theta\log\theta \tag{16.33}$$

*Remarks* 1. $g(\theta) \in [0, \log 2]$ (with $g(\frac{1}{2}) = \log 2$), so one can replace $g(\theta)$ by $\log 2$ for a $\theta$-independent bound.

2. As the Notes explain, this implies that infinite-volume entropy per unit volume is affine.

*Proof* If $\theta = 0$ or $1$, there is nothing to prove, and if $\theta \in (0,1)$, $S(1-\theta)\mu_0 + \theta\mu_1 \mid \nu) = -\infty$, (16.32) is immediate. Thus, we can suppose $(1-\theta)\mu_0 + \theta\mu_1$ is $\nu$-a.c. and $\theta \in (0,1)$, so $\mu_0, \mu_1$ are both $\nu$-.a.c. Therefore,

$$d\mu_j = \eta_j\, d\nu \tag{16.34}$$

and, by (16.2),

$$S(\mu_j \mid \nu) = -\int \eta_j \log \eta_j\, d\nu \tag{16.35}$$

We can also use (16.2) for the left side of (16.32):

$$\begin{aligned} S((1-\theta)\mu_0 + \theta\mu_1 \mid \nu) &= -\int [(1-\theta)\eta_0 + \theta\eta_1] \log[(1-\theta)\eta_0 + \theta\eta_1]\, d\nu \\ &\le -\int \{(1-\theta)\eta_0 \log[(1-\theta)\eta_0] + \theta\eta_1 \log[\theta\eta_1]\}\, d\nu \qquad (16.36) \\ &= \text{RHS of (16.32)} \end{aligned}$$

(16.36) comes from monotonicity of log on $[0, \infty)$. □

# 17
# Notes

This final chapter explores the history of convexity and provides comments on some of the themes discussed earlier in the book. There are varied historical roots to the study of convexity with input from both applied and pure sources and, sometimes, long delays between seminal work and its absorption into the mainstream.

One of the earliest discoverers of the wonders of multidimensional convex functions was Josiah Willard Gibbs in three remarkable papers [128, 129, 130] published from 1873 to 1878 in an obscure American journal. These papers on thermodynamics predated his later celebrated work in statistical mechanics. The content of the papers and their reception is discussed in detail in an historical overview on the use of convexity in thermal physics by Wightman [388].

To Gibbs, thermodynamic stability implied that internal energy of a system is a function of entropy and volume had to be convex, and this persisted to convexity in additional variables in multicomponent systems. For Gibbs, coexistence of phases corresponded to the convex function having a flat piece on its graph – or in modern parlance, to its Legendre transform having a multidimensional set of tangents. Gibbs also understood the role of certain Legendre transforms in thermodynamics and understood some other relations between multiple supporting hyperplanes for the Legendre transform and flat sections in the graph of the original function. Many of his deepest ideas lay dormant for about seventy-five years!

In terms of capturing the imagination of mathematicians, a key early paper was Jensen's 1906 paper [177] that focused on what we call midpoint convexity and proved what is usually called Jensen's inequality only in the discrete case (i.e. (1.3)). As we will discuss, work of Grolous, Henderson, Hölder, and Stolz predated Jensen, but it was the latter that became the key work for analytic researchers during the following thirty years.

Jensen and Gibbs focused on convex functions. It was Minkowski who developed the theory of general convex sets during the period 1895–1909 (Minkowski suddenly died of a ruptured appendix at age 44, a casualty of the state of medicine in his day!). Among the notions he first understood were gauges, polars, and extreme points. Because much of his important work was unpublished at the time of his untimely death and only appeared when Hilbert pulled together Minkowski's collected works [262], this has become the standard reference for his work, and we will follow the tradition – although the reader should realize it predated its publication in 1911. Before Minkowski, there was extensive work on convex polytopes (polyhedra in arbitrary dimensions) and it was in this context that Carathéodory made his contribution discussed below.

An important trend in the development of convexity was the work following up the codification by Banach of the theory of normed linear spaces. Starting about 1930, with a huge impetus due to Schwartz's development of distribution theory after the Second World War, a host of workers developed the theory we now call locally convex spaces. Some of the key people in these developments were Orlicz, Kolmogorov, Krein, Mazur, Köthe, Mackey, and Bourbaki (presumably influenced by Dieudonné and Schwartz). We will discuss some of the detailed history later.

Simultaneous with the development of locally convex spaces, the importance of convexity in proving useful inequalities was emphasized by Hardy, Littlewood, and Pólya in a paper [148] and very influential book [149].

After the Second World War, Gibbs' themes were taken up (without knowing of Gibbs' work) by mathematical economists and statisticians – the names of Kuhn, Tucker, and Blackwood come to mind.

Young [391] developed the theory of conjugate convex functions in 1912. It is surprising that the general theory of Legendre transforms, even in $\mathbb{R}^n$, was not discussed until Fenchel's great 1949 paper [114].

Choquet developed the fundamentals of the theory named after him in the mid 1950s. Over the next fifteen years, Bauer, Bishop, de Leeuw, Meyer, Klee, and others developed his and related ideas to high art.

That completes our brief summary of the high points in the early (pre-1960) history. Before turning to a chapter-by-chapter history, I mention several books on the subject. Eggleston's delightful tract on the finite-dimensional case [111] is full of interesting results. Lay [221] also focuses on the geometry of the finite-dimensional case. Roberts–Varberg [315] is a comprehensive look at some of the infinite-dimensional aspects of convexity. Rockafellar's readable book [319] focuses on duality theory and those elements of use in convex programming. The standard texts on locally convex spaces are Bourbaki [47] and Köthe [206]. Phelps [290] and Niculescu–Persson [272] are books on various aspects of Choquet theory. Pečarić *et al.* [287] focus on inequalities associated to convexity. For two books on specialized aspects of convexity, see Coppel [87] and Fuchssteiner–Lusky [120].

Proposition 1.2 is often called Jensen's inequality after Jensen [177]. It is so important (or maybe Denmark is sophisticated!) that for a time it was part of the Danish post office postmark (Jensen was Danish). Proposition 1.3 is also from Jensen's papers. While named after Jensen, (1.3) for $\theta_j = 1/n$ appeared first in

Grolous [134] and the general $\theta$ result for such $f$'s is in Hölder [164] and Henderson [156].

If $f$ is not assumed continuous, midpoint convexity does not imply convexity. View $\mathbb{R}$ as an infinite-dimensional vector space over $\mathbb{Q}$ and let $f$ be a $\mathbb{Q}$-linear functional on $\mathbb{R}$ which is not real linear. By using a $\mathbb{Q}$-basis in $\mathbb{R}$, it is easy to construct such $f$. $f$ is even midpoint-affine, but not convex. Its square is strictly midpoint-convex, but not convex.

The connection between $f'' \geq 0$ and convexity goes back to Newton. The arithmetic-geometric mean inequality (1.11) was known to the Greeks. (1.12) has the following generalization due to Muirhead [270]:

**Proposition 17.1** (Muirhead's Theorem) *Let* $\alpha_1 \geq \alpha_2 \geq \cdots \geq \alpha_n \geq 0$, $\beta_1 \geq \beta_2 \geq \cdots \geq \beta_n \geq 0$ *and define for* $a_1, a_2, \dots, a_n \geq 0$,

$$f_\alpha(a_1, \dots, a_n) = \frac{1}{n!} \sum_{\pi \in \Sigma_n} a_{\pi(1)}^{\alpha_1} \cdots a_{\pi(n)}^{\alpha_n} \tag{17.1}$$

*Then* $f_\alpha(a) \leq f_\beta(a)$ *for all such* $a$ *if and only if*

$$\sum_{j=1}^n \alpha_j = \sum_{j=1}^n \beta_j \quad \textit{and} \quad \sum_{j=1}^k \alpha_j \leq \sum_{j=1}^k \beta_j, \qquad k = 1, \dots, n-1 \tag{17.2}$$

*Remarks* 1. (1.12) is the special case $\alpha = (\frac{1}{n}, \frac{1}{n}, \dots, \frac{1}{n})$ and $\beta = (1, 0, \dots, 0)$.

2. The proof below is based on Hardy–Littlewood–Pólya [149] and Rado [301]. Rado generalizes the theorem to consider symmetrization over subgroups of $\Sigma_n$.

*Proof* Write $a_i = e^{x_i}$ and note

$$a_{\pi(1)}^{\alpha_i} \cdots a_{\pi(n)}^{\alpha_n} = \exp\left( \sum_{j=1}^n \alpha_j x_{\pi(j)} \right)$$

is convex in $\alpha_1, \dots, \alpha_n$ (since, e.g., its Hessian is a rank one positive definite matrix). Thus,

$$\Phi(\alpha_1, \dots, \alpha_n) \equiv f_\alpha(a) = \frac{1}{n!} \sum_{\pi \in \Sigma_n} \exp\left( \sum_{j=1}^n x_j \alpha_{\pi(j)} \right)$$

is convex and symmetric, that is, in $C_\Sigma(\mathbb{R}^\nu)$ in the notation of the HLP theorem (Theorem 15.5). (17.2) is condition (i) of that theorem, so (iv) of that theorem shows that (17.2) implies $f_\alpha(a) \leq f_\beta(a)$ for all $a$.

For the converse, note first that

$$f_\alpha(\lambda a_j) = \lambda^{\sum_i^n \alpha_j} f(a_j)$$

so if $\sum_i^n \alpha_j \neq \sum_i^n \beta_j$, then either for $\lambda$ very large or $\lambda$ very small, $f_\alpha(\lambda a_j) \leq f_\beta(\lambda a_j)$ has to fail. We conclude that

$$\sum_{j=1}^{n} \alpha_j = \sum_{j=1}^{n} \beta_j \tag{17.3}$$

Next, take $a_1 = a_2 = \cdots = a_k = x$ and $a_{k+1} = \cdots = a_n = 1$. Then $f_\alpha(x)$ is a finite sum of powers of $x$ which for large $x$ obeys

$$f_\alpha(x) = c_{\alpha,k,n} x^{\sum_{j=1}^{k} \alpha_j}(1 + o(1)) \tag{17.4}$$

where $c_{\alpha,k,n}$ depends on $\alpha$, $k$, and $n$ but is strictly positive (if $\alpha_{j+1} < \alpha_j$, $c_{\alpha,k,n} = k!(n-k)!/n!$). Thus, by taking $x \to \infty$, we see if $f_\alpha(x) \leq f_\beta(x)$ for all $x$, then $\sum_{j=1}^{k} \alpha_j \leq \sum_{j=1}^{k} \beta_j$. This and (17.3) imply (17.2). □

The notion of gauge of a convex set, given by (1.18), is due to Minkowski [262], at least for convex subsets of $\mathbb{R}^3$. In particular, he used the gauge to describe the closure, interior, and boundary of the convex sets (see Remark 5 after Corollary 1.10). Among Minkowski's results is the inequality named after him for sums, namely,

$$\left(\sum_{j=1}^{n} |a_j + b_j|^p\right)^{1/p} \leq \left(\sum_{j=1}^{n} |a_j|^p\right)^{1/p} + \left(\sum_{j=1}^{p} |b_j|^p\right)^{1/p}$$

Hölder's equality for $p = 2$ is, of course, the often named Schwarz inequality, due for sequences to Cauchy [70] in 1821 and for integrals to Buniakowski [58] in 1859, and independently to Schwarz [345] in 1885, but outside the former Soviet Union, "Schwarz" or "Cauchy–Schwarz" has stuck. Hölder's inequality for sums is due to Hölder [164] and for integrals to F. Riesz [306].

The "usual" proof of the Minkowski inequality follows from Hölder's inequality. One first shows that if $p$ and $q$ are conjugate indices,

$$\|f\|_p = \sup\left(\left|\int fg\,d\mu\right| \;\middle|\; \|g\|_q = 1\right) \tag{17.5}$$

by using Hölder's inequality and the special case $g = |f|^{p-1}|f|/f$. From (17.5), one easily obtains

$$\|f + g\|_p \leq \|f\|_p + \|g\|_p$$

That a bounded convex function is continuous is a result of Jensen [177]. The existence of one-sided derivatives goes back to Stolz [364].

A *supporting hyperplane* to a closed convex set $K \subset \mathbb{R}^\nu$ is an affine hyperplane $P = \{x \mid \ell(x) = \alpha\}$ for some $\ell \not\equiv 0$ with $P \cap K \neq \emptyset$ and $K$ is in the half-space $\{x \mid \ell(x) \geq \alpha\}$. One often supposes that $K \not\subset P$, which is automatic if $\dim(K) = \nu$. Tangent functionals to a convex function are the same as supporting hyperplanes to the set $\{(x,t) \in K \times \mathbb{R} \mid t \geq f(x)\}$. Tangent planes as defined in Example 1.43 and the definition after it are the same as supporting hyperplanes. The notion of supporting hyperplanes is due to Minkowski [262].

The special case (1.74) of Jensen's inequality, as we have seen, implies Hölder's inequality. In turn, the special case can be proven if one knows Hölder's inequality as follows. Since adding a constant $c$ to $f$ changes both $\int f(x)\,d\mu(x)$ and $\log(\int e^{f(x)}\,d\mu(x))$ by $c\mu(M)$, we can suppose $f \geq 0$. We can also suppose $\mu(M) = 1$. In that case, Hölder's inequality implies

$$\int f(x)\,d\mu(x) = \|f\|_1 \leq \|f\|_n \, \|1\|_{n/n-1} = \left(\int f^n \, d\mu\right)^{1/n}$$

Thus,

$$\begin{aligned} \exp\left(\int f\,d\mu\right) &= \sum_{n=0}^{\infty} (n!)^{-1} \left(\int f(x)\,d\mu\right)^n \\ &\leq \sum_{n=1}^{\infty} \int (n!)^{-1} f^n(x)\,d\mu(x) \\ &= \int \exp(-f(x))\,d\mu(x) \end{aligned}$$

which is (1.74).

For the significance of tangents to the pressure and Theorem 1.26 in statistical mechanics, see the book of Israel [175], Ruelle [324, 325], and Simon [351].

The Hahn–Banach theorem (Theorem 1.38) is due to Hahn [141] and Banach [24] after early work by Helly [154, 155] and M. Riesz [311], based in part on earlier work of his brother, F. Riesz [306, 307]. In particular, M. Riesz developed the theory in his solution to the Hamburger moment problem.

Theorem 1.42 on Baire generic existence of tangents to a convex function on a separable Banach space is due to Mazur [257]. It has a wonderful generalization due to Israel [175]. Let $F$ be a continuous convex function on $K$, an open convex subset of $V$, a separable Banach space. For each $x \in K$, let $T_x$ denote the set of tangent functionals to $F$ at $x$, that is, the set of $\ell$'s in $X^*$ with

$$F(y) - F(x) \geq \ell(y - x)$$

Let

$$G^m = \{x \in K \mid \dim(T_x) \geq m\} \tag{17.6}$$

so the Hahn–Banach theorem says $G^0 = K$ and Mazur's theorem says $K \backslash G^1$ is a dense $G_\delta$. Israel makes the set, $\mathcal{G}^k$, of $k$-dimensional subspaces of $X$ into a complete metric space under a natural topology (it is, of course, a Grassmannian manifold). He says $M \in \mathcal{G}^k$ obeys the Gibbs phase rule if $M \cap G^m = \emptyset$ for $m > k$ and $M \cap G^m$ has Hausdorff dimension at most $k - m$ if $m \leq k$. Then Israel proves that $\{M \in \mathcal{G}^k \mid M \text{ obeys the Gibbs phase rule}\}$ is a dense $G_\delta$ in $\mathcal{G}^k$.

To understand why this is called the Gibbs phase rule, Gibbs generalized the familiar fact that water, whose state space has two parameters (temperature and pressure), has a single triple point and lines of double points. Gibbs considered situations with $k$ parameters (typically temperature, pressure, and $k-2$ fraction ratios for mixtures of several substances) and claimed you had no points with $k+2$ phases, points with $k+1$ phases, etc. The number of phases is related to the dimensions of $T_x$ as follows (see Israel [175], Ruelle [324, 325], and Simon [351] for further discussion): In statistical mechanics contexts, pure phases correspond to extreme points in $T_x$ and $T_x$ is a simplex – hence, if $\dim(T_x) = m$, there are $(m+1)$ extreme points. It should be clear then why Israel defines the condition that $M$ obeys the Gibbs phase rule as he does.

Israel relied on earlier work of Anderson–Klee [9] who studied the finite-dimensional case. They proved:

**Theorem 17.2** ([9]) *Let $K \subset \mathbb{R}^\nu$ be an open convex set and $F$ a convex function on $K$. Let $G^m$ be given by* (17.6). *Then for any $m \le \nu$, $G^m$ has Hausdorff dimension at most $\nu - m$.*

*Remark* They prove the stronger statement that $G^m$ is a countable union of a family of sets of finite $\nu - m$-dimensional Hausdorff measure.

The construction of conjugate convex functions and the inequality (1.99) are due to Young [391]. Young's wife was also a mathematician and there are persistent stories that she was a crucial silent partner in an era when society frowned on women playing a professional role.

The theory of Orlicz spaces was initiated by Orlicz [277], who assumed the $\Delta_2$-condition in his definition. Orlicz defined the norm via

$$\|f\|'_F = \sup\biggl\{\biggl|\int f(x)g(x)\,d\mu(x)\biggr| \biggm| Q_G(g) \le 1\biggr\}$$

where $G$ is the conjugate function. The Luxemburg norm is due to Luxemburg [247].

The importance of the $\Delta_2$-condition in the study of convex functions on $\mathbb{R}$ was isolated by Birnbaum–Orlicz [37] even before Orlicz defined his spaces. Two other fundamental works on Orlicz spaces are Orlicz [278] and Zygmund's work on Fourier series [395].

The duality theorem for $L^p$ is due to F. Riesz [306]. Orlicz's original paper [277] showed $(L^F)^* = L^G$ if $F$ obeys the $\Delta_2$-property; see also Zaanen [393]. The general version $(E^F)^* = L^G$ is due to Krasnosel'skĭ–Rutickiĭ [210] and Luxemburg [247]. The books on the theory of Orlicz space and their use in the study of certain integral equations are Zaanen [394] and Krasnosel'skiĭ–Rutickiĭ [211]. Krasnosel'skiĭ–Rutickiĭ [211] discuss in general the connection of lack of separability and failure of the $\Delta_2$-condition.

We developed the theory of Orlicz spaces when $\mu(M) < \infty$, so we only needed the $\Delta_2$-condition at $\infty$. The opposite extreme are Orlicz spaces of sequences, and the issue is behavior of $F$ near zero. A whole chapter of Lindenstrauss–Tzafriri [235] is devoted to Orlicz sequence spaces. Not surprisingly, a key role is played by the "$\Delta_2$-condition at 0," defined by $\lim_{t\downarrow 0} F(2t)/F(t) < \infty$.

For an interesting application of Orlicz spaces to a problem in mathematical physics, see Rosen [320].

The notions of abstract spaces came rather late. Hilbert – and specifically, his student, Schmidt – discussed the explicit Hilbert spaces $\ell_2$ and $L^2(0,1)$ in the period 1905–1910, but von Neumann only gave the formal abstract definition in 1927 ([377]; see also [381]) in connection with his studies in quantum mechanics. From 1905–1920, several mathematicians, most notably F. Riesz, discussed concrete, complete, normed linear spaces like $C([0,1])$, its dual, and $L^p([0,1],dx)$. But it was only in the period 1920–1922 that Banach [23], Hahn [140], Helly [155], and Wiener [386] presented an axiomatic point of view. Helly only discussed sequence spaces and the others, the general setting. Due to his later work, his definitive book, and the development of an active school of researchers, the name Banach space, or B-space, has stuck. But the discoveries were essentially simultaneous so that, for example, Wiener called them BW-spaces in his lectures at MIT in the 1950s. One has to understand the impact of political developments on the history of mathematics. Helly, an Austrian mathematician, was close to his later work in 1912 [154], but then the First World War intervened – he was seriously wounded and captured, and became a prisoner of war in Russia until 1920! Helly was Jewish and fled Vienna in 1938, and died in Chicago (where he was writing training manuals for the US Signal Corps) in 1943. Banach's school was decimated by the Second World War. The Nazis suppressed Polish intellectuals by dismissing them from jobs and Banach, in difficult economic straits, took ill and died in 1945.

What are now called Fréchet spaces were only formally defined by Mazur and Orlicz in 1948 [258], although a close notion appeared in Banach's book as spaces of type (F) named after a basic paper of Fréchet [116].

Topological vector spaces seem to have been introduced around 1935 by Kolmogorov [201] and von Neumann [382]. In particular, Kolmogorov proved Theorem 3.14 in this paper. At about the same time, Köthe and Toeplitz [207] introduced a class of non-Banach spaces, whose further study saw the introduction of many of the ideas relevant to the general theory. Locally convex spaces were codified in the book of Bourbaki [47].

Kolmogorov defined the notion of bounded sets in that paper as follows: $A$ is bounded if and only if for any sequence $x_n \in A$ and any sequence $\alpha_n \in \mathbb{K}$ with $\alpha_n \to 0$, we have $\alpha_n x_n \to 0$. The reader should check that this agrees with the definition following Proposition 3.1.

Proposition 3.2 is called the Banach–Steinhaus principle after a joint paper [26]. Earlier special cases were found by Lebesgue [222], Hahn [139], Steinhaus [361], Saks–Tamarkin [336], Hellinger–Toeplitz [153], Landau [218], Toeplitz [372], Helly [154], and F. Riesz [306]. Proposition 3.2 was first proven by Hahn [139] with the generalization to operators between spaces by Hildebrandt [161], and Banach–Steinhaus [26].

For a discussion of uniform spaces, see the books of Kelley [191], Choquet [83], or Willard [390].

Theorem 3.7 is due to Tychonoff [375]. Theorem 3.10 is due to F. Riesz [308]. The proof we give is due to Choquet [83]. That $L^p$ for $0 < p < 1$ has no continuous linear functionals (Example 3.15) is a result of Day [93]; our discussion is related to the proof of Robertson [317]. That balls in the natural metric on $H^p$ $(0 < p < 1)$ have no open convex subsets (Example 3.16) is a result of Livingston [239], whose proof we follow, and Landsberg [219]. The failure of the Hahn–Banach theorem in $H^p$ $(0 < p < 1)$ mentioned in Example 3.16 is due to Duren, Romberg, and Shields [107]. It and other aspects of the $H^p$ space are discussed in Chapter 7 of Duren's book [106].

Distribution theory has its roots in work of Leray [224] and Friedrichs [117] on weak solutions of PDEs, and Sobolev [354] who defined what we would now call $\mathcal{D}'$ as linear functionals on $C_0^\infty(\Omega)$. It was Laurent Schwartz who systematized the theory, beginning in a 1945 paper [342] and then in a groundbreaking book [343] and, in particular, introduced $\mathcal{S}(\mathbb{R}^\nu)$ and $\mathcal{S}'(\mathbb{R}^\nu)$. Schwartz's work totally changed the theory of PDEs; see Gårding [121, Chap. 12] for historical reminiscences.

It is true that $\mathcal{D}$ is nonmetrizable, but this is only because we allow distributions of arbitrary growth. If one restricts growth as one typically can in specific problems, spaces can normally be taken metrizable. Gel'fand, in a series of books [122, 123, 124, 125, 126], especially pushed the notion of tailoring the test function space to the problem. These books have been very influential on the spread of distribution theory.

The notion of a barreled space (and the term "barrel") is due to Bourbaki [47]. Bourbaki is a pseudonym for a group of French mathematicians and it is generally believed that the main authors of the pre-1955 Bourbaki books were Dieudonné and Weil, so it is likely that much of the fundamental book on locally convex spaces is the work of Dieudonné.

The name Montel space comes from the theorem of Montel [269] that a sequence of holomorphic functions on $\Omega \subset \mathbb{C}$, uniformly bounded on compact subsets of $\Omega$, has a subsequence uniformly convergent on compact subsets. In the language of Example 3.22, his theorem says $\mathcal{H}(\Omega)$ is a Montel space.

The geometric interpretation of Hahn–Banach theorems as separation theorems exploited in Chapter 4 is due to Ascoli [15], who considered separable Banach spaces, and Mazur [257], who considered general Banach spaces. Both these authors considered separating a point from a convex set with nonempty interior. The general result, Theorem 4.1, is due to Dieudonné [98]. An extensive study of separation theorems is due to Klee [195, 196, 197, 198, 199, 200].

The notion of weakly convergent sequences in $\ell_2$ was used extensively in the work of Hilbert and Riesz starting in 1905, but despite the introduction of the general theory of topological spaces by Hausdorff in 1914 [150], it was not until 1930 that von Neumann [379] explicitly introduced the weak topology on Hilbert spaces and noted that sequences did not suffice to define the topology.

Proposition 5.1 is due to Phillips [291] and independently by Dieudonné [99], although for the special case $Y = X^*$ with $X$ a Banach space, it appears already in Banach [25] in the separable case and Alaoglu [3] in the general case. The formal notion of dual pair was introduced by Dieudonné [99] and Mackey [251] and the Mackey–Arens theorem (Theorem 5.23) was proven by Mackey [252] and Arens [12].

The notion of polar set for bounded convex sets in $\mathbb{R}^\nu$ goes back to Minkowski [262] and for convex cones to Steinitz [362], who understood the bipolar theorem in those cases. In particular, Minkowski understood the dual relationship between the gauge and indicator functions described in Example 5.20 and between gauge functions and maximums over polars described by (5.7).

Compactness theorems like Theorem 5.12 have a long history. For the special case of measures on $[0, 1]$, in the context of functions of bounded variation, the result is known as the Helly selection theorem after Helly [155]. The abstract theorem for the unit ball of the dual of a separable Banach space is due to Banach [25] and in general dual Banach spaces to Alaoglu [3], Bourbaki [46], and Kakutani [181]. The result we call the Bourbaki–Alaoglu theorem is due to Bourbaki [47] and named thus by Köthe [206] in his book.

The notion of what we call regular convex function (defined via (i)–(iv) in Proposition 5.13) is due to Fenchel [114], who proved what we call Fenchel's theorem (Theorem 5.17). As we have noted, the theory for monotone increasing convex functions on $[0, \infty)$ goes back to Young [391]. Mandelbrojt [253] discussed the general one-dimensional case, but until Fenchel, even the $\mathbb{R}^\nu$ case had not been considered.

Köthe calls Lemma 5.26 the Banach–Mackey theorem, presumably because it is a theorem of Banach (essentially a variant of the Banach–Steinhaus theorem) if $X$ is a Banach space and because Mackey [252] has a general variant.

Loewner [240] was the first to consider the issue of matrix monotone functions, to understand its connection to Loewner matrices, and to prove the deep Theorem 6.5. His result was not well known for many years so, for example, Heinz [152] and Kato [188] found proofs that for $0 < \alpha < 1$, $A \mapsto A^\alpha$ is in $M_\infty(0, \infty)$ even though the result is immediate from the "easy" half of Loewner's theorem.

Over the years, a number of rather different proofs of Loewner's theorem have appeared. Among them are

(1) Loewner's original proof [240], which studies interpolation by Herglotz functions (i.e., functions analytic in $\mathbb{C}_+ \cup \mathbb{C}_- \cup (a,b)$ with $\pm \operatorname{Im} f > 0$ if $\pm \operatorname{Im} z > 0$). We will not present his proof.
(2) The proof of Bendat–Sherman [31] presented in Chapter 7 using the Bernstein–Boas theorem and solubility of the moment problem.
(3) A continued fraction proof of Wigner–von Neumann [389].
(4) A proof of Korányi [204] on interpolation of Herglotz functions.
(5) A proof of Sparr [356] using the Hahn–Banach theorem, which we will not discuss further.
(6) A proof of Hansen–Pedersen [145] using the Krein–Milman theorem and discussed in Chapter 9.
(7) A proof of Rosenblum and Rovnyak [321] that concerns interpolation of Herglotz functions for measurable sets.

Proofs (1), (2), and (4) are discussed in detail in a lovely book of Donoghue [101].

Proposition 6.3 and the elementary approximation argument we use to prove it are due to Bendat–Sherman. Theorem 6.9(i) appeared first in Rosenblum–Rovnyak [322]; see also Donoghue [103]. Theorem 6.9(ii) appeared first in Chandler [72]; see also Donoghue [102].

Divided differences as defined by (6.13) go back at least to Cauchy's work on polynomial interpolation and were systematized by Nörlund [275]. The calculation in Theorem 6.13 is due to Daleckii and S. G. Krein [90]; Mark Krein pointed out its usefulness in studying Loewner matrices. Our proof appears in Horn–Johnson [171].

Schur products and Theorem 6.12 are due to Schur [340]. Our proof is very natural from the point of view of convexity. The extreme rays in the cone of positive definite $n \times n$ matrices are the rays through rank one projections, so it suffices to prove the result for rank one projections. For such matrices, the Schur product is a positive multiple of a rank projection by the direct calculation there.

The Schur product is often called the *Hadamard product*. Horn [170], who reviews the theory of Schur products, has some interesting historical remarks. So far as he can tell, the name "Hadamard product" first appeared in Halmos' 1948 book on "Finite Dimensional Vector Spaces" written thirty-seven years after Schur's work. Halmos told Horn that he got the name from a talk von Neumann gave at the Institute for Advanced Study. Horn is unable to find any place that Hadamard considered the matrix $a_{ij}b_{ij}$ (!) although he did discuss the sum $\sum_{i,j=1}^{n} a_{ij}b_{ij}$ $(= \operatorname{tr}(A^*B))$.

Propositions 6.14, 6.15, and 6.16 are folk theorems. At least some of them must date back to work of Kramer and Sturm in the early 1800s. Proposition 6.15 goes back at least as far as Frobenius in 1894.

Theorem 6.17 is due to Loewner, who proved (i) $\Rightarrow$ (ii),(iii) directly (the other direction required his full proof). His proof was to first prove Theorem 6.20 and take limits. All proofs of Loewner's theorem begin from the positivity of the Loewner matrix.

Rank one perturbations and the algebra behind Lemma 6.18 are discussed in Simon [350, 2nd edn.]. The proof of Theorem 6.20 using rank one perturbations follows from Donoghue [101] with some changes.

The differential inequality (6.41) and the fact that it implies $(f')^{-1/2}$ is concave is due to Dobsch [100], a student of Loewner. The other direction of Theorem 6.26 is from Donoghue's book [101]. The method of going from a positive Loewner matrix to positivity of the matrix (6.49) is also from Dobsch [100].

Proposition 6.32 for $n = \infty$ is due to Kraus [214], also a student of Loewner. The finite $n$-result and the proof we use is from Davis [92]. Theorem 6.33 is due to Bendat–Sherman [31]. Theorem 6.38 and its proof are due to Hansen–Pedersen [145], but a very closely related result appeared earlier in Davis [91, 92] in connection with what he calls the Sherman condition. Proposition 6.39 is from Hansen–Pedersen [145], but the proof we give of going through conformal maps to $(0, \infty)$ and using Kraus' theorem is new.

What we have called the Bernstein–Boas theorem (Theorem 7.1) was proven by Bernstein [33, 34] with later developments, including a simplified proof, by Boas [41], who proved the following generalization:

**Theorem 17.3** ([41]) *Let $f$ be a $C^\infty$ function on $(-1, 1)$ so that for a sequence $1 < n_1 < n_2 < \dots$ of integers with $\sup_j n_{j+1}/n_j < \infty$, we have that $f^{(n_j)}(x)$ has a definite sign (or vanishes) on $(-1, 1)$. Then $f$ is real analytic on $(-1, 1)$.*

Boas also shows if $\lim_{j\to\infty} n_{j+1}/n_j = \infty$, there are nonanalytic $C^\infty$ functions with $f^{(n_j)}(x) \geq 0$ on $(-1, 1)$. Notice that Boas does not assert analyticity in $D$, but only in a neighborhood of $(-1, 1)$. The neighborhood is only dependent on the $n_j$'s, but it is not $D$. Bendat–Sherman use this weaker version in their proof.

Donoghue [101] replaces the appeal to the Bernstein–Boas theorem as we state it with the weaker theorem that if $f$ is $C^\infty$ on $(0, \infty)$ with $(-1)^n f^{(n)}(x) \geq 0$ there, then $f$ is analytic on $\{x \mid \operatorname{Re} z > 0\}$. This result, of course, follows from Bernstein's theorem (Theorem 9.10), but also from an argument like the one we use for Theorem 7.1 without the need for Proposition 7.3. Donoghue's idea is that if $f \in M_\infty(0, \infty)$, then $(-1)^n f^{(n+1)}(x) \geq 0$, so we get analyticity in $\{x \mid \operatorname{Re} z > 0\}$. By conformal mapping, if $f \in M_\infty(-1, 1)$, we conclude $f$ is analytic in $D$.

For references to the history of the moment problem, see Akhiezer [2]. Bendat–Sherman appeal to the Hamburger moment problem rather than the simpler Herglotz moment problem.

The notions of intrinsic interior, dimension, etc. for a convex subset of $\mathbb{R}^\nu$ are due to Steinitz [362]. Minkowski [262] proved the result that $A = \operatorname{ch}(\mathcal{E}(A))$ for a closed bounded compact subset of $\mathbb{R}^\nu$. Carathéodory [63] proved that if

$x_1, \ldots, x_m \in \mathbb{R}^\nu$ and $y = \sum_{j=1}^m \theta_j x_j$ with $\theta_j \geq 0$ and $\sum \theta_j = 1$, then we can find $w_1, \ldots, w_{\nu+1}$ among the $x_j$'s and $\{\varphi_j\}_{j=1}^{\nu+1}$ with $\varphi_j \geq 0$ and $\sum_{j=1}^{\nu+1} \varphi_j = 1$ so $y = \sum_{j=1}^{\nu+1} \varphi_j x_j$. Combining the two gives Theorem 8.11, which we have given both names.

The standard proof of Carathéodory's theorem is very different and more combinatoric than the geometric proof we give. Here is a sketch. Without loss (by translation), suppose $x_m = 0$, in which case we are writing $y = \sum_{j=1}^{m-1} \theta_j x_j$ with $\theta_j \geq 0$ and $\sum_{j=1}^m \theta_j \leq 1$, and we seek $w_1, \ldots, w_\nu$ among the $\{x_j\}_{j=1}^{m-1}$ and $\{\varphi_j\}_{j=1}^\nu$, so $\varphi_j \geq 0$, $\sum_{j=1}^\nu \varphi_j \leq 1$, and $y = \sum_{j=1}^\nu \varphi_j x_j$. We may as well start by assuming $\theta_j > 0$, or else we drop those $x$'s with $\theta_j = 0$. If $m - 1 > \nu$, there is some nontrivial dependency relation $\sum_{j=1}^{m-1} \alpha_j x_j = 0$. By flipping signs, we can suppose some of $\alpha_j$'s are negative; indeed, we can suppose that $\sum_{j=1}^{m-1} \alpha_j \leq 0$. We have for any $t$,

$$y = \sum_{j=1}^{m-1} (\theta_j + t\alpha_j) x_j$$

Increase $t$ from 0, until the first point that some $\theta_{j_0} + t\alpha_{j_0} = 0$. Since some of the $\alpha_j$'s are negative, that will happen eventually. Since it is the first point, all the other $\theta_j + t\alpha_{j_0} \geq 0$. Thus, we have written $y$ as a convex combination of fewer than $m - 1$ $x$'s. Repeat the process until we get to $\nu$ or fewer $x$'s.

The Krein–Milman theorem is from [213]. The now standard proof we give is due to Kelley [190]. For an example of a compact convex subset of a topological vector space with no extreme points, see Roberts [316]. The underlying space has a topology given by a metric, but obviously, the space is not locally convex.

As Phelps [289] points out, any compact convex subset, $K$, of a separable locally convex space is affinely homeomorphic to a compact convex subset of $\ell^2$ with the weak topology, so the underlying space plays no central role. To see Phelps' remark, let $\{\ell_j\}_{j=1}^\infty$ be a sequence of continuous linear functionals on $K$ that separate points normalized so $\sup_{x \in K} |\ell_j(x)| \leq 1$. Map $K$ to $\ell^2$ by $\varphi(x_j) = 2^{-j}\ell_j(x)$. $\varphi(x) \in \ell^2$ and $\varphi[K]$ is a compact convex subset affinely homeomorphic to $K$.

The idea that ergodicity is associated to extreme points is due to Wiener [387]. (8.15) in the context of ergodic maps is due to von Neumann [380].

For a discussion of invariant means on groups, see Greenleaf [133]. It is a fascinating subject.

The proof we give of the Stone–Weierstrass theorem as Theorem 8.22 is from de Branges [96].

That the Krein–Milman theorem can be understood as an integral representation theorem (i.e., Theorem 9.2) was discussed by Choquet [78] in work that predated – and presumably motivated – his work on "Choquet theory."

The notion of barycenter codified in Theorem 9.1 is an example of a *weak integral*, that is, defined by linear functionals. An early paper on these ideas is Phillips [291]. Theorem 9.3 is due to Bauer [28, 29]. Theorem 9.5, proven by very different means, is due to Milman [261].

The fact that every $f \in L^p$, $p \in (1, \infty)$, with $\|f\|_p = 1$ is an extreme point of the unit ball (mentioned in Example 9.5) can be seen by defining

$$L(g) \equiv \int_{\{x \mid |f(x)|>0\}} g \left[ \frac{|f|^p}{f^{-1}} \right] d\mu$$

By Hölder's inequality, $|L(g)| \le \|g\|_p$ and for all $\|g\|_p \le 1$, $L(g) = 1$ if and only if $g = f$. Thus, every such $f$ is an exposed point and so, an extreme point by Theorem 8.3. One can also use uniform convexity (discussed later in these Notes) to see that $\mathcal{E}(\{f \mid \|f\|_p \le 1\}) = \{f \mid \|f\|_p = 1\}$.

The construction of Example 9.7 is from Israel's book [175]. It is further discussed in Israel–Phelps [176]. As we discuss in Chapter 11, it is a simplex. In 1961, Poulsen [297] constructed the first example of a metrizable simplex, $K$, with $\mathcal{E}(K)$ dense in $K$. In 1978, Lindenstrauss, Olsen, and Sternfeld [234] proved a number of truly remarkable properties of the Poulsen simplex including:

(i) Any two metrizable simplexes with $\mathcal{E}(K)$ dense in $K$ are equivalent under an affine homeomorphism.
(ii) Any metrizable simplex is affinely homeomorphic to a face of the Poulsen simplex.
(iii) Any two faces of the Poulsen simplex that are affinely homeomorphic are homeomorphic under an affine automorphism of the Poulsen simplex.

Bernstein's theorem is due to Bernstein [34]. The analog of Bernstein's theorem referred to in Remark 3 after Theorem 9.10 is:

**Theorem 17.4** *A bounded $C^\infty$ function $f$ on $[0, 1]$ has $f^{(n)}(x) \ge 0$ for all $n = 0, 1, 2, \dots$ and all $x$ if and only if*

$$f(x) = \sum_{n=0}^{\infty} a_n x^n \tag{17.7}$$

*with*

$$\sum_{n=1}^{\infty} a_n < \infty \tag{17.8}$$

*and*

$$a_n \ge 0 \tag{17.9}$$

*Proof* Clearly, if $f$ is given by (17.7) with (17.8) and (17.9), $f$ is $C^\infty$ on $(0, 1)$, $\sup_{0<x<1} |f(x)| = \sum_{n=0} a_n < \infty$, so $f$ is bounded and $f$ is $C^\infty$ with $f^{(n)}(x) \ge 0$. So, we need only prove the converse.

Note first if $f, f' \geq 0$ and $f$ is bounded on $(0,1)$, then

$$\|f\|_\infty = \lim_{x \uparrow 1} f(x) \tag{17.10}$$

Let

$$\mathcal{F} = \{f \text{ on } (0,1) \mid f \text{ is } C^\infty,\ f^{(n)}(x) \geq 0,\ \sup_x f(x) \leq 1\}$$

We claim if $f \in \mathcal{F}$, then

$$0 \leq f^{(n)}(x) \leq n^{2(n-1)/2}(1-x)^{-n} \tag{17.11}$$

For this clearly holds if $n = 0$. Moreover, since $f^{(n)}$ is monotone,

$$f^{(n)}(1-y)\,\frac{y}{2} \leq \int_{y/2}^{y} f^{(n)}(1-z)\,dz \leq f^{(n-1)}\left(1 - \frac{y}{2}\right)$$

so if

$$f^{(n-1)}(1-y) \leq c_{n-1} y^{-(n-1)}$$

then

$$f^{(n)}(1-y) \leq 2^n c_{n-1} y^{-n}$$

which leads to (17.11) by induction.

It follows by applying Theorem 1.20 to the convex functions $f^{(n)}$ that given any sequence, $g_m$, in $\mathcal{F}$, we can find a subsequence, $h_\ell$, so $h_\ell^{(n)}$ converges uniformly on each $[0,\alpha)$, $\alpha < 1$, for each $n$. Writing $h_\ell^{(n)}(x) = h_\ell^{(n)}(0) + \int_0^x h_\ell^{(n+1)}(y)\,dy$, we see that the limit is $C^\infty$. Thus, $\mathcal{F}$ is compact in the topology of uniform convergence.

By (17.10), Proposition 9.14 applies, since $\ell(f) = \|f\|_\infty = \lim_{x\uparrow 1} f(x)$ is linear. Let $\lambda \in (0,1)$ and let $f_\lambda(x) = f(\lambda x)$. Then

$$(f_\lambda)^{(n)}(x) = \lambda^n (f^{(n)})_\lambda(x)$$

It follows that if $f \in \mathcal{F}$, both $f_\lambda$ and $f - f_\lambda \in \mathcal{F}$ so, by Proposition 9.14, if $f \in \mathcal{E}(\mathcal{F})$, $f \neq 0$, then $f_\lambda(x) = c_\lambda f(x)$. This implies $c_{\lambda\lambda_1} = c_\lambda c_{\lambda_1}$ which yields $c_\lambda = \lambda^\alpha$ for some $\alpha$. Since $f_\lambda \leq f$, $\alpha \geq 0$ and thus $f_\lambda(x) = \gamma x^\alpha$. If $\alpha$ is not an integer and $n < \alpha < n+1$, then $f^{(n+1)}(x) < 0$, so $\alpha$ must be one of $0, 1, 2, \dots$. Moreover, $\|f\|_\infty = 1$ implies $\gamma = 1$. Thus, the only possible extreme points in $\mathcal{F}$ are $x^n$.

We next claim that each $x^n$ is an extreme point. For suppose $g \in \mathcal{F}$ and $x^n - g \in \mathcal{F}$. This implies for any $\ell$, $0 \leq g^{(\ell)} \leq (x^n)^{(\ell)}$ so that taking $\ell = n+1$, we see $g^{(n+1)} = 0$, that is, $g$ is a polynomial of degree $n$. Since $g(x) \leq x^n$, it must be that $g$ has no lower-order terms. Thus, $g = \lambda x^n$, that is, $x^n$ is extreme.

We have thus proven $\mathcal{E}(\mathcal{F}) = \{x^n\}_{n=0}^{\infty} \cup \{0\}$ and this set is discrete and closed, so the Strong Krein–Milman theorem implies any $f \in \mathcal{E}(\mathcal{F})$ has the form (17.7) where (17.9) holds and $\sum_{n=1}^{\infty} a_n \le 1$. $\square$

Extensions of Bernstein's theorem to $\mathbb{R}^n$ are discussed in Choquet's book [83] and even to certain Banach spaces; see Welland [384]. Bratteli–Kishimoto–Robinson [55] have a result related to both Bernstein's theorem and Loewner's theorem. If $e^{-tH}$ and $e^{-tK}$ are self-adjoint, positivity-preserving semigroups, then $f$ obeys $e^{-tH} \ge e^{-tK} \Rightarrow e^{-tf(H)} \ge e^{-tf(K)}$ if and only if $f'$ is completely monotone.

The proof of Bochner's theorem via the Strong Krein–Milman theorem is due to Bucy–Maltese [57], although we have borrowed from the presentation in Choquet's book [83].

As noted earlier, the Krein–Milman proof of Loewner's theorem is due to Hansen–Pedersen [145]. But they do not directly prove that the functions $\varphi_\alpha(x) = x/(1-\alpha x)$ are extreme points, but instead use the uniqueness of the representation to a posteriori conclude they are. The argument we give using Dobsch's theorem is new.

With regard to Example 9.21, there is an analog for finite intervals. If $f$ is convex on $[a,b]$ with $f(a) = f(b) = 0$ and $(D^- f)(b) - (D^+ f)(a) = 1$, then $f$ has a representation

$$f(\cdot) = \int g_y(\cdot)\, d\mu(y) \tag{17.12}$$

where

$$g_y(x) = \frac{[\max(x,y) - b][\min(x,y) - a]}{(b-a)} \tag{17.13}$$

and $\int d\mu(y) = 1$. The $g_y$'s, which obey $g_y''$ is a point measure at $y$, are the extreme points in the set of such $f$ and (17.12) is the Strong Krein–Milman representation.

There are other integral representation theorems that are examples of the Strong Krein–Milman theorem. If $\mathcal{C}$ is any class of functions (or other elements of a locally convex space) so that any $f \in \mathcal{C}$ has a representation

$$f(x) = \int_K g_\lambda(x)\, d\mu_f(\lambda)$$

where $g_\lambda \in \mathcal{C}$, $K$ is some separable, compact parameter space, $\mu_f$ is a unique probability measure, and every such integral with $\mu \in \mathcal{M}_{+,1}(K)$ defines an element of $\mathcal{C}$, then $\mathcal{C}$ is a compact convex set (in the topology induced by $\mathcal{M}_{+,1}(K)$!) and $K$ are the set of extreme points (by the uniqueness of $\mu_f$, every $g_\lambda \in \mathcal{E}(\mathcal{C})$).

Here are three other explicit examples:

(i) *The Herglotz representation theorem.*

**Theorem 17.5** *Let $\mathcal{C}$ be the set of real harmonic functions, $f$, on $D$ with $f \geq 0$ and $f(0) = 1$. Then*

$$f(re^{i\theta}) = \int_0^{2\pi} P_r(\theta - \varphi)\, d\mu(\varphi)$$

*where $d\mu$ is the measure on $\partial D$ and $P_r(\theta - \varphi)$ is the Poisson kernel given by* (17.14) *below. In particular, the extreme points of $\mathcal{C}$ are precisely the functions $u_\varphi$ given by*

$$u_\varphi(re^{i\theta}) = P_r(\theta - \varphi) = \operatorname{Re}\left[\frac{e^{i\varphi} + re^{i\theta}}{e^{i\varphi} - re^{i\theta}}\right] \tag{17.14}$$

A proof from the point of view of the Krein–Milman theorem can be found in Holland [165] and Armitage [13]. In this case, the traditional proof is so natural, simple, and direct that I have little sympathy for a proof by these methods.

(ii) *The Levy–Khintchine formula on* $[0, \infty)$. A bounded function, $f$, on $[0, \infty)$ is called *infinitely divisible* if it is positive, and for any $\alpha > 0$, $f_\alpha \equiv f^\alpha$ is completely monotone in the sense of (9.27). The name comes from the fact that for any $n$, $f = g_n^n$ ($g_n = f_{1/n}$) with $g_n$ completely monotone, and it is not hard to see this is equivalent to $f$ being infinitely divisible; we will have more to say about the definition and history below. Clearly, since $f$ is bounded and positive, $f(x) = \exp(-\log(f(x))$ so we are equivalently interested in $h$'s with $f_\alpha(x) = \exp(-\alpha h(x))$ completely monotone for all $\alpha$. We will normalize $h$ first, so $\lim_{x \downarrow 0} h(x) = 0$ and then so $h(1) = 1$. Then

**Theorem 17.6** *Let $Q$ be the set of bounded infinitely divisible functions, $f$, on $[0, \infty)$ with $\lim_{x \downarrow 0} f(x) = 1$ and $f(1) = e^{-1}$. Let $\mathcal{C} = \{-\log f \mid f \in Q\}$. Let $h_\alpha(x) = (1 - e^{-\alpha x})/(1 - e^{-\alpha})$ for $\alpha \in [0, \infty) \equiv K$ (with $h_0 \equiv x = \lim_{\alpha \downarrow 0} h_\alpha(x)$ and $h_\infty \equiv 1$). Then any $h \in \mathcal{C}$ has a unique representation*

$$h(x) = \int h_\alpha(x)\, d\mu(\alpha)$$

*for a measure $\mu$ in $\mathcal{M}_{+,1}(K)$. In particular, $\mathcal{C}$ is a convex set and $K = \mathcal{E}(\mathcal{C})$.*

The formula for an infinitely divisible function, namely,

$$f(x) = \exp\left(-\int_K \frac{1 - e^{-\alpha x}}{1 - e^{-\alpha}}\, d\mu(\alpha)\right) \tag{17.15}$$

is called the *Levy–Khintchine formula* on $[0, \infty)$. For a proof from the point of view of the Strong Krein–Milman theorem, see Kendall [192].

(iii) *The Levy–Khintchine formula on* $\mathbb{R}^\nu$. A bounded complex-valued function, $f$, on $\mathbb{R}^\nu$ is called *infinitely divisible* if for any $n = 2, 3, \dots$, there is a function, $f_n$, with $f = (f_n)^n$ and so that $f_n$ is positive definite in the sense of Bochner (in the sense of (9.57)/(9.58)). Since $f$ is complex-valued, there is a potential issue of nonuniqueness of $n$-th roots on $\mathbb{C}$. But since $f_n$ is continuous with $f_n(0) > 0$, the root is uniquely determined. By using

$$\log f = \lim_{n\to\infty} n[f_n - 1]$$

one can define a continuous value of $\log f$ with $\log f(0)$ real.

*Remark* Even if $f > 0$ on $\mathbb{R}$, it is not true that $f$ infinitely divisible on $\mathbb{R}$ implies $f \restriction [0,\infty)$ is infinitely divisible in the sense of (17.15). We will see that the prototypical infinitely divisible function on $\mathbb{R}$ is $\exp(-x^2)$. This function is positive definite in the sense of Bochner, but $\exp(-x^2) \restriction (0,\infty)$ is not completely monotone. Nonetheless, the same term "infinitely divisible" is used in both cases.

Here is the integral representation theorem:

**Theorem 17.7** (Levy–Khintchine Formula) *Any function, $f$, on $\mathbb{R}^\nu$ is infinitely divisible if and only if $f = e^{-g}$ where $g$ has a unique representation*

$$g(x) = \alpha + i\beta\cdot x + x\cdot(Ax) - \int_{\mathbb{R}^\nu} \left[e^{ix\cdot y} - 1 - \frac{ix\cdot y}{1+y^2}\right] \frac{1+y^2}{y^2}\, d\nu(y) \quad (17.16)$$

*where $\alpha \in \mathbb{R}$, $\beta \in \mathbb{R}^\nu$, $A$ is a $\nu\times\nu$ positive semidefinite matrix, and $\nu$ is a finite measure on $\mathbb{R}^\nu$ with $\nu(\{0\}) = 0$.*

For a proof of this from Bochner's theorem, see Reed–Simon [305, Sect. XIII.12]. For a proof from the point of view of the Strong Krein–Milman theorem, see Johansen [178] and Urbanik [376].

Infinitely divisible functions arise in two ways. The first concerns the following: Let $X_1, \dots, X_n$ be independent identically distributed random variables (iidrv), that is, the functions $x_1, \dots, x_\nu$ on $\mathbb{R}^\nu$ with the product measure $d\eta^{(n)}(x) = d\eta_1(x_1)\dots d\eta(x_\nu)$ where $d\eta \in \mathcal{M}_{+,1}(\mathbb{R})$. Let $\int d\eta^{(n)} = \mathbb{E}(\cdot)$ ($\mathbb{E}$ for expectation). Then

$$\mathbb{E}(e^{i\alpha(X_1+\cdots+X_n)}) = \mathbb{E}(e^{i\alpha X})^n \quad (17.17)$$

Suppose now $\{X_j\}_{j=1}^\infty$ is an infinite family of iidrv so that for some numbers $N_n$,

$$Y = \lim_{n\to\infty} \frac{(X_1+\cdots+X_n)}{N_n} \quad (17.18)$$

where we are vague about what "lim" means, but assume it in a sense that implies convergence of suitable expectations. If

$$Y_\ell = \lim_{n\to\infty} \frac{(X_1+\cdots+X_n)}{N_{n\ell}}$$

and $Y_{\ell 1}, \dots, Y_{\ell\ell}$ are $\ell$ iidrv copies of $Y_\ell$, then $Y_{\ell 1} + \cdots + Y_{\ell\ell} \to Y$ so by (17.17),

$$f_Y(\alpha) = \mathbb{E}(e^{i\alpha Y}) = \mathbb{E}(e^{i\alpha Y_\ell})^\ell$$

and $f_Y(\alpha)$ is an infinitely divisible distribution on $\mathbb{R}$. If $X \geq 0$, we can replace $i\alpha$ by $\alpha > 0$ and get an infinitely divisible distribution on $[0, \infty)$. Thus, infinitely divisible distributions describe the kinds of limits you can get for normalized sums of iidrv. If $X$ is "normal," for example, $\mathbb{E}(X) = 0$, $\mathbb{E}(X^2) = 1$, then the central limit theorem says $N_n = \sqrt{n}$ and the limit is $\mathbb{E}(e^{i\alpha x}) = \exp(-\frac{1}{2}\alpha^2)$, the prototypical infinitely divisible function. Other infinitely divisible functions parametrize the possible singular limits; see, for example, Bertoin [35].

In the above, (17.18), the $X$'s can also be $n$-dependent and the limit will still be infinitely divisible. An example are the Poisson variables. A Poisson random variable of mean $X$ has values $0, 1, 2, \dots$ with distribution

$$\text{Prob}(X = j) = \frac{e^{-\alpha}\alpha^j}{j!}$$

If $X$ and $Y$ are independent Poisson variables and $X$ has mean $\alpha$ and $Y$ has mean $\beta$, it is easy to see (by the binomial theorem) that $X + Y$ is also Poisson with mean $\alpha + \beta$. It follows that a Poisson variable of mean $\alpha$ is the sum of $n$ Poisson variables of mean $\alpha/n$ and so infinitely divisible. For $\alpha = 1$, we have

$$E(e^{-\lambda X}) = \sum_{j=0}^{\infty} \frac{e^{-1}e^{-\lambda j}}{j!} = \exp(-(1 - e^{\lambda}))$$

corresponding to $h_\alpha = 1$ in the Levy–Khintchine formula (17.15). The other extreme points in (17.15) are just scalings of this simple Poisson.

The second way the Levy–Khintchine formula enters is in answering the following question: Let $F$ be a function on $\mathbb{R}^\nu$ and let $H_0$ be the operator $G(-i\vec{\nabla})$ on $L^2(\mathbb{R}^\nu)$. For the special case, $G(y) = y^2$, $e^{-tH_0}$ has an integral kernel $(4\pi t)^{-\nu/2}\exp(-(x - y)^2/4t)$ and a key property of this kernel is its positivity. One can ask for which $G$, $e^{-tH_0}$ has a positive integral kernel. Since the integral kernel is essentially the Fourier transform of $e^{-tG(y)}$, the positivity condition asks for which $G$, $e^{-tG(y)}$ is positive definite in the sense of Bochner. The answer is given by Theorem 17.7, precisely those $G$ of the form (17.16). Among such $G$'s are $G(p) = |p|^\alpha$ $(0 < \alpha \leq 2)$ and $G(p) = (p^2 + m^2)^{1/2}$; see Reed–Simon [305, Sec. XIII.12]. Some aspects of Schrödinger operators with $-\Delta$ replaced by such $G(-i\nabla)$ are found in Carmona, Masters, and Simon [67].

The link between the two Levy–Khintchine theorems is the construction of Markov processes, called Lévy processes, associated to infinitely divisible functions; these generalize Brownian motion, the process associated to $\exp(-\frac{1}{2}x^2)$; see Bertoin [35].

The basics of Choquet theory were announced by Choquet in 1956 [79, 80, 81] with a full paper in 1960 [82]. Many of the main themes are already in this work: integrals of extreme points, the use of an order like the Choquet order, and the relation of uniqueness to geometric structure – originally, Choquet defined a simplex by requiring $K \cap (aK + b)$ to be of the form for $cK + d$.

During the sixties and seventies, Choquet theory was a hot topic with dozens of papers. A major theme was an extension to the nonseparable, that is, nonmetrizable, case since most of Choquet's work assumed metrizability. Two leaders in this effort were Bishop–de Leeuw [38] and Mokobodzki [267, 268]. In particular, the generalization of Theorem 10.7 to the nonmetrizable case is often called the Choquet–Bishop–de Leeuw theorem.

For other presentations of Choquet theory and further developments, see Choquet [83], Phelps [289], and Alfsen [4]. Edwards [110] has a sketch of Choquet's life and work.

That $\mathcal{E}(K)$ can fail to be a Baire set is seen by the following example of Bishop–de Leeuw [38]. Let $Y$ be your favorite compact set (e.g., $D$). Let $Y_0 \subset Y$ and $Y_1 = Y \backslash Y_0$. $X \subset Y \times \{0, \pm 1\}$, three copies of $Y$, namely, if $y \in Y_0, (y,0), (y,+1), (y,-1) \in X$ and if $y \in Y_1$, only $(y,0) \subset X$. $X$ is topologized as follows: A set $x_\alpha$ converges to $x \in X$, if and only if

(i) If $x = \{y,+1\}$ or $\{y,-1\}$, then eventually $x_\alpha = x$ (i.e., $\{x\}$ is open in this case!).

(ii) If $x = \{y, 0\}$, then $x_\alpha = \{y_\alpha, s_\alpha)$ with $y_\alpha \to y$, and eventually, $x_\alpha \neq \{y,+1\}$ or $\{y,-1\}$ ($y$, not $y_\alpha$). It is not hard to see that $X$ is compact, but not metrizable if $Y_0$ is uncountable. Bishop–de Leeuw call this the porcupine topology.

Let $Q \subset C(X)$ be the set of functions which obey

$$f(y,0) = \tfrac{1}{2}\, f(y,+1) + \tfrac{1}{2}\, f(y,-1)$$

for all $y \in Y_0$. Let $Q^*$ be the dual of $Q$, which is easily seen to be the quotient space $\mathcal{M}(X)/Q^\perp$ where $Q^\perp$ is the closure of the span of $\{\delta_{(y,0)} - \frac{1}{2}\delta_{y,+1)} - \frac{1}{2}\delta_{(y,-1)}\}$. Let $K = \{\mu \in Q^* \mid \mu \geq 0, \|\mu\| = 1\}$. Then $K$ is compact and one can show $\mathcal{E}(K) = \{\delta_{(y,0)} \mid y \in Y_1\} \cup \{\delta_{(y,+1)} \mid y \in Y_0\} \cup \{\delta_{(y,-1)} \mid y \in Y_0\}$. If $Y_0$ is chosen non-Baire, then it can be seen $\mathcal{E}(K)$ is not Baire either.

In many ways, the replacement for "concentrated on $\mathcal{E}(K)$" becomes maximal in the Choquet order. Nevertheless, Bishop–de Leeuw examine ways in which a maximal measure is sort of concentrated on $\mathcal{E}(K)$.

Choquet order, at least in finite-dimensional contexts, had been studied prior to Choquet's work. As we discussed in Chapter 15 (see Theorem 15.37), the HLP theorem can be viewed as a statement about Choquet order. In terms of dilation of measures (discussed shortly), the notion was used by statisticians; see, for example, Blackwell [39].

There are several equivalent definitions of Choquet order of some interest and considerable illuminative value. Loomis [242] introduced an order as follows: Given two probability measures, $\mu, \nu$, on a compact convex subset, $K$, of locally convex spaces, we say $\mu \prec_{\text{Loo}} \nu$ if and only if whenever $\mu = \sum_{j=1}^n \theta_j \mu_j$ with $\mu_j \in \mathcal{M}_{+,1}(K)$ and $\theta_j \geq 0$, $\sum_{j=1}^n \theta_j = 1$, there exist $\nu_j \in \mathcal{M}_{+,1}(K)$ with $R(\nu_j) = R(\mu_j)$ and $\nu = \sum_{j=1}^n \theta_j \nu_j$. Loomis showed $\mu \prec_{\text{Loo}} \nu$ implies $\mu \prec \nu$ in Choquet sense (if $\mu$ is a finite point measure, that is obvious and a limiting argument handles general $\mu$) and used his order to discuss the uniqueness part of Choquet theory. Cartier, Fell, and Meyer [69] then proved the Loomis order is equivalent to Choquet order.

Related to Loomis order is the notion of dilation of measures. Given two measures $\mu$ and $\nu$ in $\mathcal{M}_{+,1}(K)$ with $K$ a compact convex subset of a locally convex space, we say $\nu$ is a *dilation* of $\mu$ if and only if for $\mu$ a.e. $x \in K$, there is a measure $\gamma_x \in \mathcal{M}_{+,1}(K)$ so that

(a) $x \mapsto \gamma_x$ is weakly Baire measurable, that is, $x \mapsto \int f(y)\, d\gamma_x(y)$ is Baire measurable for each $f \in C(K)$.

(b)
$$R(\gamma_x) = x \tag{17.19}$$

(c)
$$\nu(f) = \int \gamma_x(f)\, d\mu(x) \tag{17.20}$$

for all $f \in C(K)$.

Thus, dilation says that $\nu$ is obtained from $\mu$ by smearing out each part of $\mu$. Dilation in an intuitive sense gets one closer to the boundary of $K$.

Suppose $a_1, \dots, a_n, b_1, \dots, b_n \in [0,1]$, when is $\nu = \frac{1}{n}\sum_{k=1}^n \delta_{b_k}$ a dilation of $\mu = \frac{1}{n}\sum_{k=1}^n \delta_{a_k}$? It must be that

$$\gamma_{a_k} = \sum_{j=1}^n \alpha_{kj}\delta_{b_j} \tag{17.21}$$

That $\gamma$ is a probability measure implies

$$\alpha_{kj} \geq 0, \qquad \sum_{j=1}^n \alpha_{kj} = 1 \tag{17.22}$$

That (17.20) holds says

$$\sum_{k=1}^n \alpha_{kj} = 1 \tag{17.23}$$

Finally, (17.19) says

$$a_k = \sum_{j=1}^{n} \alpha_{kj} b_j \tag{17.24}$$

so $\nu$ is a dilation of $\mu$ if and only if $a = Db$ for a doubly stochastic matrix $D$.

Returning to the general case, since $\gamma(f) \geq f(R(\gamma))$ for every continuous convex function, (17.19) and (17.20) immediately show that if $\nu$ is a dilation of $\mu$, then $\nu \succ \mu$ in the Choquet order. In fact, they are equivalent if $K$ is separable.

**Theorem 17.8** (Cartier's Theorem) *Let $K$ be a metrizable, compact convex subset of a locally convex space, $X$. Let $\mu, \nu \in \mathcal{M}_{+,1}(K)$. Then $\nu$ is a dilation of $\mu$ if and only if $\nu \succ \mu$ in the Choquet order.*

This theorem is from Cartier–Fell–Meyer [69], but has been attributed to Cartier alone by Meyer, so it has come to be called Cartier's theorem. There is also a proof in Alfsen's book [4].

Notice, by our analysis above, Cartier's theorem is a vast generalization of the HLP theorem that $a \prec_{\text{HLP}} b$ if and only if $a = Db$ for a doubly stochastic matrix $D$. This is made clearer if we use $\gamma_x$ to define

$$(Tf)(x) = \int f(y)\, d\gamma_x(y)$$

Then $\gamma_x \in \mathcal{M}_{+,1}(X)$ implies $\|Tf\|_\infty \leq \|f\|_\infty$, while (17.20) becomes $\|Tf\|_{L^1(d\mu)} \leq \|f\|_{L^1(d\nu)}$ with equality if $f \geq 0$.

An interesting reinterpretation of Choquet's theorem as the fact that measures maximal in the Choquet order are concentrated on the extreme points is due to Niculescu–Persson [271, 272]. In 1883, Hermite [158] proved an inequality, rediscovered by Hadamard [136] in 1893, that for $f$ a convex function on $[a, b] \subset \mathbb{R}$, continuous at the endpoints, one has that

$$f\left(\frac{a+b}{2}\right) \leq \frac{1}{b-a} \int_a^b f(x)\, dx \leq \frac{f(a)+f(b)}{2} \tag{17.25}$$

which is easy to see from $f(\frac{a+b}{2}) \leq \frac{1}{2} f(\frac{a+b+c}{2}) + \frac{1}{2} f(\frac{a+b-c}{2})$ for $0 < c < b - a$ and $f(\theta a + (1-\theta) b) \leq \theta f(a) + (1-\theta) f(b)$. The point is that the existence of maximal measure implies

**Theorem 17.9** (Generalized Hermite–Hadamard Inequality [271, 272]) *Let $K$ be a metrizable compact convex subset of a locally convex space, and let $\mu \in \mathcal{M}_{+,1}(K)$. Then there exist $x \in K$ and $\nu \in \mathcal{M}_{+,1}(K)$ with $\nu(\mathcal{E}(K)) = 1$ so that for all continuous convex functions, $f$, on $K$, we have that*

$$f(x) \leq \int f(y)\, d\mu(y) \leq \int f(y)\, d\nu(y) \tag{17.26}$$

To prove this, one lets $x$ be the barycenter of $\mu$ and $\nu$ a Choquet maximal measure with $\mu \prec \nu$.

Corollary 10.6 characterizing $\mathcal{E}(A)$ as the set of points where $\hat{f}(x) = f(x)$ for all $f \in C(A)$ is due to Hervé [160], although the idea appeared implicitly earlier in Kadison [179]. Hervé [160] also proved that if $A$ has a strictly convex function, $A$ is metrizable!

Choquet's original presentation of unicity [81] defined the notion in terms of intersections of $a + \lambda K \cap K$ being of the form $b + \mu K$. The refined Theorem 11.5 and method of proof are due to Choquet–Meyer [84]. Theorem 11.13 is due to Bauer [29] and for that reason, simplexes, $A$, with $\mathcal{E}(A)$ closed are usually called *Bauer simplexes*.

Ordered vector spaces and vector lattices have an enormous literature that predates the work of Choquet and Choquet–Meyer. The Riesz decomposition property, Proposition 11.11, was emphasized by Riesz in 1940 [310]. For this reason, vector lattices are often called Riesz spaces. Other critical early contributions to the theory include Freudenthal [118], Kantorovitch [184, 185], Stone [365], Krein [212], and Kakutani [182, 183]. Book presentations of the modern theory can be found in Luxemburg–Zaanen [249] and Schaefer [337]. For other book discussions of vector lattices and convex cones, see Aliprantis–Burkinshaw [5] and Aliprantis–Tourky [6].

Two heavily studied Banach lattices are the $M$- and $L$-spaces. An *$M$-space* is a Banach space with an order making it into a lattice, so if $x, y \geq 0$, then $\|x \vee y\| = \max(\|x\|, \|y\|)$. The canonical examples are $C(X)$ and $L^\infty(\Omega, d\mu)$. An *$L$-space* has a norm obeying if $x, y \geq 0$, then $\|x+y\| = \|x\| + \|y\|$. The canonical examples are $\mathcal{M}(X)$ and $L^1(\Omega, d\mu)$. An interesting theorem is that the dual of an $M$-space is an $L$-space and vice-versa.

Example 11.16 is not as specialized as it looks. It is a theorem of Downarowicz [104] that given any metrizable Choquet simplex, $K$, there is a compact metric space, $X$, and continuous bijection $\tau\colon X \to X$ so $K$ is affinely homeomorphic to $\mathcal{M}^I_{+,1}(T)$. Another general structure theorem is the result of Choquet [83] and Haydon [151] that any complete separable metric space is homeomorphic to the set of extreme points of some separable Bauer simplex.

There is considerable literature on extensions of the Krein–Milman theorem and Choquet's existence theorem to noncompact, bounded, closed subsets of certain Banach spaces. The theory is fairly complete for spaces obeying what is called the Radon–Nikodym property. Existence is due to Edgar [109] and uniqueness to Bourgin–Edgar [49] and Saint Raymond [333]. For further discussions, see Fonf–Lindenstrauss–Phelps [115] and Bourgin [48].

The Hadamard three-line theorem for bounded functions on a strip is attributed to Hadamard by Bohr–Landau [42], but Hadamard never published a proof. Bernstein's lemma (Theorem 12.2) is due to [32]. The extensions are associated to the work of Phrágmen and Lindelöf.

In 1927, M. Riesz first found interpolation theorems between $L^p$ space mappings in Riesz [312] in connection with the argument to extend his theorem on $L^p$ boundedness of harmonic conjugation from $p = 2n$ to all of $(2, \infty)$. He did not use complex methods and could only interpolate maps from $L^p$ to $L^q$ with $q \geq p$. The complex method and general theorem is due to Thorin [371], a student of Riesz, who went into the insurance business after writing his initial 1938 paper, only returning to it for a thesis ten years later [371].

The idea of varying the operator also is due to Stein [359], who found the application in Theorem 12.10.

Abstract versions of the Riesz–Thorin and Stein interpolation theorem are due to Calderón [62] and Lions [236, 237, 238], who set up families of spaces $X_t$, $0 \leq t \leq 1$, which include $\mathcal{I}_p$ spaces in addition to $L^p$. ($\mathcal{I}_p$ is discussed in Simon [350].)

Young's inequality goes back to Young [392]. The optimal constants in Young's inequality on $\mathbb{R}^\nu$ are known, namely,

$$\|f(g * h)\|_1 \leq (C_p C_q C_r)^\nu \|f\|_p \|g\|_q \|h\|_r \tag{17.27}$$

where

$$C_p^2 = \frac{p^{1/p}}{(p')^{1/p'}}$$

with $p' = (1 - p^{-1})^{-1}$. This is due to Beckner [30] and Brascamp–Lieb [52]. The latter prove that there is equality if and only if $f, g, h$ are Gaussians about some point multiplied by plane waves. For a generalization of (17.27) to optimal bounds for integrals of products of Gaussians with products of the form $f_i(B_i x)$ with $B_i$ a linear map, see Lieb [230].

Theorem 12.8 goes back to Hardy and Littlewood [146, 147] in case $\nu = 1$ and Sobolev [355] in the general case. Best constants are known in case $q = s$ $(= 2\nu/(2\nu - \sigma))$ where

$$C = \pi^{\sigma/2} \frac{\Gamma(\frac{\nu}{2} - \frac{\sigma}{2})}{\Gamma(\nu - \frac{\sigma}{2})} \left\{ \frac{\Gamma(\frac{\nu}{2})}{\Gamma(\nu)} \right\}^{-1+\sigma/\nu}$$

This result is due to Lieb [229]. Pedagogical presentations of the best constant results in both Young's and Sobolev's inequality can be found in Lieb–Loss [231]. It follows from Theorem 14.8 that (12.36) implies (12.35).

One reason that Sobolev inequalities are important is because they provide borderline control on Sobolev embedding theorems. For $2k < \nu$, $(-\Delta + 1)^{-k}$ applied to functions in $\mathcal{S}(\mathbb{R}^\nu)$ is convolution by a function $f_{\nu,k}(x)$ which obeys

$$|f_{\nu,k}(x)| \leq C_{\alpha,\nu} \exp(-\alpha|x|), \qquad |x| \geq 1 \text{ and } \alpha < 1 \tag{17.28}$$

and

$$|f_{\nu,k}(x)| \leq D_{\nu,k} |x|^{(-\nu+2k)}, \qquad |x| \leq 1 \tag{17.29}$$

If $p_{\rm crit} = \nu/(\nu - 2k)$, then $f \in L^1 \cap L_w^{p_{\rm crit}}$ so $(-\Delta + 1)^{-k}$ map $L^q$ into $L^r$ for any $r$ with $q \le r \le r_{\rm crit} = (q^{-1} + p_{\rm crit}^{-1} - 1)^{-1}$ so long as $1 < q < (1 - p_{\rm crit}^{-1})^{-1}$. This only holds at the endpoint $r_{\rm crit}$ because of the Sobolev inequality. These mapping results show $\{f \in L^q \mid D^\alpha f \in L^q, |\alpha| \le 2k\}$ lies in $L^q \cap L^{r_{\rm crit}}$, known as a *Sobolev embedding theorem*.

The Strichartz inequality was proven by Strichartz [366]. By Theorem 14.23, it is equivalent to $|x|^{-\nu/s}|\Delta|^{-\nu/2s} \colon L^p \to L^p$ for $1 < p < s$ and the best constants are determined by that special case. Since $|x|^{-\nu/s}|\Delta|^{-\nu/2\nu}$ is scaling invariant, one can partly understand it using Mellin transforms; see Herbst [157].

One application of the Strichartz theorem is that it implies if $\nu \ge 5$, $|x|^{-2}$ is $-\Delta$-bounded in the sense of Kato [189]. The relative bound is not zero. This is further discussed in [89].

The Brunn–Minkowski inequality (Theorem 13.1) goes back to Brunn [56] in 1887 and Minkowski [262] in the convex case. The general case presented at the end of Chapter 13 is due to Lusternik [244]. The proof we give is due to Hadwiger and Ohman [138]. For further discussion of this result, extensions to other geometries, and to other geometric inequalities, see Burago–Zalgaller [59] and Federer [113].

Prékopa's theorem (Theorem 13.9) is due to Prékopa [299] based on his earlier work in [298]. It is a corollary of a more general result dubbed the Prékopa–Leindler theorem by Brascamp–Lieb [53] after this work of Prékopa and Leindler [223]. Closely related ideas were developed and published by others at about the same time, notably Rinott [314] and Borell [44, 45]. Brascamp–Lieb [50] discovered Theorem 13.9 independently, but they did not publish their preprint after learning of Prékopa's paper. This is unfortunate since they have several proofs, including the one we use! Brascamp–Lieb apply these and related ideas in [51, 52, 53].

Special cases of Theorem 13.11 predated the general result (which appeared in Brascamp–Lieb [50]). In one dimension, it is a result of Schoenberg [338]. Anderson [8] and Sherman [347] proved results about convolutions of the characteristic functions of convex sets that follow from the log concavity of this convolution. Corollary 13.14 was first proven by different means by Gross [135]. The applications to Brunn–Minkowski theorems for Lebesgue and Gauss measures are taken from Brascamp–Lieb [50]. Since $\nu$ does not appear in Theorem 13.13, it extends to Gaussian processes.

*Isoperimetric inequalities* are a major theme not only in the applications in Chapters 13 and 14 but also in their history; in particular, Brunn–Minkowski and Steiner symmetrization are rooted in attempts at proving the classical isoperimetric inequalities, so it makes sense to say something about the subject. More can be found in books of Pólya and Szegő [296], Burago–Zalgaller [59], Bandle [27], and Chavel [74], or the review articles of Osserman [279], Payne [282, 283], Hersch [159], and

Ashbaugh [16]. In terms we will frame momentarily, some of these focus on the geometric inequalities and some on the analytic inequalities.

One could claim that none of the isoperimetric results that we prove in Chapters 13 and 14 are "true" isoperimetric inequalities, for we only show for various quantities $q$ associated to a region, $\Omega$, that $q(\Omega) \geq q(\Omega^*)$ where $\Omega^*$ is the ball of the same volume. "True" isoperimetric inequalities also prove the inequality is strict if $\Omega \neq \Omega^*$. These sharper results are very interesting and often harder to prove.

The classical isoperimetric problem is to find the region with maximal area for given perimeter and the geometric problems are the analogs of this in higher dimension and on suitable homogeneous surfaces (like a sphere and the Lobachevsky plane). There is a closely related set of analytic problems where some quantity is associated to a region – for example, a lowest Dirichlet eigenvalue – and one wants to minimize this for a given volume.

The geometric problem was known to the Greeks who knew the answer in two and three dimensions. The problem was known in ancient times as Dido's problem after Queen Dido, the legendary founder of Carthage. As told in Virgil's Aeneid, Queen Dido was offered the amount of land a bull hide could encompass. She cut the hide into strips, tied them together, and used this "cord" to enclose a maximal area.

The Greeks knew the solution was a disk in two dimensions and a ball in three. But formal proofs were only sought in the nineteenth century. The earliest geometric proofs were by Steiner [360] in 1838, using the idea of repeated symmetrization as we do in Chapter 14, but without realizing there was any kind of convergence issue to prove. This idea was pushed to a careful conclusion by Schwarz [344] in 1884. Minkowski [262] used the Brunn–Minkowski inequality. Other important early approaches are due to Weierstrass, Edler, and Hurwitz [174].

The two-dimensional case has several factors that make it easier. If $\Omega$ is a region, $\mathrm{ch}(\Omega)$ has a larger volume and, in two dimensions, a shorter perimeter (but not in three dimensions; let $\Omega$ be a ball plus a very long spike), so in two dimensions, it is clear that for the isoperimetric ratio, one always does better with convex sets.

Secondly, in two dimensions, there are some remarkable inequalities that go back to Bonnesen [43], the most famous of which is

$$P^2 \geq 4\pi A + \pi^2(R - r)^2 \qquad (17.30)$$

where

$P =$ perimeter of a simple closed curve, $\gamma$

$A =$ area of enclosed region

$R =$ out radius, the radius of the smallest circle enclosing $\gamma$

$r =$ in radius, the radius of the largest circle inside $\gamma$

$R = r$ only if $\gamma$ is a circle, so (17.30) implies the strong form of the two-dimensional isoperimetric inequality.

The analytic side of the inequalities goes back to nineteenth-century conjectures proven in the twentieth century. The first such conjecture was in 1856 by Saint-Venant [334] that torsional rigidity was maximized by a circular cross-section. This result, Theorem 14.20, was proven by Pólya in 1948 [293] and discussed further in the book of Pólya and Szegő [296].

The most famous analytic result is Theorem 14.19 conjectured by Lord Rayleigh in 1877 in Section 210 of his "Theory of Sound" [302]. It was proven independently by Faber [112] and Krahn [208] in 1923–1925.

There are a number of other interesting isoperimetric results involving eigenvalues of the Laplacian. In 1952, Kornhauser–Stakgold [205] conjectured that for the Neumann Laplacian where the lowest eigenvalue, $\mu_0^N(\Omega) = 0$, one has an isoperimetric inequality on the first eigenvalue

$$\mu_1^N(\Omega^*) \geq \mu_1^N(\Omega) \tag{17.31}$$

(notice the opposite direction from the Faber–Krahn inequality that $e_D(\Omega^*) \leq e_D(\Omega)$). This was proven in two dimensions for simply connected $\Omega$ by Szegő [368] in 1954 and in general by Weinberger [383] in 1956.

Returning to the Dirichlet Laplacian, if $\mu_j^D(\Omega)$ is the $j$-th eigenvalue of the Dirichlet Laplacian (so $e_D(\Omega) = \mu_1^D(R)$), then Payne, Pólya, and Weinberger conjectured in 1955–1956 [284, 285] that in two dimensions,

$$\frac{\mu_2^D(\Omega)}{\mu_1^D(\Omega)} \leq \frac{\mu_2^D(\Omega^*)}{\mu_1^D(\Omega^*)} \tag{17.32}$$

This was proven not only in two dimensions but in $n$-dimensions by Ashbaugh and Benguria [17, 18, 19]. They also proved that [20]

$$\frac{\mu_4^D(\Omega)}{\mu_2^D(\Omega)} \leq \frac{\mu_2^D(\Omega^*)}{\mu_1^D(\Omega^*)} \tag{17.33}$$

which implies for $m = 2, 3$ that

$$\frac{\mu_{m+1}^D(\Omega)}{\mu_m^D(\Omega)} \leq \frac{\mu_2^D(\Omega^*)}{\mu_1^D(\Omega^*)} \tag{17.34}$$

It is an open problem that (17.34) holds for all $m$ and also an open conjecture of Payne, Pólya, and Weinberger [285] that in two dimensions,

$$\frac{\mu_2^D(\Omega) + \mu_3^D(\Omega)}{\mu_1^D(\Omega)} \leq \frac{\mu_2^D(\Omega^*) + \mu_3^D(\Omega)}{\mu_1^D(\Omega^*)}$$

For a discussion of additional isoperimetric inequalities on eigenvalues (e.g., the fourth-order clamped plate problem), see the review of Ashbaugh [16].

Isoperimetric inequalities connected to the Coulomb energy go back to Poincaré [292] in 1902, who proved Theorem 14.21 on Coulomb energy and conjectured a result on the capacity (Theorem 14.22). Actually, he claimed the result on capacity but his proof was incomplete. In 1918, Carleman [65] proved Poincaré's capacity conjecture in two dimensions using conformal mapping. Szegő [367] proved the general three-dimensional result in 1930. In 1945, Pólya–Szegő [295] proved this using Steiner symmetrization, a technique extended in their book and behind our discussion. A lovely presentation of the capacity result – without the shortcuts in presentation that we take – is in Lieb–Loss [231].

It is intuitively obvious that repeated Steiner symmetrization in different directions should result in a set converging to a ball of the same volume, so much so that proofs of this fact (nontrivial, it turns out!) were late in coming. Convergence in the Hausdorff metric was proven in 1909 by Carathéodory and Study [64]. Convergence in measure (our Theorem 14.13) seems to have only been proven by Brascamp, Lieb, and Luttinger [54] in 1974. Our proof closely follows their ideas; Lieb–Loss [231] have some alternates of the proof, using the more "elementary" Helly selection theorem rather than the theorem of M. Riesz.

There are two threads in Chapters 14 and 15 centered around the BLL inequality (Theorem 14.8) and the HLP theorem (Theorem 15.5). Both themes were central in a remarkable paper of Hardy–Littlewood–Pólya [148] and in their book [149]. There are important precursors to their work, especially on the HLP theorem. Muirhead [270] proved Proposition 17.1 in 1903 using the idea in Lemma 15.9 to prove his result. This is a precursor to the HLP idea of doubly stochastic matrices. Also, Schur's work on Schur convex functions [341], which we will discuss below, predated the work of HLP.

With regard to the first theme, HLP proved the following theorem which is a discrete analog of Riesz's rearrangement inequality (Theorem 14.1):

**Theorem 17.10** *Let $x, y, c$ be nonnegative sequences of size $2k+1$ indexed by $j \in \{-k, -k+1, \dots, k-1, k\}$. Suppose $c$ has the property that it takes its maximum value an odd number of times and every other value an even number of times. Let $c^*$ be the rearrangement of $c$ to a sequence $\{c^*_\ell\}_{\ell=-k}^{k}$ with*

$$c_0^* \geq c_1^* = c_{-1}^* \geq c_2^* = c_{-2}^* \geq \cdots \geq c_k^* = c_{-k}^*$$

*Let $x^+$ be the rearrangement with*

$$x_0^+ \geq x_1^+ \geq x_{-1}^+ \geq x_2^+ \geq x_{-2}^+ \geq \cdots \geq x_k^+ \geq x_{-k}^+$$

*and ${}^+y$ the rearrangement with*

$${}^+y_0 \geq {}^+y_{-1} \geq {}^+y_1 \geq {}^+y_{-2} \geq {}^+y_2 \geq \cdots \geq {}^+y_{-k} \geq {}^+y_k$$

*Then*

$$\sum_{j=-k}^{k} \sum_{\ell=-k}^{k} c_{j-\ell} x_j y_\ell \le \sum_{j=-k}^{k} \sum_{\ell=-k}^{k} c_{j-\ell}^{*} x_j^{++} y_\ell$$

Motivated by this work, F. Riesz proved the one-dimensional Theorem 14.1 [309]. Riesz remarks: "The extension of the inequality to the case of several variables is immediate." Since he refers to the paper of HLP, it seems evident he had in mind integrals of the form

$$\int f_1(x_1) f_2(x_2) \dots f_\ell(x_\ell) g(x_1 + x_2 + x_3 + \cdots + x_\ell)\, dx_1 \dots dx_\ell$$

rather than higher-dimensional $x$'s, but he is not explicit, and there is some confusion about the history of the higher-dimensional result so that the Theorem 14.1 with $x$ and $y$ in $\mathbb{R}^\nu$ and $\nu$-dimensional rearrangements is sometimes mistakenly attributed to Riesz. It is interesting to note that Riesz's "paper" is not a conventional publication but a letter Riesz sent to Hardy, which Hardy read into the minutes of the London Mathematical Society!

Special cases of Riesz's inequality, but in higher dimensions, had appeared in Blaschke [40], Carleman [65], and Lichtenstein [225]. Interest in these inequalities was revived by Luttinger [245], who wished to apply the inequality to reorderings of potentials in path integrals with applications like our Theorem 14.17 in mind. Together with Friedberg [246], he proved a one-dimensional inequality with multiple convolutions. They conjectured the one-dimensional BLL inequalities.

Brascamp–Lieb–Luttinger, in a masterful paper [54], not only proved this conjecture with a simple, elegant, and new use of the Brunn–Minkowski inequality (essentially, the proof we present of the $\nu = 1$ version of Theorem 14.8), but also, as noted above, settled the issue of higher dimensions by a careful proof of the requisite convergence theorem for repeated Steiner symmetrization.

Corollary 14.6 that $f \mapsto f^*$ is a contraction on $L^p$ is due to Chiti [76] and Crandall–Tartar [88]. We showed in Theorem 14.18 that $f \mapsto f^*$ is a contraction in the Sobolev space $W_1^2$, and it is more generally known that

$$\int |\nabla f^*|^p \, d^\nu x \le \int |\nabla f|^p \, d^\nu x \tag{17.35}$$

(discovered at about the same time by Aubin [21], Duff [105], Lieb [228], Sperner [357, 358], and Talenti [369]). It is therefore surprising that $f \mapsto f^*$ may be discontinuous in Sobolev norm ((14.91) only implies continuity at $f = 0$ since $^*$ is a nonlinear map); see Almgren–Lieb [7]. Burchard [60] studied when equality holds in the $\nu$-dimensional Riesz convolution inequality. Almgren–Lieb [7] and Lieb [228] have extensions of the Riesz theorem.

Doubly substochastic matrices were introduced in the context of the HLP theorem by HLP, who proved (i) $\Rightarrow$ (iii) in Theorem 15.5 not by using the structure of the doubly stochastic matrices but, following Muirhead, by using Lemma 15.9. The term doubly stochastic was only introduced systematically in the probability literature around 1950.

Given that HLP did consider this family of matrices starting in 1929, it is surprising that their extreme points were only found by Birkhoff [36] in 1946. Since then, many proofs of his theorem (Theorem 15.4) have been found; see the review article of Mirsky [265] for references. The proof we give is due to Hoffman–Wielandt [163]. Safarov [332] has an extension of Birkhoff's theorem to certain infinite-dimensional situations (and discusses other such extensions).

There is an enormous literature on HLP majorization (i.e., what we denote $a \prec_{\text{HLP}} b$). There is even a 500-page book (Marshall–Olkin [256]) on the subject, which is essentially a long love poem to the idea, so much so that it includes thumbnail biographies and pictures of the major figures, including Muirhead, Schur, Hardy, Littlewood, and Pólya! A short book on majorization is Arnold [14]. Ando [11] is a later review article on some aspects of majorization.

What we call the HLP theorem (Theorem 15.5) includes more than in the original HLP paper [148] and book [149]. They only proved the equivalence of (i), (iii), (v), and (vi). That (ii) (convex hull of $\{M_\pi a\}$) is involved was noted first by Rado [301] twenty years after, using the separating hyperplane theorem as we do. That (iv) is involved is an earlier result of Schur [341], as we will explain.

Shortly after HLP, unaware of their work, Karamata [186] found the results (i) $\Leftrightarrow$ (v) in the HLP theorem. For an alternate proof of (i) $\Rightarrow$ (v), see Fuchs [119].

Fuchs [119] actually proves a more general result than (i) $\Rightarrow$ (v) in HLP. Namely, he shows if $x = x^*$, $y = y^*$, and for fixed $p_j$'s, one has $\sum_{j=1}^k p_j y_j \le \sum_{j=1}^k p_j x_j$ (with equality if $k = \nu$), then $\sum_{j=1}^\nu p_j \varphi(y_j) \le \sum_{j=1}^\nu p_j \varphi(x_j)$ for all convex $\varphi$. Pečarić [286] proved the converse of $p_j \ge 0$ and found a counterexample of the converse if $p_j$ are arbitrary reals.

Extensions to majorization with respect to groups other than permutation groups are discussed by Eaton–Perlman [108] and Niezgoda [273, 274].

There has been study, given $b, a \in \mathbb{R}_+^\nu$ with $b \prec_{\text{HLP}} a$, of $\{D \in \mathbb{D}_\nu \mid b = Da\}$. Cheon–Song [75] have described its extreme points and Chao–Wong [73] have proven that if $b = b^*$, $a = a^*$, then there is always a symmetric doubly stochastic matrix that takes $a$ to $b$.

The term Schur convex function has become standard, even though what is involved is a monotonicity condition, not a convexity condition. In his paper on the subject, Schur [341] called them "convex functions" and what we (and everyone else now) call convex functions, he called Jensen convex functions. So the name "Schur convex" has stuck.

Schur's main result is that a $C^1$ function, $\Phi$, is Schur convex if and only if $\Phi$ is permutation invariant and the differential equality (15.31) holds. Most discussions since have stated this version rather than the form we state in Theorem 15.10, which is more general and easier to check many cases (!). Schur found the examples of the elementary symmetric functions (Example 15.11) and was motivated by Theorem 15.39 and its application to Hadamard's determinantal inequality (Theorem 15.40).

Schur's original paper dealt with the case where $I = [0, \infty)$ and $a \in \mathbb{R}^\nu_+$. It was Ostrowski [280] who noted the extension to $\mathbb{R}^\nu$. While it is easy to derive the $\mathbb{R}^\nu$ result from the $\mathbb{R}^\nu_+$ result, Theorem 15.10 is often called the Schur–Ostrowski theorem.

Going through the many variants of the HLP theorem, the reader may be tempted to quote Yogi Berra's "it's déjà vu all over again." While the theorems are similar with similar proofs, each is useful in somewhat different contexts, and each variant occurred in a different time frame (Theorem 15.16 around 1950, Theorem 15.17 around 1960, and Theorem 15.36 around 1970) with occasional rediscovery of the analog of an equivalence in the new context.

Theorem 15.12 is due to Ando [10]; see also Ando [11]. Equation (15.146) has a converse (Horn [168], Mirsky [263]; see also Chan–Li [71] and Carlen–Lieb [66]): if $a \prec \lambda$ for two sequences in $\mathbb{R}^n$, then there exists a Hermitian matrix, $A$, with diagonal element $a_i$ and eigenvalues $\lambda_i$.

Motivated by Weyl's work (Theorem 15.20; see below), Pólya [294] noted the core of Theorem 15.16 that is that majorization with equality at the top level replaced by an inequality implies (15.52) for all monotone convex $\varphi\colon \mathbb{R} \to \mathbb{R}$. For he remarked that one could add a $(\nu + 1)$-st component to $a$ and $b$ for which $(b, b_{\nu+1}) \prec_{\text{HLP}} (a, a_{\nu+1})$, and then the HLP theorem implied the needed result.

Extensions to the complex case (Theorem 15.17) were driven by applications to operator ideals, especially among the Russians. Mitjagin [266] is credited by both Gohberg–Krein [131] and Simon [350] with the proof of (i) $\Rightarrow$ (ii) in Theorem 15.17, and while it is correct that he presented this proof in this context, Mitjagin essentially rediscovered Rado's proof [301] in the HLP context. A key paper in the complex version (Theorem 15.17) is Markus [255].

Lemma 15.19 and Weyl's inequality (Theorem 15.20) are due to Weyl [385]. Horn's inequality (Theorem 15.21) is from Horn [167]. In [169], Horn found a "converse" to Weyl's lemma, namely, a necessary and sufficient condition for $n$ complex numbers $|\lambda_1| \geq |\lambda_2| \geq \cdots \geq |\lambda_n|$ and $n$ nonnegative numbers $\mu_1 \geq \mu_2 \geq \cdots \geq \mu_n$ to be the eigenvalues and singular values of an $n \times n$ matrix is

$$
\begin{aligned}
|\lambda_1 \dots \lambda_k| &\leq \mu_1 \dots \mu_k, \qquad k = 1, 2, \dots, n-1 \\
|\lambda_1 \dots \lambda_n| &= \mu_1 \dots \mu_n
\end{aligned}
$$

For another proof, see Mirsky [264].

Prior to Weyl's work, special cases of (15.60) were known. In 1909, Schur [339] proved the $p = 2$ result in the form

$$\sum_{j=1}^{n} |\lambda_j(A)|^2 \leq \operatorname{tr}(A^*A)$$

After a partial result by Lalesco [217], Gheorghiu [127], and Hille–Tamarkin [162] proved the $p = 1$ case in the form

$$\sum_{j=1}^{n} |\lambda_j(AB)| \leq \operatorname{tr}(A^*A)^{1/2} \operatorname{tr}(B^*B)^{1/2}$$

The discrete infinite-dimensional case (Theorem 15.22) has been discussed by Markus [255] and Mirsky [265].

HLP already noted that for nonnegative functions $f$ and $g$ on $[0,1]$, $\int_0^1 F(g(s))\,ds \leq \int_0^1 F(f(s))\,ds$ for all convex functions $F$ if and only if $\int_0^t g^*(s)\,ds \leq \int_0^t f^*(s)\,ds$ for all $t$ and $\int_0^1 g(s)\,ds = \int_0^1 f(s)\,dr$. The themes of general measure spaces and the continuum analogs of doubly stochastic matrices were only taken up systematically after 1963 in a series of papers by Ryff [326, 327, 328, 329, 330, 331] and Day [94, 95]. Precursors and related work include Burkholder [61], Rota [323], Luxemburg [248], and Chong [77].

The definition (15.87) of $S_\lambda(f)$ in the presence of atoms and variational principles, (vi) and (vii) in Proposition 15.26 are due to Hundertmark–Simon [173].

Theorem 15.28 is a special case of the fact that the set of values of a measure without atoms and with values in $\mathbb{R}^\nu$ is a convex subset of $\mathbb{R}^\nu$. This result is due to Lyapunov [250]. Our proof follows Halmos [142]. Marcus [254] discusses what happens when there are atoms. He refers to the result as "the Darboux property" in honor of Darboux who emphasized the intermediate value theorem for functions. Marcus has extensive references to other results in this area.

What we call the Lorentz–Ryff lemma (Theorem 15.32 and the following Theorem 15.33) appeared in Ryff [331] for the case of functions on $[0,1]$ but, as noted by Day [95], his proof works for general measure spaces. Lorentz [243, p. 60] had the result earlier than Ryff, although Ryff seems to have been unaware of his earlier work.

Theorem 15.36 is due to Day [95], with earlier work by Ryff [327] for the case $M = [0,1]$ and $d\mu = dx$. In Ryff [328], it is proven that the extreme points of $\operatorname{cch}(\{h \in L^p \mid h \text{ is measurable with } f\})$ is precisely the set of $h$ which are equimeasurable with $f$. For a related result, see Horsley–Wrobel [172]. On the other hand, it seems difficult to find the proper analog of Birkhoff's theorem, that is, to find all the extreme points of the set of maps $\zeta\colon L^1((0,1),dx) \to L^1((0,1),dx)$ which are contractions of $L^1$ and $L^\infty$, positivity preserving, and obeying $\zeta 1 = 1$, $\int_0^1 (\zeta f)(x)\,dx = \int_0^1 f(x)\,dx$. Ryff [326] notes this if $T\colon [0,1] \to [0,1]$ is measure preserving, $(\zeta f)(x) = f(Tx)$ is an extreme point in this set, but he shows $(\zeta f)(x) = \frac{1}{2}f(\frac{1}{2}x) + \frac{1}{2}f(\frac{1}{2}(x+1))$ is also an extreme point and not of this form.

The interpolation argument used in our proof of Proposition 15.38 is from Simon [348]. The basis $\{\varphi_j\}_{j=1}^n$ used in the proof of Theorem 15.39 is called a *Schur basis* after Schur [341]. Schur also found the application to the Hadamard's inequality (our proof of (Theorem 15.40).

Hadamard's determinantal inequality (Theorem 15.40) is due to Hadamard [137] and Minkowski's determinantal inequality (Corollary 15.42) is due to Minkowski [262]. Here is an alternate proof of Hadamard's inequality in the form of Theorem 15.41, close to Hadamard's:

*Direct proof of Theorem 15.41* If the rows of $A$ are dependent, $\det(A) = 0$ and the inequality is trivial. So suppose the rows $r_j = (a_{j1}, \dots, a_{jn})$ are independent. Use the Gram–Schmidt process to write

$$r_j = \sum_{i=1}^{j} \alpha_{ji} e_i$$

where $e_i$ is an orthonormal basis. Since the $e_i$ are orthonormal, $\|r_j\|^2 = \sum_{i=1}^{j} |\alpha_{ji}|^2$. In particular,

$$|\alpha_{ji}| \le \|r_j\|$$

On the other hand, writing the determinant in the $e_i$ basis, we see

$$\det(A) = \prod_{j=1}^{n} \alpha_{jj} \le \prod_{j=1}^{n} \|r_j\|$$

which is Hadamard's inequality. □

Majorization has been applied to probability theory, queueing theory, reliability testing, and other areas. There are many applications in the book of Marshall–Olkin [256]; see also Proschan [300], Sakata–Nomakuchi [335], Kästner–Zylka [187], Lonc [241], Tong [373], and Towsley [374].

Entropy originated in the development of thermodynamics in the first half of the nineteenth century. Key names associated with these ideas are Carnot and Clausius, with critical later contributions by Kelvin and Gibbs. For an axiomatic approach to entropy and the second law, see Lieb–Yngvason [232, 233]. It was Boltzmann in the 1870s who first realized entropy as an expectation of the log of a counting function. Shannon's realization in 1948 [346] that the negative of entropy measured information was a significant goad to further developments. Other critical developments in the applicability of entropy beyond physics include Kolmogorov's introduction [202, 203] of the metric entropy of dynamical systems (also called Kolmogorov–Sinai entropy) and the notion of topological entropy of a map by Adler, Konheim, and McAndrew [1]. Entropy is almost a religion and a deep set of ideas, and we cannot plumb its depths here.

Variational principles for the entropy go back to Gibbs; in their modern guise, they appear in Robinson–Ruelle [318] and Lanford–Robinson [220]; see the discussion in the books of Israel [175], Ruelle [324, 325], and Simon [351]. Upper semicontinuity of $S(\mu \mid \nu)$ in $\mu$ is discussed in many places. Upper semicontinuity in $\nu$ is discussed by Cohen *et al.* [86], Kullback–Leibler [216], and Ohya–Petz [276]. Theorem 16.1, in the form we state it, appears in Killip–Simon [193] whose proof we follow. There are changes to accommodate $F$'s, which are other than $F(x) = \log x$. Stating the general $F$ result both simplifies and illuminates the theorems.

The almost convexity of $S(\mu \mid \nu)$ in $\mu$ is a result of Robinson–Ruelle [318]. It is especially interesting because $|S(\mu \mid \nu)|$ can be very large and dwarf $g$. For example, for product measure $S(\mu_1 \otimes \mu_2 \mid \nu_1 \otimes \nu_2) = S(\mu_1 \mid \nu_1) + S(\mu_2 \mid \nu_2)$, so one expects that, for large statistical mechanical systems, entropy grows as the volume, and that for infinite systems, one needs to restrict to finite volume, compute entropy, divide by the volume, and take the limit as the volume goes to infinity. This entropy per volume is both concave and convex, so affine.

That completes the notes on topics discussed in this book. Here is a brief discussion of some issues related to convexity that are not discussed in this book.

(1) **Uniform convexity.** A Banach space, $B$, is called *uniformly convex* if and only if for all $\varepsilon \in (0, 2)$,

$$\delta(\varepsilon) \equiv \inf\left\{1 - \left\|\frac{x+y}{2}\right\| \;\middle|\; \|x\| = \|y\| = 1,\ \|x-y\| > \varepsilon\right\} > 0$$

Uniformly convex spaces are known to be reflexive. This is a theorem of Milman [260] and Pettis [288], proven also by Kakutani [180]. Ringrose [313] has a half-page proof. $L^p$ for $1 < p < \infty$ is uniformly convex. This follows from inequalities of Clarkson [85] and Hanner [144]; for the case of trace ideals, see McCarthy [259] and Ball–Carlen–Lieb [22].

(2) **Points at infinity.** We have not said much about unbounded convex sets in $\mathbb{R}^\nu$. One can add a sphere at infinity, $S^{\nu-1}$ to $\mathbb{R}^\nu$. Given $A \subset \mathbb{R}^\nu$ a convex set, we say $e \in S^{\nu-1}$ is a *point at infinity* affiliated to $A$ if and only if $x \in A$ implies $x + \lambda e \in A$ for all $\lambda > 0$. Adding these points allows one to extend some of the theory to unbounded sets. See Eggleston's book [111]. There are various results extending the Krein–Milman theorem to cases with points at infinity. This theory, largely due to Klee [195, 196, 197, 198], is discussed in Rockafellar [319].

(3) **Convex programming** is a generalization of linear programming and involves minima of convex functions with constraints. It is the subject of Part VI of Rockafellar's book [319]. A seminal paper is by Kuhn and Tucker [215].

(4) **Helly's theorem.** Let $\{C_\alpha\}_{\alpha \in I}$ be a family of compact convex subsets of $\mathbb{R}^\nu$. Helly's theorem asserts that if any $\nu + 1$ subsets in the family have a nonempty intersection, then $\bigcap_{\alpha \in I} C_\alpha \neq \emptyset$. This is discussed, for example, in Eggleston [111] and Rockafellar [319]. In particular, Eggleston discusses the close relation to Carathéodory's theorem.

(5) **Generalizations of convexity.** In an interesting book, Hörmander [166] discusses notions like subharmonicity as analogs of convex functions. See also Krantz [209].

(6) **Convexity inequalities for matrices.** We have already seen that there are several results about matrices that follow from convexity, for example, Theorems 15.20, 15.21, 15.40, 15.41, and Corollary 15.42. But there is a lot more; some references on these issues are the books of Gohberg–Krein [131] and Simon [350]. Among the important further results are Lieb concavity [226] that for $t \in [0, 1]$ and $X$ fixed, $f(A, B) = \text{tr}(X^* A^t X B^{1-t})$ is joint concave on pairs $(A, B)$ of positive matrices, a result of Lieb [227] that for certain functions $F$ on matrices $|F(\sum_{i=1}^m C_i)|^2 \leq F(\sum_{i=1}^m |C_i|) F(\sum_{i=1}^m |C_i^*|)$, and the Golden [132]–Thompson [370] inequality that $\text{tr}(e^{A+B}) \leq \text{tr}(e^A e^B)$ for $A, B$ self-adjoint.

(7) **A minimax principle.** There are interesting results about min-max and max-min being equal for functions convex in one set of variables and concave in the others. The earliest results are due to von Neumann [378]. For a simple proof of a general infinite-dimensional result, see Kindler [194]. The result is important in game theory and mathematical economics.

# References


[1] R. L. Adler, A. G. Konheim, and M. H. McAndrew, Topological entropy, *Trans. Amer. Math. Soc.* **114** (1965), 309–319.

[2] N. I. Akhiezer, *The Classical Moment Problem and Some Related Questions in Analysis*, Hafner, New York, 1965; Russian original, 1961.

[3] L. Alaoglu, Weak topologies of normed linear spaces, *Ann. of Math. (2)* **41** (1940), 252–267.

[4] E. M. Alfsen, *Compact Convex Sets and Boundary Integrals*, Ergebnisse der Mathematik und ihrer Grenzgebiete 57, Springer, Berlin-Heidelberg-New York, 1971.

[5] C. D. Aliprantis and O. Burkinshaw, *Locally Solid Riesz Spaces With Applications to Economics*, 2nd edition, Mathematical Surveys and Monographs 105, American Mathematical Society, Providence, RI, 2003.

[6] C. D. Aliprantis and R. Tourky, *Cones and Duality*, Graduate Studies in Mathematics 84, American Mathematical Society, Providence, RI, 2007.

[7] F. J. Almgren, Jr. and E. H. Lieb, Symmetric decreasing rearrangement is sometimes continuous, *J. Amer. Math. Soc.* **2** (1989), 683–773.

[8] T. W. Anderson, The integral of a symmetric unimodal function over a symmetric convex set and some probability inequalities, *Proc. Amer. Math. Soc.* **6** (1955), 170–176.

[9] R. D. Anderson and V. L. Klee, Convex functions and upper semi-continuous collections, *Duke Math. J.* **19** (1952), 349–357.

[10] T. Ando, Majorization, doubly stochastic matrices, and comparison of eigenvalues, *Linear Algebra Appl.* **118** (1989), 163–248.

[11] T. Ando, Majorizations and inequalities in matrix theory, *Linear Algebra Appl.* **199** (1994), 17–67.

[12] R. F. Arens, *Duality in linear spaces*, Duke Math. J. **14** (1947), 787–794.

[13] D. H. Armitage, The Riesz–Herglotz representation for positive harmonic functions via Choquet's theorem, in *Potential Theory – ICPT 94* (Kouty, 1994), pp. 229–232, de Gruyter, Berlin, 1996.

[14] B. C. Arnold, *Majorization and the Lorenz Order: A Brief Introduction*, Lecture Notes in Statistics 43, Springer, Berlin, 1987.

[15] G. Ascoli, Sugli spazi lineari metrici e le loro varietà lineari, *Ann. Mat. Pura Appl. (4)* **10** (1932), 33–81, 203–232.

[16] M. S. Ashbaugh, Isoperimetric and universal inequalities for eigenvalues, in *Spectral Theory and Geometry* (Edinburgh, 1998), pp. 95–139, London Math. Soc. Lecture Note Series 273, Cambridge University Press, Cambridge, 1999.

[17] M. S. Ashbaugh and R. D. Benguria, Proof of the Payne–Pólya–Weinberger conjecture, *Bull. Amer. Math. Soc.* **25** (1991), 19–29.

[18] M. S. Ashbaugh and R. D. Benguria, A second proof of the Payne–Pólya–Weinberger conjecture, *Comm. Math. Phys.* **147** (1992), 181–190.

[19] M. S. Ashbaugh and R. D. Benguria, A sharp bound for the ratio of the first two eigenvalues of Dirichlet Laplacians and extensions, *Ann. of Math.* **135** (1992), 601–628.

[20] M. S. Ashbaugh and R. D. Benguria, More bounds on eigenvalue ratios for Dirichlet Laplacians in $n$ dimensions, *SIAM J. Math. Anal.* **24** (1993), 1622–1651.

[21] T. Aubin, Problèmes isopérimétriques et espaces de Sobolev, *C. R. Acad. Sci. Paris Sér. A-B* **280** (1975), A279–A281.

[22] K. Ball, E. Carlen, and E. H. Lieb, Sharp uniform convexity and smoothness inequalities for trace norms, *Invent. Math.* **115** (1994), 463–482.

[23] S. Banach, Sur les opérations dans les ensembles abstraits et leur application aux équations intégrales, *Fund. Math.* **3** (1922), 133–181.

[24] S. Banach, Sur les fonctionnelles linéaires, I, II, *Studia Math.* **1**, (1929), 211–216, 223–239.

[25] S. Banach, Théorie des opérations linéaires, *Monografje Matematyczne* **1**, Warsaw, 1932.

[26] S. Banach and H. Steinhaus, Sur le principe de la condensation de singularités, *Fund. Math.* **9** (1927), 50–61.

[27] C. Bandle, *Isoperimetric Inequalities and Applications*, Monographs and Studies in Mathematics 7, Pitman, Boston-London, 1980.

[28] H. Bauer, Un problème de Dirichlet pour la frontière de Silov d'un espace compact, *C. R. Acad. Sci. Paris* **247** (1958), 843–846.

[29] H. Bauer, Schilowscher Rand und Dirichletsches Problem, *Ann. Inst. Fourier Grenoble* **11** (1961), 89–136.

[30] W. Beckner, Inequalities in Fourier analysis, *Ann. of Math.* **102** (1975), 159–182.

[31] J. Bendat and S. Sherman, Monotone and convex operator functions, *Trans. Amer. Math. Soc.* **79** (1955), 58–71.

[32] S. Bernstein, Sur l'ordre de la meilleure approximation des fonctions continues par des polynomes de degré donné, *Mém. Acad. Royale Belgique* (2) **4** (1912), 1–104.

[33] S. Bernstein, Leçons sur les propriétes extrémales et la meilleure approximation des fonctions analytiques d'une variable réelle, in *Collection de monographies sur la théorie des fonctions*, pp. 196–197, Gauthier–Villars, Paris, 1926.

[34] S. Bernstein, Sur les fonctions absolument monotones, Acta Math. **52** (1928), 1–66.

[35] J. Bertoin, *Lévy Processes*, Cambridge Tracts in Mathematics 121, Cambridge University Press, Cambridge, 1996.

[36] G. Birkhoff, Three observations on linear algebra, *Univ. Nac. Tucumán, Revista Ser. A* **5** (1946), 147–150. [Spanish]

[37] Z. Birnbaum and W. Orlicz, Über die Verallgemeinerung des Begriffes der zueinander konjugierten Potenzen, *Studia Math.* **3** (1931), 1–67.

[38] E. Bishop and K. de Leeuw, The representations of linear functionals by measures on sets of extreme points, *Ann. Inst. Fourier Grenoble* **9** (1959), 305–331.

[39] D. Blackwell, Equivalent comparisons of experiments, *Ann. Math. Stat.* **24** (1953), 265–272.

[40] W. Blaschke, Eine isoperimetrische Eigenschaft des Kreises, *Math. Z.* **1** (1918), 52–57.

[41] R. P. Boas, Jr., Functions with positive derivatives, *Duke Math. J.* **8** (1941), 163–172.

[42] H. Bohr and E. Landau, Beiträge zur Theorie der Riemannschen Zetafunktion, *Math. Ann.* **74** (1913), 3–30.

[43] T. Bonnesen, Sur une amélioration de l'inégalité isopérimetrique du cercle et la démonstration d'une inégalite de Minkowski, *C. R. Acad. Sci. Paris* **172** (1921), 1087–1089.

[44] C. Borell, Convex measures on locally convex spaces, *Ark. Mat.* **12** (1974), 239–252.

[45] C. Borell, Convex set functions in $d$-space, *Period. Math. Hungar.* **6** (1975), 111–136.

[46] N. Bourbaki, Sur les espaces de Banach, *C. R. Acad. Sci.* **206** (1938), 1701–1704.

[47] N. Bourbaki, Eléments de mathématique, Livre V, Espaces vectoriels topologiques, *Act. Sci. et Ind.* **1189** Herman & Cie, Paris, 1953; **1229** Hermann & Cie, Paris, 1955.

[48] R. D. Bourgin, *Geometric Aspects of Convex Sets with Radon–Nikodým Property*, Lecture Notes in Mathematics 993, Springer, Berlin, 1983.

[49] R. D. Bourgin and G. A. Edgar, Non-compact simplexes in Banach spaces with the Radon–Nikodým property, *J. Funct. Anal.* **23** (1976), 162–176.

[50] H. J. Brascamp and E. H. Lieb, A logarithmic concavity theorem with some applications, unpublished, 1974.

[51] H. J. Brascamp and E. H. Lieb, Some inequalities for Gaussian measures and the long-range order of the one-dimensional plasma, in *Functional Integration and its Applications*, Clarendon Press, Oxford, 1975.

[52] H. J. Brascamp and E. H. Lieb, Best constants in Young's inequality, its converse, and its generalization to more than three functions, *Advances in Math.* **20** (1976), 151–173.

[53] H. J. Brascamp and E. H. Lieb, On extensions of the Brunn–Minkowski and Prékopa–Leindler theorems, including inequalities for log concave functions, and with an application to the diffusion equation, *J. Funct. Anal.* **22** (1976), 366–389.

[54] H. J. Brascamp, E. H. Lieb, and J. M. Luttinger, A general rearrangement inequality for multiple integrals, *J. Funct. Anal.* **17** (1974), 227–237.

[55] O. Bratteli, A. Kishimoto, and D. W. Robinson, Positivity and monotonicity properties of $C_0$-semigroups. I, *Comm. Math. Phys.* **75** (1980), 67–84.

[56] H. Brunn, *Über Ovale und Eiflächen*, Inaug. Diss., München, 1887.

[57] R. S. Bucy and G. Maltese, Extreme positive definite functions and Choquet's representation theorem, *J. Math. Anal. Appl.* **12** (1965), 371–377.

[58] V. Buniakowski, Sur quelques inégalites concernant les intégrales ordinaires et les intégrales aux différences finies, *Mém. Acad. St. Petersburg (7)* **1** (1859), 1–18.

[59] Yu. D. Burago and V. A. Zalgaller, *Geometric Inequalities*, Grundlehren der Mathematischen Wissenschaften 285, Springer Series in Soviet Mathematics, Springer, Berlin, 1988.

[60] A. Burchard, Cases of equality in the Riesz arrangement inequality, *Ann. of Math.* **143** (1996), 499–527.

[61] D. L. Burkholder, Sufficiency in the undominated case, *Ann. Math. Statist.* **32** (1961), 1191–1200.

[62] A.-P. Calderón, Intermediate spaces and interpolation, the complex method, *Studia Math.* **24** (1964), 113–190.

[63] C. Carathéodory, Über den Variabilitätsbereich der Fourier'schen Konstanten von positiven harmonischen Funktionen, *Rend. Circ. Mat. Palermo* **32** (1911), 193–217.

[64] C. Carathéodory and E. Study, Zwei Beweise des Satzes daß der Kreis unter allen Figuren gleichen Umfanges den größten Inhalt hat, *Math. Ann.* **68** (1909), 133–140.

[65] T. Carleman, Über ein Minimalproblem der mathematischen Physik, *Math. Z.* **1** (1918), 208–212.

[66] E. A. Carlen and E. H. Lieb, Short proofs of theorems of Mirsky and Horn on diagonals and eigenvalues of matrices, *Electron. J. Linear Algebra* **18** (2009), 438–441.

[67] R. Carmona, W. Masters, and B. Simon, Relativistic Schrödinger operators: Asymptotic behavior of the eigenfunctions, *J. Funct. Anal.* **91** (1990), 117–142.

[68] N. L. Carothers, *Real Analysis*, Cambridge University Press, Cambridge, 2000.

[69] P. Cartier, J. M. G. Fell, P. A. Meyer, Comparaison des mesures portées par un ensemble convexe compact, *Bull. Soc. Math. France* **92** (1964), 435–445.

[70] A. L. Cauchy, *Cours d'analyse de l'École Royale polytechnique. I. Analyse algébrique*, Debure, Paris, 1821.

[71] N. N. Chan and K. H. Li, Diagonal elements and eigenvalues of a real symmetric matrix, *J. Math. Anal. Appl.* **91** (1983), 562–566.

[72] J. D. Chandler, Jr., Extensions of monotone operator functions, *Proc. Amer. Math. Soc.* **54** (1976), 221–224.

[73] K.-M. Chao and C. S. Wong, Applications of $M$-matrices to majorization, *Linear Algebra Appl.* **169** (1992), 31–40.

[74] I. Chavel, *Isoperimetric Inequalities. Differential Geometric and Analytic Perspectives*, Cambridge Tracts in Mathematics 145, Cambridge University Press, Cambridge, 2001.

[75] G.-S. Cheon and S.-Z. Song, On the extreme points on the majorization polytope $\Omega_3(y \prec x)$, *Linear Algebra Appl.* **269** (1998), 47–52.

[76] G. Chiti, Rearrangements of functions and convergence in Orlicz spaces, *Applicable Anal.* **9** (1979), 23–27.

[77] K. M. Chong, Doubly stochastic operators and rearrangement theorems, *J. Math. Anal. Appl.* **56** (1976), 309–316.

[78] G. Choquet, Theory of capacities, *Ann. Inst. Fourier Grenoble* **5** (1955), 131–295.

[79] G. Choquet, Existence des représentations intégrales au moyen des points extrémaux dans les cônes convexes, *C. R. Acad. Sci. Paris* **243** (1956), 699–702.

[80] G. Choquet, Existence des représentations intégrales dans les cônes convexes, *C. R. Acad. Sci. Paris* **243** (1956), 736–737.

[81] G. Choquet, Unicité des représentations intégrales au moyen des points extrémaux dans les cônes convexes réticulés, *C. R. Acad. Sci. Paris* **243** (1956), 555–557.

[82] G. Choquet, Le théorème de représentation intégrales dans les ensemble convexes compacts, *Ann. Inst. Fourier Grenoble* **10** (1960), 333–344.

[83] G. Choquet, *Lectures on Analysis, Volume 1: Integration and Topological Vector Spaces; Volume II: Representation Theory; Volume III: Infinite Dimensional Measures and Problem Solutions*, W. A. Benjamin, New York-Amsterdam, 1969.

[84] G. Choquet and P.-A. Meyer, Existence et unicité des représentations intégrales dans les convexes compacts quelconques, *Ann. Inst. Fourier Grenoble* **13** (1963), 139–154.

[85] J. A. Clarkson, Uniformly convex spaces, *Trans. Amer. Math. Soc.* **40** (1936), 396–414.

[86] J. E. Cohen, J. H. B. Kemperman, and Gh. Zbăganu, *Comparisons of Stochastic Matrices. With Applications in Information Theory, Statistics, Economics, and Population Sciences*, Birkhäuser, Boston, 1998.

[87] W. A. Coppel, *Foundations of Convex Geometry*, Australian Mathematical Society Lecture Series 12, Cambridge University Press, Cambridge, 1998.

[88] M. G. Crandall and L. Tartar, Some relations between nonexpansive and order preserving mappings, *Proc. Amer. Math. Soc.* **78** (1980), 358–390.

[89] H. Cycon, R. Froese, W. Kirsch and B. Simon, *Schrödinger Operators with Application to Quantum Mechanics and Global Geometry*, Texts and Monographs in Physics. Springer Study Edition, Springer-Verlag, Berlin, 1987; corrected and extended 2nd printing, 2008.

[90] Ju. L. Daleckii and S. G. Krein, Formulas of differentiation according to a parameter of functions of Hermitian operators, *Dokl. Akad. Nauk SSSR* **76** (1951), 13–16 [Russian].

[91] C. Davis, A Schwarz inequality for convex operator functions, *Proc. Amer. Math. Soc.* **8** (1957), 42–44.

[92] C. Davis, *Notions generalizing convexity for functions defined on spaces of matrices*, Proc. Symp. Pure Math. 7, pp. 187–201, American Mathematical Society, Providence, RI, 1963.

[93] M. M. Day, The spaces $L^p$ with $0 < p < 1$, *Bull. Amer. Math. Soc.* **46** (1940), 816–823.

[94] P. W. Day, Rearrangement inequalities, *Canad. J. Math.* **24** (1972), 930–943.

[95] P. W. Day, Decreasing rearrangements and doubly stochastic operators, *Trans. Amer. Math. Soc.* **178** (1973), 383–392.

[96] L. de Branges, The Stone–Weierstrass theorem, *Proc. Amer. Math. Soc.* **10** (1959), 822–824.

[97] S. A. Denisov and A. Kiselev, Spectral properties of Schrödinger operators with decaying potentials, in *Spectral Theory and Mathematical Physics: A Festschrift in Honor of Barry Simon's 60th birthday*, pp. 565–589, Proc. Sympos. Pure Math. 76.2, American Mathematical Society, Providence, RI, 2007.

[98] J. Dieudonné, Sur le théorème de Hahn–Banach, *Revue Sci.* **79** (1941), 642–643.

[99] J. Dieudonné, La dualité dans les espaces vectoriels topologiques, *Ann. Sci. École Norm. Sup.* (3) **59** (1942), 107–139.

[100] O. Dobsch, Matrixfunktionen beschränkter Schwankung, *Math. Z.* **43** (1938), 353–388.

[101] W. F. Donoghue, Jr., *Monotone Matrix Functions and Analytic Continuation*, Die Grundlehren der mathematischen Wissenschaften 207, Springer, New York-Heidelberg-Berlin, 1974.

[102] W. F. Donoghue, Jr., Monotone operator functions on arbitrary sets, *Proc. Amer. Math. Soc.* **78** (1980), 93–96.

[103] W. F. Donoghue, Jr., Another extension of Loewner's theorem, *J. Math. Anal. Appl.* **110** (1985), 323–326.

[104] T. Downarowicz, The Choquet simplex of invariant measures for minimal flows, *Israel J. Math.* **74** (1991), 241–256.

[105] G. F. D. Duff, A general integral inequality for the derivative of an equimeasurable rearrangement, *Canad. J. Math.* **28** (1976), 793–804.

[106] P. L. Duren, *Theory of $H^p$ Spaces*, Pure and Applied Mathematics 38, Academic Press, New York-London 1970.

[107] P. L. Duren, B. W. Romberg, and A. L. Shields, Linear functionals on $H^p$ spaces with $0 < p < 1$, *J. Reine Angew. Math.* **238** (1969), 32–60.
[108] M. L. Eaton and M. D. Perlman, Reflection groups, generalized Schur functions, and the geometry of majorization, *Ann. Probability* **5** (1977), 829–860.
[109] G. A. Edgar, A non-compact Choquet theorem, *Proc. Amer. Math. Soc.* **49** (1975), 354–358.
[110] D. A. Edwards, Gustave Choquet, 1915–2006, *Bull. London Math. Soc.* **42** (2010), 341–370.
[111] H. G. Eggleston, *Convexity*, Cambridge Tracts in Mathematics and Mathematical Physics 47, Cambridge University Press, New York, 1958.
[112] C. Faber, Beweiss, dass unter allen homogenen Membranen von gleicher Fläche und gleicher Spannung die kreisförmige die tiefsten Grundton gibt, *Sitzungsber. Bayer. Akad. Wiss. München, Math.-Phys. Kl.* (1923), 169–172.
[113] H. Federer, *Geometric Measure Theory*, Die Grundlehren der mathematischen Wissenschaften 153, Springer, New York, 1969.
[114] W. Fenchel, On conjugate convex functions, *Canad. J. Math.* **1** (1949), 73–77.
[115] V. P. Fonf, J. Lindenstrauss, and R. R. Phelps, Infinite dimensional convexity, in *Handbook of the Geometry of Banach Spaces, Vol. I*, pp. 599–670, North-Holland, Amsterdam, 2003.
[116] M. Fréchet, Sur quelques points du calcul fonctionnel, *Rend. Circ. Mat. Palermo* **22** (1906), 1–74.
[117] K. Friedrichs, On differential operators in Hilbert spaces, *Amer. J. Math.* **61** (1939), 523–544.
[118] H. Freudenthal, Teilweise geordnete Moduln, *Nederl. Akad. Wetensch. Proc.* **39** (1936), 641–651.
[119] L. Fuchs, A new proof of an inequality of Hardy–Littlewood–Pólya, *Mat. Tidsskr. B.* **1947** (1947), 53–54.
[120] B. Fuchssteiner and W. Lusky, *Convex Cones*, North-Holland Mathematics Studies 56; Notas de Matemática [Mathematical Notes], 82, North-Holland, Amsterdam-New York, 1981.
[121] L. Gårding, *Some Points of Analysis and Their History*, University Lecture Series 11, American Mathematical Society, Providence, RI; Higher Education Press, Beijing, 1997.
[122] I. M. Gel'fand and G. E. Shilov, *Generalized Functions. Volume 1: Properties and Operations*, Academic Press, New York-London, 1964.
[123] I. M. Gel'fand and G. E. Shilov, *Generalized Functions. Volume 2: Spaces of Fundamental and Generalized Functions*, Academic Press, New York-London, 1968.
[124] I. M. Gel'fand and G. E. Shilov, *Generalized Functions. Volume 3: Theory of Differential Equations*, Academic Press, New York-London, 1967.
[125] I. M. Gel'fand and N. Ya. Vilenkin, *Generalized Functions. Volume 4: Applications of Harmonic Analysis*, Academic Press, New York-London, 1964.
[126] I. M. Gel'fand, M. I. Graev, and N. Ya. Vilenkin, *Generalized Functions. Volume 5: Integral Geometry and Representation Theory*, Academic Press, New York-London, 1966.
[127] S. A. Gheorghiu, Sur l'équation de Fredholm, Thesis, Paris, 1928.
[128] J. W. Gibbs, Graphical methods in the thermodynamics of fluids, *Trans. Connecticut Acad.* **2** (1873), 309–342.

[129] J. W. Gibbs, A method of geometrical representation of the thermodynamic properties of substances by means of surfaces, *Trans. Connecticut Acad.* **2** (1873), 382–404.

[130] J. W. Gibbs, On the equilibrium of heterogeneous substances, *Trans. Connecticut Acad.* **3** (1875–1876), 108–248; (1877–1878), 343–524.

[131] I. C. Gohberg and M. G. Krein, *Introduction to the Theory of Linear Non-Selfadjoint Operators*, Transl. Math. Monographs 18, American Mathematical Society, Providence, RI, 1969.

[132] S. Golden, Lower bounds for the Helmholtz function, *Phys. Rev.* (2) **137** (1965), B1127–B1128.

[133] F. P. Greenleaf, *Invariant Means on Topological Groups and Their Applications*, Van Nostrand Math. Studies 16, Van Nostrand Reinhold, New York-Toronto-London, 1969.

[134] J. Grolous, Un théorème sur les fonctions, *Inst. (2)* **3** (1875), 401.

[135] L. Gross, Measurable functions on Hilbert space, *Trans. Amer. Math. Soc.* **105** (1962), 372–390.

[136] J. Hadamard, Étude sur les propriétés des fonctions entières et en particulier d'une fonction considérée par Riemann, *J. Math. Pures Appl.* (4) **9** (1893), 171–215.

[137] J. Hadamard, Résolution d'une question relativ aux déterminants, *Bull. Sci. Math.* (2) **17** (1893), 240–246.

[138] H. Hadwiger and D. Ohman, Brunn–Minkowskischer Satz und Isoperimetrie, *Math. Z.* **66** (1956), 1–8.

[139] H. Hahn, Über die Darstellung gegebener Funktionen durch singuläre Integrale, II, *Denkschriften der K. Akad. Wien. Math.-Naturwiss. Kl.* **93** (1916), 657–692.

[140] H. Hahn, Über Folgen linearer Operationen, *Monatsh. f. Math.* **32** (1922), 3–88.

[141] H. Hahn, Über lineare Gleichungssysteme in linearen Räumen, *J. Reine Angew. Math.* **157** (1927), 214–229.

[142] P. R. Halmos, The range of a vector measure, *Bull. Amer. Math. Soc.* **54** (1948), 416–421.

[143] P. R. Halmos, *Measure Theory*, Van Nostrand, New York, 1950.

[144] O. Hanner, On the uniform convexity of $L^p$ and $l^p$, *Ark. Mat.* **3** (1956), 239–244.

[145] F. Hansen and G. K. Pedersen, Jensen's inequality for operators and Löwner's theorem, *Math. Ann.* **258** (1981/1982), 229–241.

[146] G. H. Hardy and J. E. Littlewood, Some properties of fractional integrals. I, *Math. Z.* **27** (1928), 565–606.

[147] G. H. Hardy and J. E. Littlewood, Notes on the theory of series (XII): On certain inequalities connected with the calculus of variations, *J. London Math. Soc.* **5** (1930), 34–39.

[148] G. H. Hardy, J. E. Littlewood, and G. Pólya, Some simple inequalities satisfied by convex functions, *Messenger of Math.* **58** (1929), 145–152.

[149] G. H. Hardy, J. E. Littlewood, and G. Pólya, *Inequalities*, Cambridge University Press, Cambridge, 1934.

[150] F. Hausdorff, *Grundzüge der Mengenlehre*, Veit, Leipzig, 1914.

[151] R. Haydon, A new proof that every Polish space is the extreme boundary of a simplex, *Bull. London Math. Soc.* **7** (1975), 97–100.

[152] E. Heinz, Beiträge zur Störunstheorie der Spektralzerlegung, *Math. Ann.* **123** (1951), 415–438.

[153] E. Hellinger and O. Toeplitz, Grundlagen für eine Theorie der unendlichen Matrizen, *Math. Ann.* **69** (1910), 289–330.

[154] E. Helly, Über lineare Funktionaloperationen, *S.-B. K. Akad. Wiss. Wien Math.-Naturwiss. Kl.* **121** (1912), 265–297.

[155] E. Helly, Über Systeme linearer Gleichungen mit unendlich vielen Unbekannten, *Monatsh. Math. Phys.* **31** (1921), 60–91.

[156] R. Henderson, Moral values, *Bull. Amer. Math. Soc.* **2** (1895), 46–51.

[157] I. W. Herbst, Spectral theory of the operator $(p^2 + m^2)^{1/2} - Ze^2/r$, *Comm. Math. Phys.* **53** (1977), 285–294.

[158] Ch. Hermite, Sur deux limites d'une intégrale dé finie, *Mathesis* **3** (1883), 82.

[159] J. Hersch, Isoperimetric monotonicity: Some properties and conjectures (connections between isoperimetric inequalities), *SIAM Rev.* **30** (1988), 551–577.

[160] M. Hervé, Sur les représentations intégrales à l'aide des points extrémaux dans un ensemble compact convexe métrisable, *C. R. Acad. Sci. Paris* **253** (1961) 366–368.

[161] T. H. Hildebrandt, On uniform limitedness of sets of functional operations, *Bull. Amer. Math. Soc.* **29** (1923), 309–315.

[162] E. Hille and J. D. Tamarkin, On the characteristic values of linear integral equations, *Acta Math.* **57** (1931), 1–76.

[163] A. J. Hoffman and H. W. Wielandt, The variation of the spectrum of a normal matrix, *Duke Math. J.* **20** (1953), 37–40.

[164] O. Hölder, Über einen Mittelwertsatz, *Nachr. Akad. Wiss. Göttingen. Math.-Phys. Kl.* (1889), 38–47.

[165] F. Holland, The extreme points of a class of functions with positive real part, *Math. Ann.* **202** (1973), 85–87.

[166] L. Hörmander, *Notions of Convexity*, Progress in Mathematics 127, Birkhäuser Boston, Boston, 1994.

[167] A. Horn, On the singular values of a product of completely continuous operators, *Proc. Nat. Acad. Sci. USA* **36** (1950), 374–375.

[168] A. Horn, Doubly stochastic matrices and the diagonal of a rotation matrix, *Amer. J. Math.* **76** (1954), 620–630.

[169] A. Horn, On the eigenvalues of a matrix with prescribed singular values, *Proc. Amer. Math. Soc.* **5** (1954), 4–7.

[170] R. A. Horn, The Hadamard product, in *Matrix Theory and Applications*, *Proc. Sympos. Appl. Math.* 40, pp. 87–169, American Mathematical Society, Providence, RI, 1990.

[171] R. A. Horn and C. R. Johnson, *Matrix Analysis*, corrected reprint of the 1985 original, Cambridge University Press, Cambridge, 1990.

[172] A. Horsley and A. Wrobel, The extreme points of some convex sets in the theory of majorization, *Nederl. Akad. Wetensch. Indag. Math.* **49** (1987), 171–176.

[173] D. Hundertmark and B. Simon, An optimal $L^p$-bound on the Krein spectral shift function, *J. Anal. Math.* **87** (2002), 199–208.

[174] A. Hurwitz, Sur le problème des isopérimétres, *C. R. Acad. Sci. Paris* **132** (1901), 401–403.

[175] R. B. Israel, *Convexity in the Theory of Lattice Gases*, Princeton Series in Physics, Princeton University Press, Princeton, NJ, 1979.

[176] R. B. Israel and R. R. Phelps, Some convexity questions arising in statistical mechanics, *Math. Scand.* **54** (1984), 133–156.

[177] J. L. Jensen, Sur les fonctions convexes et les inégalités entre les valeurs moyennes, *Acta Math.* **30** (1906), 175–193.

[178] S. Johansen, An application of extreme point methods to the representation of infinitely divisible distribution, *Z. Wahr. Verw.* Gebiete **5** (1966), 304–316.

[179] R. V. Kadison, A representation theory for commutative topological algebra, *Mem. Amer. Math. Soc.* **7** (1951).

[180] S. Kakutani, Weak topologies and regularity of Banach spaces, *Proc. Imp. Acad. Tokyo* **15** (1939), 169–173.

[181] S. Kakutani, Weak topology, bicompact set and the principle of duality, *Proc. Imp. Acad. Tokyo* **16** (1940), 169–173.

[182] S. Kakutani, Concrete representation of abstract $L$-spaces and the mean ergodic theorem, *Ann. of Math.* **42** (1941), 63–67.

[183] S. Kakutani, Concrete representation of abstract $M$-spaces. (A characterization of the space of continuous functions), *Ann. of Math. (2)* **42** (1941), 994–1024.

[184] L. V. Kantorovitch, Lineare halbgeordnete Räume, *Receuil Math. Moscow* **2** (1937), 121–168.

[185] L. V. Kantorovitch, Linear operators in semi-ordered spaces, *Mat. Sbornik* **7 (49)** (1940), 209–284.

[186] J. Karamata, Sur une inégalité rélative aux fonctions convexes, *Publ. Math. Univ. Belgrade* **1** (1932), 145–148.

[187] J. Kästner and C. Zylka, An application of majorization in the theory of dynamical systems, in *Nonlinear Dynamics and Quantum Dynamical Systems* (Gaussig, 1990), Math. Res. 59, pp. 42–51, Akademie-Verlag, Berlin, 1990.

[188] T. Kato, Notes on some inequalities for linear operators, *Math. Ann.* **125** (1952), 208–212.

[189] T. Kato, *Perturbation Theory for Linear Operators*, 2nd edition, Grundlehren der Mathematischen Wissenschaften 132, Springer, Berlin-New York, 1976.

[190] J. L. Kelley, Note on a theorem of Krein and Milman, *J. Osaka Inst. Sci. Tech. Part I*, **3** (1951), 1–2.

[191] J. L. Kelley, *General Topology*, reprint of the 1955 edition, Graduate Texts in Mathematics 27, Springer-Verlag, New York-Berlin, 1975.

[192] D. G. Kendall, Extreme-point methods in stochastic analysis, *Z. Wahr. Verw. Gebiete* **1** (1962/1963), 295–300.

[193] R. Killip and B. Simon, Sum rules for Jacobi matrices and their applications to spectral theory, *Ann. of Math.* **158** (2003), 253–321.

[194] J. Kindler, A simple proof of Sion's minimax theorem, *Amer. Math. Monthly* **112** (2005), 356–358.

[195] V. L. Klee, Convex sets in linear spaces, *Duke Math. J.* **18** (1951), 443–466.

[196] V. L. Klee, Convex sets in linear spaces, II, *Duke Math. J.* **18** (1951), 875–883.

[197] V. L. Klee, Convex sets in linear spaces, III, *Duke Math. J.* **20** (1953), 105–111.

[198] V. L. Klee, Separation properties of convex cones, *Proc. Amer. Math. Soc.* **6** (1955), 313–318.

[199] V. L. Klee, Strict separation of convex sets, *Proc. Amer. Math. Soc.* **7** (1956), 735–737.

[200] V. L. Klee, Maximal separation theorems for convex sets, *Trans. Amer. Math. Soc.* **134** (1968), 133–147.

[201] A. N. Kolmogorov, *Zur Normierbarkeit eines allgemeinen topologischen linearen Raumes*, Studia Math. **5** (1934), 29–33.

[202] A. N. Kolmogorov, A new metric invariant of transient dynamical systems and automorphisms in Lebesgue spaces, *Dokl. Akad. Nauk SSSR (N.S.)* **119** (1958), 861–864 [Russian].

[203] A. N. Kolmogorov, Enropy per unit time as a metric invariant of automorphisms, *Dokl. Akad. Nauk SSSR* **124** (1959), 754–755 [Russian].

[204] A. Korányi, On a theorem of Löwner and its connections with resolvents of self-adjoint transformations, *Acta Sci. Math. Szeged* **17** (1956), 63–70.

[205] E. T. Kornhauser and I. Stakgold, A variational theorem for $\nabla^2 u + \lambda u = 0$ and its application, *J. Math. Phys.* **31** (1952), 45–54.

[206] G. Köthe, *Topological Vector Spaces I*, Die Grundlehren der mathematischen Wissenschaften 159, Springer-Verlag, New York, 1969.

[207] G. Köthe and O. Toeplitz, Lineare Räume mit unendlich vielen Koordinaten und Ringe unendlicher Matrizen, *J. Reine Angew. Math.* **171** (1934), 193–226.

[208] E. Krahn, Über eine von Rayleigh formulierte Minimaleigenschafte des Kreises, *Math. Ann.* **94** (1925), 97–100.

[209] S. G. Krantz, Convexity in complex analysis. Several complex variables and complex geometry, Part 1, (Santa Cruz, CA, 1989), *Proc. Sympos. Pure Math.* 52.1, pp. 119–137, American Mathematical Society. Providence, RI, 1991.

[210] M. Krasnosel'skiĭ and Ya. Rutickiĭ, On linear functionals in Orlicz spaces, *Dokl. Akad. Nauk SSSR (N.S.)* **97** (1954), 581–584 [Russian].

[211] M. Krasnosel'skiĭ and Ya. Rutickiĭ, *Convex Functions and Orlicz Spaces*, P. Noordhoff, Groningen, 1961.

[212] M. Krein and S. Krein, On an inner characteristic of the set of all continuous functions defined on a bicompact Hausdorff space, *C. R. (Doklady) Acad. Sci. URSS (N.S.)* **27** (1940), 427–430.

[213] M. Krein and D. Milman, On extreme points of regularly convex sets, *Studia Math.* **9** (1940), 133–138.

[214] F. Kraus, Über konvexe Matrixfunktionen, *Math. Z.* **41** (1936), 18–42.

[215] H. W. Kuhn and A. W. Tucker, Nonlinear programming, in *Proc. Second Berkeley Symp. on Mathematical Statistics and Probability*, pp. 481–492, University of California Press, Berkeley and Los Angeles, 1951.

[216] S. Kullback and R. A. Leibler, On information and sufficiency, *Ann. Math. Statistics* **22** (1951), 79–86.

[217] T. Lalesco, Une thèoréme sur les noyaux composés, *Bull. Soc. Sci. Acad. Roumania* **3** (1914–1915), 271–272.

[218] E. Landau, Über einen Konvergensatz, *Nachr. Akad. Wiss. Göttingen. Math.-Phys. Kl.* (1907), 25–27.

[219] M. Landsberg, Lineare topologische Räume, die nicht lokalkonvex sind, *Math. Z.* **65** (1956), 104–112.

[220] O. E. Lanford and D. W. Robinson, Statistical mechanics of quantum spin systems, III, *Comm. Math. Phys.* **9** (1968), 327–338.

[221] S. R. Lay, *Convex Sets and Their Applications*, revised reprint of the 1982 original, Robert E. Krieger Publishing, Malabar, FL, 1992.

[222] H. Lebesgue, Sur les intégrales singulières, *Ann. Fac. Sci. Toulouse Sci. Math. Sci. Phys. (3)* **1** (1909), 25–117.

[223] L. Leindler, On a certain converse of Hölder's inequality, II, *Acta Sci. Math. (Szeged)* **33** (1972), 217–223.

[224] J. Leray, Sur le mouvement d'un fluide visqueux emplissant l'espace, *Acta Math.* **63** (1934), 193–248.

[225] L. Lichtenstein, Über eine isoperimetrische Aufgabe der mathematischen Physik, *Math. Z.* **3** (1919), 8–28.

[226] E. H. Lieb, Convex trace functions and the Wigner–Yanase–Dyson conjecture, *Advances in Math.* **11** (1973), 267–288.

[227] E. H. Lieb, Inequalities for some operator and matrix functions, *Advances in Math.* **20** (1976), 174–178.

[228] E. H. Lieb, Existence and uniqueness of the minimizing solution of Choquard's nonlinear equation, *Studies in Appl. Math.* **57** (1976/1977), 93–105.

[229] E. H. Lieb, Sharp constants in the Hardy–Littlewood–Sobolev and related inequalities, *Ann. of Math.* **118** (1983), 349–374.

[230] E. H. Lieb, Gaussian kernels have only Gaussian maximizers, *Invent. Math.* **102** (1990), 179–208.

[231] E. H. Lieb and M. Loss, *Analysis*, Graduate Studies in Math. 14, American Mathematical Society, Providence, RI, 1997.

[232] E. H. Lieb and J. Yngvason, The physics and mathematics of the second law of thermodynamics, *Phys. Rep.* **310** (1999), 1–96.

[233] E. H. Lieb and J. Yngvason, A fresh look at entropy and the second law of thermodynamics, *Physics Today* **53** (2000), 32–37.

[234] J. Lindenstrauss, G. H. Olsen, and Y. Sternfeld, The Poulsen simplex, *Ann. Inst. Fourier Grenoble* **28** (1978), 91–114.

[235] J. Lindenstrauss and L. Tzafriri, *Classical Banach Spaces I: Sequence Spaces*, Ergebnisse der Mathematik und ihrer Grenzgebiete 92, Springer-Verlag, Berlin-New York, 1977.

[236] J. Lions, Théorèmes de trace et d'interpolation, I, *Ann. Scuola Norm. Sup. Pisa* (3) **13** (1959), 389–403.

[237] J. Lions, Théorèmes de trace et d'interpolation, II, *Ann. Scuola Norm. Sup. Pisa* (3) **15** (1960), 317–331.

[238] J. Lions, Théorèmes de trace et d'interpolation, III, *J. Math. Pures Appl.* (9) **42** (1963), 195–203.

[239] A. E. Livingston, The space $H^p$, $0 < p < 1$, is not normable, *Pacific J. Math.* **3** (1953), 613–616.

[240] K. Loewner, Über monotone Matrixfunktionen, *Math. Z.* **38** (1934), 177–216.

[241] Z. Lonc, Majorization, packing, covering and matroids, Graph Theory (Niedzica Castle, 1990), *Discrete Math.* **121** (1993), 151–157.

[242] L. H. Loomis, Unique direct integral decompositions on convex sets, *Amer. J. Math.* **84** (1962), 509–526.

[243] G. G. Lorentz, *Bernstein polynomials*, Mathematical Expositions, no. 8, University of Toronto Press, Toronto, 1953.

[244] L. A. Lusternik, Die Brunn–Minkowskische Ungleichung für beliebige messbare Mengen, *C. R. (Doklady) Acad. Sci.* URSS **8** (1935), 55–58.

[245] J. M. Luttinger, Generalized isoperimetric inequalities, I, II, III, *J. Math. Phys.* **14** (1973), 586–593, 1444–1447, 1448–1450.

[246] J. M. Luttinger and R. Friedberg, A new rearrangement inequality for multiple integrals, *Arch. Rational Mech. Anal.* **61** (1976), 45–64.

[247] W. A. J. Luxemburg, Banach Function Spaces, Thesis, Technische Hogeschool te Delft, 1955.

[248] W. A. J. Luxemburg, Rearrangement invariant Banach function spaces, *Queen's Paper in Pure and Applied Math.* **10** (1967), 83–144.

[249] W. A. J. Luxemburg and A. C. Zaanen, *Riesz Spaces, Vol. I*, North-Holland Mathematical Library, North-Holland, Amsterdam-London; American Elsevier Publishing, New York, 1971.

[250] A. A. Lyapunov, On completely additive vector functions, *Izv. Akad. Nauk SSSR* **4** (1940), 465–478

[251] G. W. Mackey, On infinite-dimensional linear spaces, *Trans. Amer. Math. Soc.* **57** (1945), 155–207.

[252] G. W. Mackey, On convex topological linear spaces, *Trans. Amer. Math. Soc.* **60** (1946), 519–537.

[253] S. Mandelbrojt, Sur les fonctions convexes, *C. R. Acad. Sci. Paris* **209** (1939), 977–978.

[254] S. Marcus, Atomic measures and Darboux property, *Rev. Math. Pures Appl.* **7** (1962), 327–332.

[255] A. S. Markus, Characteristic numbers and singular numbers of the sum and product of linear operators, *Soviet Math. Dokl.* **3** (1962), 104–108; Russian original in Dokl. Akad. Nauk SSSR **146** (1962), 34–36.

[256] A. W. Marshall and I. Olkin, *Inequalities: Theory of Majorization and Its Applications*, Mathematics in Science and Engineering 143, Academic Press, New York-London, 1979.

[257] S. Mazur, Über konvexe Mengen in linearen normierten Räumen, *Studia Math.* **4** (1933), 70–84.

[258] S. Mazur and W. Orlicz, Sur les espaces métriques linéaires, I, *Studia Math.* **10** (1948), 184–208.

[259] C. A. McCarthy, $c_p$, *Israel J. Math.* **5** (1967), 249–271.

[260] D. P. Milman, On some criteria for the regularity of spaces of type (B), *C. R. (Doklady) Acad. Sci.* URSS **20** (1938), 243–246.

[261] D. P. Milman, Characteristics of extremal points of regularly convex sets, *Dokl. Akad. Nauk SSSR* **57** (1947), 119–122 [Russian].

[262] H. Minkowski, *Theorie der Konvexen Körper, insbesondere Begründung ihres Oberflächenbegriffs*, Gessamelte Abhandlungen 2, Leipzig, 1911.

[263] L. Mirsky, Matrices with prescribed characteristic roots and diagonal elements, *J. London Math. Soc.* **33** (1958), 14–21.

[264] L. Mirsky, Remarks on an existence theorem in matrix theory due to A. Horn, *Monatsh. Math.* **63** (1959), 241–243.

[265] L. Mirsky, Results and problems in the theory of doubly-stochastic matrices, *Z. Wahr. Verw. Gebiete* **1** (1962/1963) 319–334.

[266] B. S. Mitjagin, Normed ideals of intermediate type, *Amer. Math. Soc. Transl.* **63** (1967), 180–194.

[267] G. Mokobodzki, Balayage défini par une cône convexe de fonctions numériques sur un espace compact, *C. R. Acad. Sci. Paris* **254** (1962), 803–805.

[268] G. Mokobodzki and D. Sibony, Cônes de fonctions continues, *C. R. Acad. Sci. Paris Sér. A-B* **264** (1967), A15–A18.

[269] P. Montel, *Leçons sur les familles normales de fonctions analytiques et leurs applications*, Gauthier-Villars, Paris, 1927.

[270] R. F. Muirhead, Some methods applicable to identities and inequalities of symmetric algebraic functions of $n$ letters, *Proc. Edinburgh Math. Soc.* **21** (1902), 144–162.

[271] C. P. Niculescu and L.-E. Persson, Old and new on the Hermite–Hadamard inequality, *Real Anal. Exchange* **29** (2003/2004), 663–685.

[272] C. P. Niculescu and L.-E. Persson, *Convex Functions and Their Applications. A Contemporary Approach*, CMS Books in Mathematics/Ouvrages de Mathématiques de la SMC 23, Springer, New York, 2006.

[273] M. Niezgoda, Group majorization and Schur type inequalities, *Linear Algebra Appl.* **268** (1998), 9–30.

[274] M. Niezgoda, On Schur–Ostrowski type theorems for group majorizations, *J. Convex Anal.* **5** (1998), 81–105.

[275] N. Nörlund, *Vorlesungen über Differenzenrechnung*, Springer, Berlin, 1924.

[276] M. Ohya and D. Petz, *Quantum Entropy and Its Use*, Texts and Monographs in Physics, Springer-Verlag, Berlin, 1993.

[277] W. Orlicz, Über eine gewisse Klasse von Räumen vom Typus B, *Bull. Intern. Acad. Pol. Ser.* A **8/9** (1932), 207–220.

[278] W. Orlicz, Über Räume $(L^M)$, *Bull. Intern. Acad. Pol. Ser. A* (1936), 93–107.

[279] R. Osserman, The isoperimetric inequality, *Bull. Amer. Math. Soc.* **84** (1978), 1182–1238.

[280] A. M. Ostrowski, Sur quelques applications des fonctions convexes et concaves au sens de I. Schur, *J. Math. Pures Appl. (9)* **31** (1952), 253–292.

[281] J. C. Oxtoby, *Measure and Category. A Survey of the Analogies Between Topological and Measure Spaces*, 2nd edition, Graduate Texts in Mathematics 2, Springer-Verlag, New York-Berlin, 1980.

[282] L. E. Payne, Isoperimetric inequalities and their applications, *SIAM Rev.* **9** (1967), 453–488.

[283] L. E. Payne, Some comments on the past fifty years of isoperimetric inequalities, in *Inequalities: Fifty Years On from Hardy, Littlewood, and Pólya*, pp. 143–162, Lecture Notes in Pure and Appl. Math. 129, Marcel Dekker, New York, 1987.

[284] L. E. Payne, G. Pólya, and H. F. Weinberger, Sur le quotient de deux fréquences propres consécutives, *C. R. Acad. Sci. Paris* **241** (1955), 917–919.

[285] L. E. Payne, G. Pólya, and H. F. Weinberger, On the ratio of consecutive eigenvalues, *J. Math. Phys.* **35** (1956), 289–298.

[286] J. E. Pečarić, On the Fuchs's generalization of the majorization theorem, *Bul. Ştiinţ. Tehn. Inst. Politehn. "Traian Vuia" Timişoara* **25(39)** (1980), 10–11 (1981).

[287] J. E. Pečarić, F. Proschan, and Y. L. Tong, *Convex Functions, Partial Orderings, and Statistical Applications*, Mathematics in Science and Engineering 187, Academic Press, Boston, 1992.

[288] B. J. Pettis, A proof that every uniformly convex space is reflexive, *Duke Math. J.* **5** (1939), 249–253.

[289] R. R. Phelps, Integral representations for elements of convex sets, in *Studies in Functional Analysis*, pp. 115–157, MAA Stud. Math. 21, Mathematical Association of America, Washington, DC, 1980.

[290] R. R. Phelps, *Lectures on Choquet's Theorem*, 2nd edition, Lecture Notes in Math. 1757, Springer-Verlag, Berlin, 2001.

[291] R. S. Phillips, Integration in a convex linear topological space, *Trans. Amer. Math. Soc.* **47** (1940), 114–145.

[292] H. Poincaré, *Figures d'équilibre d'une masse fluide*, Gauthier-Villars, Paris, 1902.

[293] G. Pólya, Torsional rigidity, principal frequency, electrostatic capacity and symmetrization, *Quart. Appl. Math.* **6** (1948), 267–277.

[294] G. Pólya, Remark on Weyl's note "Inequalities between the two kinds of eigenvalues of a linear transformation", *Proc. Nat. Acad. Sci. USA* **36** (1950), 49–51.

[295] G. Pólya and G. Szegő, Inequalities for the capacity of a condenser, *Amer. J. Math.* **67** (1945), 1–32.

[296] G. Pólya and G. Szegő, *Isoperimetric Inequalities in Mathematical Physics*, Annals of Mathematics Studies 27, Princeton University Press, Princeton, NJ, 1951.

[297] E. T. Poulsen, A simplex with dense extreme boundary, *Ann. Inst. Fourier Grenoble* **11** (1961), 83–87.

[298] A. Prékopa, Logarithmic concave measures with application to stochastic programming, *Acta Sci. Math. (Szeged)* **32** (1971), 301–316.

[299] A. Prékopa, On logarithmic concave measures and functions, *Acta Sci. Math. (Szeged)* **34** (1973), 335–343.

[300] F. Proschan, *Applications of majorization and Schur functions in reliability and life testing*, Reliability and Fault Tree Analysis (Berkeley, CA, 1974), pp. 237–258, Soc. Indust. Appl. Math., Philadelphia, 1975.

[301] R. Rado, An inequality, *J. London Math. Soc.* **27** (1952), 1–6.

[302] (Lord) J. S. W. Rayleigh, *The Theory of Sound*, Macmillan, New York, 1877. Reprinted: Dover Publications, New York, 1945.

[303] M. Reed and B. Simon, *Methods of Modern Mathematical Physics, I. Functional Analysis*, Academic Press, New York-London, 1972.

[304] M. Reed and B. Simon, *Methods of Modern Mathematical Physics, II. Fourier Analysis, Self-Adjointness*, Academic Press, New York, 1975.

[305] M. Reed and B. Simon, *Methods of Modern Mathematical Physics, IV. Analysis of Operators*, Academic Press, New York-London, 1978.

[306] F. Riesz, Untersuchungen über Systeme integrierbarer Funktionen, *Math. Ann.* **69** (1910), 449–497.

[307] F. Riesz, *Les systèmes d'équations linéaires à une infinité d'inconnues*, Gauthier-Villars, Paris, 1913.

[308] F. Riesz, Über lineare Funktionalgleichungen, *Acta Math.* **41** (1916), 71–98.

[309] F. Riesz, Sur une inégalité intégrale, *J. London Math. Soc.* **5** (1930), 162–168.

[310] F. Riesz, Sur quelques notions fondamentales dans la théorie générale des opérations linéaires, *Ann. of Math.* (2) **41** (1940), 174–206.

[311] M. Riesz, Sur le problème des moments. I, II, III, *Ark. Mat. Astr. Fys.* **16** (1922) no. 12, no. 19; **17** (1923), no. 16.

[312] M. Riesz, Sur les maxima des fonctions bilinéaires et sur les fonctionelles linéaires, *Acta Math.* **49** (1927), 465–497.

[313] J. R. Ringrose, A note on uniformly convex spaces, *J. London Math. Soc.* **34** (1959), 92.

[314] Y. Rinott, On convexity of measures, *Ann. Probability* **4** (1976), 1020–1026.

[315] A. W. Roberts and D. E. Varberg, *Convex Functions*, Pure and Applied Mathematics 57, Academic Press, New York-London, 1973.

[316] J. W. Roberts, A compact convex set with no extreme points, *Studia Math.* **60** (1977), 255–266.

[317] W. J. R. Robertson, Contributions to the General Theory of Linear Topological Spaces, Thesis, Cambridge, 1954.

[318] D. W. Robinson and D. Ruelle, Mean entropy of states in classical statistical mechanics, *Comm. Math. Phys.* **5** (1967), 288–300.

[319] R. T. Rockafellar, *Convex Analysis*, Princeton Mathematical Series 28, Princeton University Press, Princeton, NJ 1970; reprinted 1997.

[320] J. Rosen, Sobolev inequalities for weight spaces and supercontractivity, *Trans. Amer. Math. Soc.* **222** (1976), 367–376.

[321] M. Rosenblum and J. Rovnyak, Two theorems on finite Hilbert transforms, *J. Math. Anal. Appl.* **48** (1974), 708–720.

[322] M. Rosenblum and J. Rovnyak, An operator-theoretic approach to theorems of the Pick–Nevanlinna and Loewner types, I, *Integral Equations Operator Theory* **3** (1980), 408–436.

[323] G.-C. Rota, An "Alternierende Verfahren" for general positive operators, *Bull. Amer. Math. Soc.* **68** (1962), 95–102.

[324] D. Ruelle, *Statistical Mechanics: Rigorous Results*, W. A. Benjamin, New York-Amsterdam, 1969.

[325] D. Ruelle, Thermodynamic formalism. The mathematical structures of classical equilibrium statistical mechanics, in *Encyclopedia of Mathematics and its Applications* 5, Addison–Wesley, Reading, MA, 1978.

[326] J. V. Ryff, On the representation of doubly stochastic operators, *Pacific J. Math.* **13** (1963), 1379–1386.

[327] J. V. Ryff, Orbits of $L^1$ functions under doubly stochastic transformations, *Trans. Amer. Math. Soc.* **117** (1965), 92–100.

[328] J. V. Ryff, Extreme points of some convex subsets of $L^1(0,1)$, *Proc. Amer. Math. Soc.* **18** (1967), 1026–1034.

[329] J. V. Ryff, On Muirhead's theorem, *Pacific J. Math.* **21** (1967), 567–576.

[330] J. V. Ryff, Majorized functions and measures, *Indag. Math.* **30** (1968), 431–437.

[331] J. V. Ryff, Measure preserving transformations and rearrangements, *J. Math. Anal. Appl.* **31** (1970), 449–458.

[332] Yu. Safarov, Birkhoff's theorem and multidimensional numerical range, *J. Funct. Anal.* **222** (2005), 61–97.

[333] J. Saint Raymond, Représentation intégrale dans certains convexes, *Séminaire Choquet, 14e année (1974/75), Initiation à l'analyse, Exp. No. 2*, Secrétariat Math., Paris, 1975.

[334] B. Saint-Venant, Mémoire sur la torsion des prismes, *Mém. présentés par divers savants à l'Academie des Sciences* **14** (1856), 233–560.

[335] T. Sakata and K. Nomakuchi, Some examples of statistical applications of majorization theory, *Mem. Fac. Gen. Ed. Kumamoto Univ. Natur. Sci.* no. 17 (1982), 15–21.

[336] S. Saks and J. D. Tamarkin, On a theorem of Hahn–Steinhaus, *Ann. of Math.* (2) **34** (1933), 595–601.

[337] H. H. Schaefer, *Banach Lattices and Positive Operators*, Die Grundlehren der mathematischen Wissenschaften 215, Springer-Verlag, New York-Heidelberg-Berlin, 1974.

[338] I. J. Schoenberg, On Pólya frequency functions, I: The totally positive functions and their Laplace transforms, *J. Anal. Math.* **1** (1951), 331–374.

[339] I. Schur, Über die charakteristischen Wurzeln einer linearen Substitution mit einer Anwendung auf die Theorie der Integralgleichung, *Math. Ann.* **66** (1909), 488–510.

[340] I. Schur, Bemerkungen zur Theorie der beschränkten Bilinearformen mit unendlich vielen Veränderlichen, *J. Reine Angew. Math.* **140** (1911), 1–29.

[341] I. Schur, Über eine Klasse von Mittelbildungen mit Anwendungen auf die Determinantentheorie, *Sitzungsber. Berlin Math. Gesellschaft* **22** (1923), 9–20.

[342] L. Schwartz, Géneralisation de la notion de fonction, de transformation de Fourier et applications mathématiques et physiques, *Ann. Univ. Grenoble Sect. Sci. Math. Phys. (N.S.)* **21** (1945), 57–74.

[343] L. Schwartz, *Théorie des distributions*, Actual. Scient. Ind. 1091 and 1122, Hermann, Paris, 1950–1951.

[344] H. A. Schwarz, Beweis des Satzes, dass die Kugel kleinere Oberfläche besitzt, als jeder andere Körper gleichen Volumens, *Nachr. Ges. Wiss. Göttingen* (1884), 1–13.

[345] H. A. Schwarz, *Gesammelte Mathematische Abhandlungen, Band I*, Springer, Berlin, 1890.

[346] C. E. Shannon, A mathematical theory of communication, *Bell System Tech. J.* **27** (1948), 379–423, 623–656,

[347] S. Sherman, A theorem on convex sets with applications, *Ann. Math. Statist.* **26** (1955), 763–767.

[348] B. Simon, Notes on infinite determinants of Hilbert space operators, *Advances in Math.* **24** (1977), 244–273.

[349] B. Simon, *Functional Integration and Quantum Physics*, Academic Press, 1979; 2nd edition, AMS Chelsea Publishing, Providence, RI, 2005.

[350] B. Simon, *Trace Ideals and Their Applications*, Cambridge University Press, Cambridge-New York, 1979; 2nd edition, Mathematical Surveys and Monographs 120, American Mathematica Society, Providence, RI, 2005.

[351] B. Simon, *The Statistical Mechanics of Lattice Gases, Vol. 1*, Princeton University Press, Princeton, NJ, 1993.

[352] B. Simon, *Representations of Finite and Compact Groups*, Graduate Studies in Mathematics 10, American Mathematical Society, Providence, RI, 1996.

[353] B. Simon, *Szegő's Theorem and Its Descendants: Spectral Theory for $L^2$ Perturbations of Orthogonal Polynomials*, Princeton University Press, Princeton, NJ, 2011.

[354] S. L. Sobolev, Méthode nouvelle à résoudre le problème de Cauchy pour les équations linéaires hyperboliques normales, *Mat. Sb. (N.S.)* **1** (1936), 39–72.

[355] S. L. Sobolev, Sur un théorème d'analyse fonctionelle, *Mat. Sb. (N.S.)* **4** (1938), 471–497. English translation in *Amer. Math. Soc. Transl. Ser. 2*, **34** (1963), 39–68.

[356] G. Sparr, A new proof of Löwner's theorem on monotone matrix functions, *Math. Scand.* **47** (1980), 266–274.

[357] E. Sperner, Jr., Symmetrisierung für Funktionen mehrerer reeler Variablen, *Manuscripta Math.* **11** (1974), 159–170.

[358] E. Sperner, Jr., Zur Symmetrisierung von Funktionen auf Sphären, *Math. Z.* **134** (1973), 317–327.

[359] E. Stein, Interpolation of linear operators, *Trans. Amer. Math. Soc.* **83** (1956), 482–492.

[360] J. Steiner, Einfache Beweise der isoperimetrischen Hauptsätze, *J. Reine Angew. Math.* **18** (1838), 281–296.

[361] H. Steinhaus, Sur les développements orthogonaux, *Bull. Int. Acad. Polon. Sci. Sér. A* (1926), 11–39.

[362] E. Steinitz, Bedingt konvergente Reihen und konvexe Systeme, I, II, III, *J. Reine Angew. Math.* **143** (1913), 128–175; **144** (1914), 1–40; **146** (1916), 1–52.

[363] T. Stieltjes, Recherches sur les fractions continues, *Ann. Fac. Sci. Univ. Toulouse* **8** (1894–1895), J76–J122; ibid. **9**, A5–A47.

[364] O. Stolz, *Grundzüge der Differential und Integralrechnung, I*, Teubner, Leipzig, 1893.

[365] M. H. Stone, Applications of the theory of Boolean rings to general topology, *Trans. Amer. Math. Soc.* **41** (1937), 375–481.

[366] R. Strichartz, Multipliers on fractional Sobolev spaces, *J. Math. Mech.* **16** (1967), 1031–1060.

[367] G. Szegő, Über einige Extremalaufgaben der Potentialtheorie, *Math. Z.* **31** (1930), 583–593.

[368] G. Szegő, Inequalities for certain eigenvalues of a membrane of given area, *J. Rational Mech. Anal.* **3** (1954), 343–356.

[369] G. Talenti, Best constant in Sobolev inequality, *Ann. Mat. Pura Appl.* (4) **110** (1976), 353–372.

[370] C. Thompson, Inequality with applications in statistical mechanics, *J. Math. Phys.* **6** (1965), 1812–1813.

[371] O. Thorin, An extension of a convexity theorem due to M. Riesz, *K. Fysiogr. Sällsk. Lund Förh.* **8** (1938), no. 14.

[372] O. Toeplitz, Über allgemeine lineare Mittelbildungen, *Prace Math.-Fiz.* **22** (1911), 113–119.

[373] Y. L. Tong, Some recent developments on majorization inequalities in probability and statistics, *Linear Algebra Appl.* **199** (1994), 69–90.

[374] D. Towsley, Application of majorization to control problems in queueing systems, in *Scheduling Theory and Its Applications*, pp. 295–311, Wiley, Chichester, 1995.

[375] A. Tychonoff, Ein Fixpunktsatz, *Math. Ann.* **111** (1935), 767–776.

[376] K. Urbanik, Extreme point method in probability theory, in *Probability–Winter School* (Proc. Fourth Winter School, Karpacz, 1975), pp. 169–194, Lecture Notes in Math. 472, Springer, Berlin, 1975.

[377] J. von Neumann, Mathematische Begründung der Quantenmechanik, *Nachr. Gesell. Wiss. Göttingen Math.-Phys. Kl.* (1927), 1–57.

[378] J. von Neumann, Zur Theorie der Gesellschaftsspiele, *Math. Ann.* **100** (1928), 295–320.

[379] J. von Neumann, Zur Algebra der Funktionaloperationen und Theorie der Normalen Operatoren, *Math. Ann.* **102** (1930), 370–427.

[380] J. von Neumann, Proof of the quasi-ergodic hypothesis, *Proc. Nat. Acad. Sci. USA* **18** (1932), 70–82.

[381] J. von Neumann, Über adjungierte Funktionaloperatoren, *Ann. of Math. (2)* **33** (1932), 294–310.

[382] J. von Neumann, On complete topological spaces, *Trans. Amer. Math. Soc.* **37** (1935), 1–20.

[383] H. F. Weinberger, An isoperimetric inequality for the $N$-dimensional free membrane problem, *J. Rational Mech. Anal.* **5** (1956), 633–636.

[384] R. Welland, Bernstein's theorem for Banach spaces, *Proc. Amer. Math. Soc.* **19** (1968), 789–792.

[385] H. Weyl, Inequalities between the two kinds of eigenvalues of a linear transformation, *Proc. Nat. Acad. Sci. USA* **35** (1949), 408–411.

[386] N. Wiener, Limit in terms of continuous transformation, *Bull. Soc. Math. France* **50** (1922), 119–134.

[387] N. Wiener, The ergodic theorem, *Duke Math. J.* **5** (1939), 1–18.

[388] A. S. Wightman, Convexity and the notion of equilibrium state in thermodynamics and statistical mechanics, introduction to *Convexity in the Theory of Lattice Gases* by R. B. Israel, Princeton University Press, Princeton, NJ, 1979.

[389] E. Wigner and J. von Neumann, Significance of Löwner's theorem in the quantum theory of collisions, *Ann. of Math.* (2) **59** (1954), 418–433.

[390] S. Willard, *General Topology*, reprint of the 1970 original, Dover Publications, Mineola, NY, 2004.

[391] W. H. Young, On classes of summable functions and their Fourier series, *Proc. R. Soc. Lond.* A **87** (1912), 225–229.

[392] W. H. Young, On the determination of the summability of a function by means of its Fourier constants, *Proc. London Math. Soc.* **12** (1913), 71–88.

[393] A. C. Zaanen, On a certain class of Banach spaces, *Ann. of Math.* (2) **47** (1946), 654–666.

[394] A. C. Zaanen, *Linear analysis. Measure and Integral, Banach and Hilbert Space, Linear Integral Equations*, Interscience Publishers, New York; North-Holland, Amsterdam; P. Noordhoff, Groningen, 1953.

[395] A. Zygmund, *Trigonometric Series, Vols. I, II*, 2nd edition, Cambridge University Press, London-New York, 1968.

# Author index

# Subject index